# Terotechnology XI

11<sup>th</sup> International Conference on Terotechnology, 27-28
September 2019, Kielce, Poland

### Editor
### Agnieszka Szczotok

Silesia University of Technology,
agnieszka.szczotok@polsl.pl,  http://www.polsl.pl

Peer review statement

All papers published in this volume of "Materials Research Proceedings" have
been peer reviewed. The process of peer review was initiated and overseen by the
above proceedings editors. All reviews were conducted by expert referees in
accordance to Materials Research Forum LLC high standards.

Published under License by **Materials Research Forum LLC**
Millersville, PA 17551, USA

Published as part of the proceedings series
**Materials Research Proceedings**
Volume 17 (2020)

ISSN 2474-3941 (Print)
ISSN 2474-395X (Online)

ISBN 978-1-64490-102-1 (Print)
ISBN 978-1-64490-103-8 (eBook)

This book contains information obtained from authentic and highly regarded sources. Reasonable efforts have been made to publish reliable data and information, but the author and publisher cannot assume responsibility for the validity of all materials or the consequences of their use. The authors and publishers have attempted to trace the copyright holders of all material reproduced in this publication and apologize to copyright holders if permission to publish in this form has not been obtained. If any copyright material has not been acknowledged please write and let us know so we may rectify in any future reprint.

Distributed worldwide by

**Materials Research Forum LLC**
105 Springdale Lane
Millersville, PA 17551
USA
http://www.mrforum.com

Manufactured in the United States of America
10 9 8 7 6 5 4 3 2 1

# Table of Contents

# Preface

Terotechnology is the technology of installation, commissioning, maintenance, replacement and removal of plant machinery and equipment, of feedback on operation and design thereof, and on related subjects and practices.

It has been eighteen years since the first Conference on Terotechnology was held. The conference is still a venture of Kielce University of Technology, Polish Maintenance Society, Kielce Trade Fairs and Polish Society for Stereology. The first conference was held in 2001 and subsequent meetings were held in 2005-2009 in the annual cycle. In 2011, a two-year conference cycle was established.

The 11th International Conference on Terotechnology was held at the Kielce Trade Fair Centre, Kielce, POLAND on 27-28 September, 2019. Kielce is situated halfway between Kraków and Warsaw. The selected papers from the 11th Conference, having been reviewed by at least two experts, were prepared for this volume of *Materials Research Proceedings*. The papers were chosen on the basis of their quality and relevancy to the conference. The volume represents the recent advances in materials, technologies and methods.

Dozens of people meet at our conference: experienced researchers, young scientists and many practitioners from industry. A special feature of our Conference is a broad formula resulting from the extent of the concept of "terotechnology".

The aim of the conference is to present results and exchange experiences of research work in the area of broadly defined use of machinery. The conference is addressed to academic staff, research institutes, manufacturing and transportation, as well as the local government administration. The conference covers all areas of exploitation, namely: the theories of exploitation, industrial management, reliability, technical diagnostics, materials science, design of experiments, tribology and technical safety.

The Organizing Committee is grateful to deans and workers of the Centre for Laser Technologies of Metals of Kielce University of Technology, Faculty of Mechanical Engineering of Cracow University of Technology who helped with the organization of the 11th Conference and the edition of this volume. The Committee would also like to express their gratitude to all who have updated and reviewed the papers submitted to the conference, the scientific secretary for her editing work, and Materials Research Forum LLC for producing the volume.

# Conference Organizers

Kielce University of Technology,
Polish Maintenance Society,
Kielce Trade Fairs,
Polish Society for Stereology

# Scientific Committee

Stanisław Adamczak, Poland
Bogdan Antoszewski, Poland
Vladimir Antonjuk, Belarus
Vyacheslav Bilousov, Ukraine
Zdzisław Bogdanowicz, Poland
Otakar Bokuvka, Slovakia
Jozef Broncek, Slovakia
Robert Cep, Czech
Ryszard Dindorf, Poland
Emil Evin, Slovakia
Peter Fabian, Slovakia
Bogusław Grabas, Poland
Ivo Hlavaty, Czech
Zbigniew Koruba, Poland
Czesław Kundera, Poland
Tomasz Lipiński, Poland
Monika Madej, Poland
Włodzimierz Makieła, Poland
Vladimir Marcinkowski, Ukraine
Jozef Mesko, Slovakia
Milos Mician, Slovakia
Edward Miko, Poland
Andrzej Niewczas, Poland
Łukasz Orman, Poland
Dariusz Ozimira, Poland
Peter Palcek, Slovakia
Ryszard Pawlak, Poland
Jana Petru, Czech
Jacek Pietraszek, Poland
Norbert Radek, Poland
Leszek Radziszewski, Poland
Mieczysław Scendo, Poland
Jacek Selejdak, Poland
Augustin Sládek, Slovakia
Tomasz Stańczyk, Poland
Michał Styp-Rekowski, Poland
Robert Ulewicz, Poland
Vyacheslav Tarelnik, Ukraine
Włodzimierz Zowczak, Poland
Wojciech Żórawski, Poland

# Organizing Committee

Bogdan Antoszewski – chairman
Mieczysław Scendo – co-chairman
Augustin Sládek – co-chairman
Ivo Hlavaty – co-chairman
Norbert Radek – scientific secretary
Dariusz Gontarski – organizing secretary
Jacek Pietraszek
Aneta Gądek-Moszczak
Agnieszka Szczotok
Justyna Kasińska
Piotr Sęk
Piotr Kurp
Izabela Pliszka
Szymon Tofil
Hubert Danielewski
Dominika Soboń
Joanna Machniewska

# Failure Prediction of Spool Control Valve using CFD Simulation

DOMAGALA Mariusz[1][a]*, BIKASS Saeed[2], FABIS-DOMAGALA Joanna[1],
MOMENI Hassan[2] and FILO Grzegorz[1]

[1] Cracow University of Technology, Institute of Applied Informatics, Al Jana Pawla II 37,31-841 Cracow, Poland

[2] Western Norway University of Applied Sciences, Department of Mechanical and Marine Engineering, Inndalsveien 28, 5063 Bergen, Norway

[a] domagala@mech.pk.edu.pl

**Keywords:** Failure Prediction, Spool Control Valve, CFD Simulation

**Abstract.** Spool control valves play an important role in fluid power systems and required high reliability. Prediction of possible failures for such valves is a key issue. According to research most of failures which may appear are caused by contaminated fluid. Contaminants of fluid may cause erosion of metallic surfaces or cavitation wear. Both types of wear may be predicted by using CFD simulation tools and multiphase flow which includes phase changes and possibility of simulation of flow with solid particles.

## Introduction

Failures prediction for hydraulic valve is very important issue due to the fact that its malfunction may cause human injuries or environmental pollution. Among all causes of hydraulic systems failures, contamination of hydraulic fluid is one of the most common. It is estimated that majority (up to 90%) of failures is caused by contamination of hydraulic fluid which might be solid particles (remaining of wear or dust) or chemical agents (liquids or gases). It is not possible to prevent hydraulic fluid against contamination even using sophisticated filtration systems. Contamination may appear in the hydraulic systems internally as a normal process of operation, as a result of wear of hydraulic components. Solid particles in very specific flow conditions, very often with high flow velocity, gain momentum from fluid and may cause erosive wear inside components. Additionally, fluid flowing inside hydraulic components, particularly inside valves, rapidly changing direction and velocity what may lead to sudden pressure drop, below saturation pressure and cavitation and wear of surface of valve components. Both type of wears: erosive and cavitation has been investigated from decades on experimental way. Recent development of simulation methods (in particular computational fluid dynamics: CFD) allows to predict wear before their occurrence [1-4]. This paper presents research on flow inside spool control valve for two cases. First one is multiphase flow with contaminants (metallic particles) while the second deals with phase change flow (cavitation). Both simulations might be used for predicting spool valve wear.

The failure prediction presented here can be used in analogous issues, both related to fluid flows (e.g. biotechnology [5, 6], heat flows [7]) and in the field of solid materials engineering (e.g. materials science [8, 9] and design of medical implants [10]) as well as enhancement of surface layers [11], where the aim is to improve tribological properties [12] and fatigue resistance [13]. It is certain that this will have an impact on related methods of management optimization [14, 15] and data analysis [16, 17] as well as CFD numerical methods [18-20] aimed at erosion modeling [21, 22] and associated failure analysis methods [23, 24].

**Modelling of erosion and cavitation**
The wear caused by impact of solid particles on walls for metals is a function of impact angle and particle velocity [25]:

$$E = kV_p^n f(\gamma) \tag{1}$$

where:

$E$ – dimensionless mass (mass of eroded wall material divided by the mass of particle),

$V_p$ – particle velocity,

$n$ – exponent value which for metals is between 2.3 and 2.5,

$f(\gamma)$ – function of impact angle.

One of the models that describes rate of wear is Finnie model [26] :

$$E = kV_p^2 f(\gamma) \tag{2}$$

where:

$$f(\gamma) = \tfrac{1}{3} cos^2(\gamma) \; if \; tan(\gamma) > \tfrac{1}{3} \tag{3}$$

$$f(\gamma) = sin(2\gamma) - 3sin^2(\gamma) \; if \; tan(\gamma) > \tfrac{1}{3} \tag{4}$$

Other model which describes erosion rate is Tabakoff and Grant [27]:

$$E = k_1 f(\gamma) V_p^2 cos^2(\gamma) \left[ 1 - R_T^2 \right] + f(V_{PN}) \tag{5}$$

where:

$$f(\gamma) = \left[ 1 + k_2 k_{12} sin \left( \gamma \frac{\frac{\pi}{2}}{\gamma_0} \right) \right]^2 \tag{6}$$

$$R_T = 1 - k_4 V_p sin(\gamma) \tag{7}$$

$$f(V_{PN}) = k_3 \left( V_p sin(\gamma) \right)^4 \tag{8}$$

$$k_2 = \begin{cases} 1.0 \; if \; \gamma \le 2\gamma_0 \\ 0 \; if \; \gamma > 2\gamma_0 \end{cases} \tag{9}$$

$k_1, k_{12}$ – constants.

Simulation of particles in continuous fluid in CFD is treated as two phases flow: discreet one (particles) and continuous one (fluid). Discreet particles are tracked into fluid domain during fluid flow. Equation of motion for a single particle according to Basset, Boussinesq and Oseen may be presented in the following way:

$$m_p \frac{dU_p}{dt} = F_D + F_B + F_R + F_{VM} + F_P + F_{BA} \tag{10}$$

where: $F_D$ – drag force, $F_B$ – buoyancy force, $F_R$ – Coriolis force, $F_{VM}$ - is inertia force of fluid occupied by particle (Virtual Mass), $F_P$ – pressure force, $F_{BA}$ – Basset force

The inertia force of fluid occupied by particle (Virtual Mass) is expressed by the following equation:

$$F_{VM} = \frac{C_{VM}}{2} m_F \left( \frac{dU_F}{dt} - \frac{dU_P}{dt} \right) \tag{11}$$

Transforming Eq. 10 we obtain:

$$\frac{dU_p}{dt} = \left( \frac{1}{m_p + \frac{C_{VM}}{2} m_F} \right) (F_D + F_B + F_R + F'_{VM} + F_P + F_{BA}) \tag{12}$$

where

$$F'_{VM} = \frac{C_{VM}}{2} m_F (U_F \nabla U_F) \tag{13}$$

$m_P = \frac{\pi}{6} d_P^3 \rho_P$ — particle mass,

$m_F = \frac{\pi}{6} d_P^3 \rho_F$ — fluid mass,

$d_p$ — the particle diameter,

$\rho_P, \rho_F$ — density of particle and fluid respectively.

Assuming that:

$$R_{VM} = \frac{m_p}{m_p + \frac{C_{VM}}{2} m_F} = \frac{\rho_P}{\rho_P + \frac{C_{VM}}{2} \rho_F} \tag{14}$$

Finally Eq. 10 has the following form:

$$\frac{dU_p}{dt} = \left( \frac{R_{VM}}{m_p} \right) (F_D + F_B + F_R + F'_{VM} + F_P + F_{BA}) \tag{15}$$

In generally, tendency of fluid flow to cavitation may be expressed as:

$$c_a = \frac{p - p_v}{0.5 \rho U^2} \tag{16}$$

One of the approaches which describes bubble dynamics is The Rayleigh-Plesset formula:

$$R_b \frac{d^2 R_b}{dt^2} + \frac{3}{2} \left( \frac{dR_b}{dt} \right)^2 + \frac{2\sigma}{\rho R_b} = \frac{p_v - p}{\rho} \tag{17}$$

After neglecting surface tension and term of second order the above eq. have the following form:

$$\frac{dR_b}{dt} = \sqrt{\frac{2}{3} \left( \frac{p_v - p}{\rho} \right)} \tag{18}$$

The rate of changes of bubble volume is:

$$\frac{dV_b}{dt} = 4\pi R_b^2 \sqrt{\frac{2}{3} \left( \frac{p_v - p}{\rho} \right)} \tag{19}$$

The rate of changes of bubble mass is:

$$\frac{dm_g}{dt} = 4\pi R_b^2 \rho_g \sqrt{\frac{2}{3} \left( \frac{p_v - p}{\rho} \right)} \tag{20}$$

The number of bubbles Nb per unit volume rg is expressed by:

$$r_g = \frac{4}{3} \pi R_b^2 N_b \tag{21}$$

Terotechnology XI                                                          Materials Research Forum LLC
Materials Research Proceedings **17** (2020) 1-8                    https://doi.org/10.21741/9781644901038-1

The total interphase mass transfer per unit volume is:

$$\dot{m}_{fg} = 3\frac{r_g \rho_g}{R_b}\sqrt{\frac{2}{3}\frac{p_v - p}{p_f}} \tag{22}$$

After including condensation to above expression have the following form:

$$\dot{m}_{fg} = 3F\frac{r_g \rho_g}{R_b}\sqrt{\frac{2}{3}\frac{p_v - p}{p_f}}\,\mathrm{sgn}(p_v - p) \tag{23}$$

And finally the vapor transport equation have form:

$$\frac{\partial}{\partial t}(\alpha \rho_v) + \nabla \cdot (\alpha \rho_v \mathbf{U_v}) = R_e - R_c \tag{24}$$

where:

$R_b$ – bubble radius, $p_v$ – vapor pressure, $p$ – pressure of liquid surrounding the bubble, $\rho$ – liquid density, $\rho_g$ – vapor density, $\sigma$ – surface tension, $U_v$ – vapour phase velocity, $\alpha$ – vapor volume fraction, $n$ – bubble number, $R$ – phase change rate, $f_v$ – vapor mass fraction, $f_g$ – noncondesable gases, $R_e$, $R_c$ – mass transfer source terms connected to the growth and collapse of the vapor bubbles, respectively.

**Spool control valve**

The spool control valve presented in Fig. 1. is used to control the direction of flow and movements speed of actuators. Position of spool is determined by proportional solenoids which allows to control both direction of flow and receivers movement speed. Return to neutral position (any receiver is supplied) is provided by two springs. The shape of the spool determines flow configuration (connections between ports: P, T, A, B).

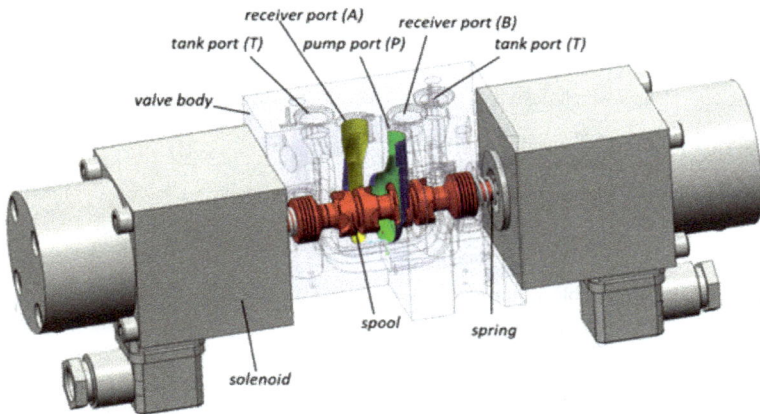

*Fig. 1. Spool control valve.*

Terotechnology XI                                              Materials Research Forum LLC
Materials Research Proceedings **17** (2020) 1-8        https://doi.org/10.21741/9781644901038-1

## CFD analysis of flow control valve

CFD simulations were conducted in Ansys CFX a general purpose CFD code, which use Finite Volume Method (FVM) for solving flow governing equations and applies multiphase flow model with phase change and Lagrangian particle tracking modelling technique. For numerical simulation path P-A was used with initial opening of 0.3 mm. Boundary conditions and grid is presented in Fig. 3. CFD simulations were performed for cavitation and flow with solid particles independently.

Outlet: receiver port (A)

Inlet: pump port (P)

*Fig. 2. CFD model: grid and boundary conditions.*

CFD simulation was performed for fixed spool position for steady state conditions and for the following assumptions:
- Fluid (hydraulic oil) has a constant properties: density 880 [kg/m$^3$], viscosity $\upsilon$=40 [mm$^2$/s];
- Flow is turbulent: k-$\omega$ turbulence model was used;
- Model is in thermodynamics equilibrium, heat transfer is not included;

Cavitation simulation included below assumptions:
- Saturation pressure is set to 0.2 [bar];
- Fluid is homogeneous without any dissolved gases;

Whereas particle flow simulation included following assumptions:
- Particles have uniform shape (spheres with diameter 20 [μm]) and constant properties, 20 particles in 1 [cm3];
- Erosion model: Finnie;
- Interaction of particles and fluid is fully coupled;
- Simulations were performed for metallic particles (steel);
- Particles are uniformly injected at inlet to the valve.

Exemplary CFD simulations for solid particle flow are presented in Fig. 3 and Fig. 4. There are presented pathlines of hydraulic oil colored by velocity values and trajectory of solid particles (grey lines).

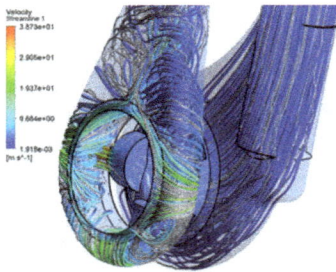
*Fig. 3. Fluid path lines (color) and particles trajectory (grey).*

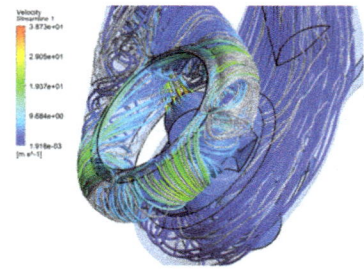
*Fig. 4. Fluid path lines (color) and particles trajectory (grey).*

Erosion rate density is presented in Fig. 5, which allows to identify areas which are exposed to erosion wear. Figure 6 presents distribution of gas fraction during fluid flow for given conditions.

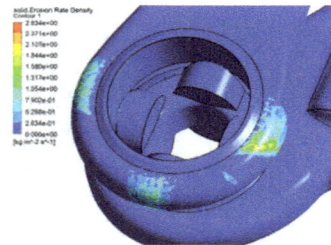
*Fig. 5. Erosion rate density.*

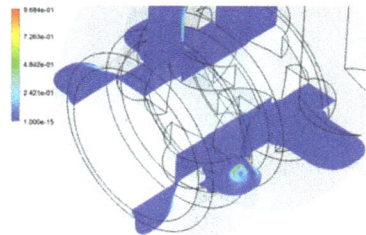
*Fig. 6. Oil vapor fraction.*

**Conclusions**

This paper presents attempt of failure prediction of spool control valve caused by contaminated fluid which may cause erosive or cavitation wear. Both are complex phenomena, however, available numerical simulation methods (CFD) allows conducting multiphase flows which includes phase changes (cavitation) and particle tracking. This paper presents simulation of flow inside spool valve for two cases, the first includes hydraulic fluids with metallic contaminants while the second flow with phase changes (cavitation). Obtained results allows to indicate areas which are mostly exposed on wear or to define flow conditions for which wear may occur. CFD simulation tools seem to be an effective tool for perdition of failures in fluid power components.

**References**

[1] Y. Yaobaoa, Y. Jiayanga, G. Shengrongb, Numerical study of solid particle erosion in hydraulic spool valves, Wear 392-393 (2017) 147-189.
https://doi.org/10.1016/j.wear.2017.09.021

[2] X. Liu, H. Ji, W. Min, Z. Zheng, J. Wang, Erosion behavior and influence of solid particles in hydraulic spool valve without notches, Eng. Fail. Anal. 108 (2020) art. 104262. https://doi.org/10.1016/j.engfailanal.2019.104262

[3] Z. Kudzma, M. Stasiak, Studies of flow and cavitation of hydraulic lift valve, Archives of Civil and Mechanical Engineering 15 (2015) 951-961. https://doi.org/10.1016/j.acme.2015.05.003

[4] R. Amirante, E. Distaso, P. Tamburrano, Experimental and numerical analysis of cavitation in hydraulic proportional directional valves, Energy Conversion and Management, 87 (2014) 208–219. https://doi.org/10.1016/j.enconman.2014.07.031

[5] E. Skrzypczak-Pietraszek, K. Reiss, P. Zmudzki, J. Pietraszek, J. Enhanced accumulation of harpagide and 8-O-acetyl-harpagide in Melittis melissophyllum L. agitated shoot cultures analyzed by UPLC-MS/MS. PLOS One. 13 (2018) art. e0202556. https://doi.org/10.1371/journal.pone.0202556

[6] E. Skrzypczak-Pietraszek, K. Piska, J. Pietraszek, Enhanced production of the pharmaceutically important polyphenolic compounds in Vitex agnus castus L. shoot cultures by precursor feeding strategy. Engineering in Life Sciences 18 (2018) 287-297. https://doi.org/10.1002/elsc.201800003

[7] L.J. Orman, Boiling heat transfer on single phosphor bronze and copper mesh microstructures. EPJ Web of Conf. 67 (2014) art. 02087. https://doi.org/10.1051/epjconf/20146702087

[8] A. Tiziani, A. Molinari, J. Kazior, G. Straffelini, Effect of vacuum sintering on the mechanical-properties of copper-alloyed stainless-steel. Powder Metall. 22 (1990) 17-19.

[9] L. Mosinska, K. Fabisiak, K. Paprocki, M. Kowalska, P. Popielarski, M. Szybowicz, A. Stasiak, Diamond as a transducer material for the production of biosensors. Przem. Chem. 92 (2013) 919-923.

[10] D. Klimecka-Tatar, Electrochemical characteristics of titanium for dental implants in case of the electroless surface modification. Arch. Metall. Mater. 61 (2016) 923-26. https://doi.org/10.1515/amm-2016-0156

[11] N. Radek, A. Szczotok, A. Gadek-Moszczak, R. Dwornicka, J. Broncek, J. Pietraszek, The impact of laser processing parameters on the properties of electro-spark deposited coatings. Arch. Metall. Mater. 63 (2018) 809-816.

[12] S. Wojciechowski, D. Przestacki, T. Chwalczuk, The evaluation of surface integrity during machining of Inconel 718 with various laser assistance strategies. MATEC Web of Conf. 136 (2017) art. 01006. https://doi.org/10.1051/matecconf/201713601006

[13] R. Ulewicz, F.R. Novy, The influence of the surface condition on the fatigue properties of structural steel. J. Balk. Tribol. Assoc. 22 (2016) 1147-1155.

[14] D. Malindzak, A. Pacana, H. Pacaiova, An effective model for the quality of logistics and improvement of environmental protection in a cement plant. Przem. Chem. 96 (2017) 1958-1962.

[15] A. Pacana, K. Czerwinska, R. Dwornicka, Analysis of non-compliance for the cast of the industrial robot basis, METAL 2019 28th Int. Conf. on Metallurgy and Materials (2019), Ostrava, Tanger 644-650. https://doi.org/10.37904/metal.2019.869

[16] J. Pietraszek, A. Gadek-Moszczak, The Smooth Bootstrap Approach to the Distribution of a Shape in the Ferritic Stainless Steel AISI 434L Powders. Solid State Phenomena 197 (2012) 162-167. https://doi.org/10.4028/www.scientific.net/SSP.197.162

[17] A. Gadek-Moszczak, J. Pietraszek, B. Jasiewicz, S. Sikorska, L. Wojnar, The Bootstrap Approach to the Comparison of Two Methods Applied to the Evaluation of the Growth Index in the Analysis of the Digital X-ray Image of a Bone Regenerate. New Trends in Comp. Collective Intell. 572 (2015) 127-136. https://doi.org/10.1007/978-3-319-10774-5_12

[18] E. Lisowski, G. Filo, CFD analysis of proportional flow control valve with an innovative opening shape, Energy Conversion and Management 123 (2016) 15-28. https://doi.org/10.1016/j.enconman.2016.06.025

[19] E. Lisowski, G. Filo, J. Rajda, Analysis of flow forces in the initial phase of throttle gap opening in a proportional control valve, Flow Measurement and Instrumentation 59 (2018) 157-167. https://doi.org/10.1016/j.flowmeasinst.2017.12.011

[20] H. Momeni, M. Domagala, CFD simulation of transport solid particles by jet pumps, Technical Transactions 112 (7) 185-191.

[21] M. Domagala, H. Momeni, J. Domagala-Fabis, G. Filo, D. Kwiatkowski, Simulation of Cavitation Erosion in a Hydraulic Valve, Materials Research Proceedings 5 (2018) 1-6. https://doi.org/10.21741/9781945291814-1

[22] M. Domagala, H. Momeni, J. Domagala-Fabis, G. Filo, M. Krawczyk, J. Rajda, Simulation of particle erosion in a hydraulic valve, Materials Research Proceedings 5 (2018) 17-24. https://doi.org/10.21741/9781945291814-4

[23] J. Fabis-Domagala, H. Momeni, G. Filo, M. Domagała, Instruments of identification of hydraulic components potential failures, MATEC Web Conf. 183 (2018) art. 03008. https://doi.org/10.1051/matecconf/201818303008

[24] J. Fabis-Domagala, H. Momeni, M. Domagała, G. Filo, Matrix FMEA analysis as a preventive method for quality design of hydraulic components, CzOTO 1(1) (2019) 684-691. https://doi.org/10.2478/czoto-2019-0087

[25] I.M. Hutchings, Mechanical and metallurgical aspects of the erosion of metals, Conf. Corrosion-Erosion of Coal Conversion System Materials, NACE, Houston, TX (1979) 393-428.

[26] S. Dosanjh, J.A.C. Humphrey, The influence of turbulence on erosion by a particle laden fluid jet, Wear 102 (1985) 309-330. https://doi.org/10.1016/0043-1648(85)90175-9

[27] W. Dosa Tabakoff, G. Grant, R. Ball, An experimental investigation of certain aerodynamics effect on erosion, AIAA Paper No. 74-639, 1974. https://doi.org/10.2514/6.1974-639

Terotechnology XI
Materials Research Proceedings **17** (2020) 9-15

Materials Research Forum LLC
https://doi.org/10.21741/9781644901038-2

# Matrix FMEA Analysis of a Spool Control Valve

FABIS-DOMAGALA Joanna[1, a *], BIKASS Saeed[2], MOMENI Hassan[2],
DOMAGALA Mariusz[1], FILO Grzegorz[1] and KWIATKOWSKI Dominik[1]

[1] Cracow University of Technology, Institute of Applied Informatics, Al J. Pawla II 37,31-841
Cracow, Poland

[2] Western Norway University of Applied Sciences, Department of Mechanical and Marine
Engineering, Inndalsveien 28, 5063 Bergen, Norway

[a] joanna.fabis-domagala@mech.pk.edu.pl

**Keywords:** FMEA, Spool Control Valve

**Abstract.** Spool control valves are one of the most common use components of fluid power systems. They play an important role in hydraulic systems controlling flow direction and movement of actuators. Therefore, high reliability of such valves is required. It may be achieved by using quality improvement tools. This paper presents implementation of matrix FMEA analysis of directional flow control valve. Results of conducted analysis allowed to indicate failure with highest probability of occurrence during valve operation.

## Introduction

Directional flow control valve which is also known as spool control valve is one of the hydraulic components which are used in almost each drive systems. It is a key component of hydraulic system which find application in wide industrial branches, from agriculture and heavy machinery to manufacturing machines. Reliability of such valves plays important role and more and more effort is put on avoiding possible failures, which may cause improper work, malfunctioning or even a damages and in consequences financial loses. Improving quality is endless process which involves using efficient tools and preventing measures. Reliability can be improved by implementation of a qualitative methods such as matrix FMEA analysis [1]. The FMEA might also be implemented when defects should be identified as early as possible due to the human health and safety as well as environmental protection. This can be achieved by using FMEA analysis during preliminary design work what allows to obtain information about strengths and weaknesses of the valve and implement appropriate changes. The matrix FMEA is relatively easy for programming and allows identify possible failures and preparing preventing measures.

This paper presents an implementation of matrix FMEA analysis [2] for directional flow control valve. There are presented three approaches of matrix FMEA analysis for the valve which allows to identify failures [3, 4] and its causes in more details [5-7]. This paper also shows that FMEA analysis strongly depends on input data and in any case should be supported by knowledge base which should include as rich as possible information about analyzed problem.

The presented approach to damage analysis should also be interesting in terms of management [8], but also for other industries such as biotechnology [9, 10], heat-loaded structures [11], materials engineering [12, 13] and biomaterials [14], and improving the properties of their surface layers directly [15, 16] or by applying coatings [17]. It may also be inspiring for related data analysis methods such as statistical corrections (adjustment calculus) of temperature fields [18, 19] or sub-sampling methods (bootstrap) for estimating statistical distribution of properties [20, 21].

**Spool control valve**
Directional flow control valve main purpose is controlling direction movement and speed of actuators in hydraulic systems. Presented in Fig. 1 valve is proportional directional flow control valve mounting on subplate is controlled by solenoids. It consists of body with receiver ports, pump and tank port, solenoids, spool and sealing rings. Direction of oil flow and flow rate is controlled by proportional solenoids which set the proper position of the spool. Return to neutral position (none of any receiver is supplied by fluid) is provided by two centering springs. Spool with controlling edges control flow directions and seals flow in selected directions.

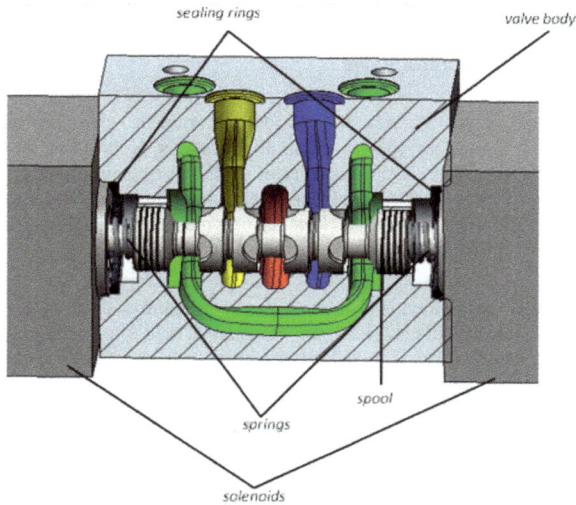

*Fig. 1. Flow control valve.*

The FMEA analysis was conducted only for mechanical system (all valve components). Analysis of control system for solenoid and solenoid itself was not included in this work.

**Matrix FMEA analysis of flow control valve**
In general, matrix FMEA analysis consists of two stages which first one is recognized as a preparation stage. During this stage valve was decomposed into individual components for which list of possible failures was prepared. Additionally, functions which those components realize were determined. At this stage two matrices CF and EC were created. The matrix CF corresponds to relation between components (C) and possible failures (F), while the second EC to relations between functions (E) and components (C). The following components of directional control valve were included in matrix FMEA analysis:
- C1: valve body,
- C2: spool,
- C3: spring,
- C4: sealing ring,

The flow control valve presented in Fig. 1 includes two springs and sealing rings, but both, springs and sealing rings have the same structure and realize the same function. Therefore, FMEA analysis was conducted for one spring and sealing ring. For above mentioned components the following failures were selected:

- F1: seizure, s
- F2: cracking,
- F3: wear,
- F4: damage.

The matrix FMEA analysis was performed in three approaches, the first one uses the simplest evaluation methods for failures using two values 0 when failure does not exists and 1 when exists.

The matrix **CF** is shown in Table 1.

Functions for selected components were set as below:

- E1: fixing: fixing position of spool;
- E2: controlling: controlling flow rate and flow direction in the valve;
- E3: transferring force;
- E4: sealing.

The matrix **EC** has the form shown in Table 2.

*Table 1. **CF** matrix for flow control valve.*

|    | F1 | F2 | F3 | F4 |
|----|----|----|----|----|
| C1 | 0  | 1  | 1  | 0  |
| C2 | 1  | 1  | 1  | 0  |
| C3 | 0  | 1  | 0  | 0  |
| C4 | 0  | 1  | 0  | 1  |

*Table 2. **EC** matrix for flow control valve.*

|    | C1 | C2 | C3 | C4 |
|----|----|----|----|----|
| E1 | 1  | 0  | 0  | 0  |
| E2 | 0  | 1  | 0  | 0  |
| E3 | 0  | 0  | 1  | 0  |
| E4 | 1  | 1  | 0  | 1  |

Multiplication of matrices **EC** and **EF** allows [1] to obtain matrix **EF** (Table 3):

$$EF = EC \times CF$$

*Table 3. **EF** matrix for flow control valve.*

|    | F1 | F2 | F3 | F4 |
|----|----|----|----|----|
| E1 | 0  | 1  | 1  | 0  |
| E2 | 1  | 1  | 1  | 0  |
| E3 | 0  | 1  | 0  | 0  |
| E4 | 1  | 3  | 2  | 1  |

Terotechnology XI
Materials Research Proceedings 17 (2020) 9-15

Materials Research Forum LLC
https://doi.org/10.21741/9781644901038-2

This matrix (**EF**) allows to identify failure which may appear with highest probability. The maximal values of matrix **EF** is for failure F2 (cracking) for function E4 (sealing).

Presenting above results do not take into consideration the probability of failures which may appear for each components, therefore the other approach was used. Instead of simple valuation methods of failure appearance it was used appropriate (weighted) factors, which allows to define how often selected failure may appear:

- 0: failure does not exist;
- 0.25: failure appears incidentally;
- 0.5: failure appears seldom;
- 0.75: failure appears often;
- 1: failure appears very often.

The new matrix Component-Failure (**CF**) is presented in Table 4.

*Table 4. Modified **CF** matrix for weighted evaluation method.*

|      | F1   | F2   | F3   | F4   |
|------|------|------|------|------|
| C1   | 0    | 0.25 | 0.5  | 0    |
| C2   | 0.75 | 0.25 | 0.75 | 0    |
| C3   | 0    | 0.25 | 0    | 0    |
| C4   | 0    | 0.25 | 0    | 0.25 |

Modified **CF** matrix multiplied by **EF** matrix allowed to obtain new Function-Failure (**EC**) (Table 5).

*Table 5. **EF** matrix for weighted evaluation method for **CF** matrix.*

|      | F1   | F2   | F3   | F4   |
|------|------|------|------|------|
| E1   | 0    | 0.25 | 0.5  | 0    |
| E2   | 0.75 | 0.25 | 0.75 | 0    |
| E3   | 0    | 0.25 | 0    | 0    |
| E4   | 0.75 | 0.75 | 1.25 | 0.25 |

Modified **CF** matrix indicates that the highest probability of occurrence is for failure wear (F3) for function sealing (E4). In comparison to the first analysis other type of wear was indicated as dominant.

Another approach in FMEA analysis was using more detailed types of failures (wear and seizure). In such case the list of potential failures is as follows:

- F1: seizure (contamination);
- F2: seizure (incorrect tolerances and operating conditions);
- F3: seizure (distortion due to excessive load);
- F4: cracking;
- F5: wear (abrasive wear);
- F6: wear (corrosive wear);
- F7: wear (erosive wear);
- F8: wear (sliding);
- F9: damage.

Matrix CF for extended failures have form presented in Table 6.

*Table 6. CF matrix for extended failures.*

|      | F1   | F2   | F3   | F4   | F5   | F6   | F7   | F8   | F9   |
|------|------|------|------|------|------|------|------|------|------|
| C1   | 0    | 0    | 0    | 0.25 | 0.25 | 0.25 | 0.5  | 0    | 0    |
| C2   | 0.75 | 0.25 | 0.25 | 0.25 | 0.5  | 0.25 | 0.75 | 0.25 | 0.25 |
| C3   | 0    | 0    | 0    | 0.25 | 0    | 0    | 0    | 0    | 0    |
| C4   | 0    | 0    | 0    | 0.25 | 0    | 0    | 0    | 0    | 0.25 |

After multiplication of **EC** with **CF** we obtain **EF** matrix which is presented in Table 7.

*Table 7. EF matrix for extended failures.*

|      | F1   | F2   | F3   | F4   | F5   | F6   | F7   | F8   | F9   |
|------|------|------|------|------|------|------|------|------|------|
| E1   | 0    | 0    | 0    | 0.25 | 0.25 | 0.25 | 0.5  | 0    | 0    |
| E2   | 0.75 | 0.25 | 0.25 | 0.25 | 0.5  | 0.25 | 0.75 | 0.25 | 0.25 |
| E3   | 0    | 0    | 0    | 0.25 | 0    | 0    | 0    | 0    | 0    |
| E4   | 0.75 | 0.25 | 0.25 | 0.75 | 0.75 | 0.5  | 1.25 | 0.25 | 0.5  |

Above matrix indicates failure F7 (erosive wear) as a type of wear with highest probability of occurrence. Recent research [3] shows that erosion is one of the typical wear for spool valve. It may also indicate that the cause of erosive wear might be contaminated hydraulic fluid what leads to conclusion that preventing measures is controlling fluid cleanness.

**Conclusions**

This paper presents implementation of matrix FMEA analysis for one of the most important and one of the most common use components in fluid power systems which is directional flow control valve. There were presented various approaches in the FMEA matrix analysis, starting with the easiest evaluation methods for possible failures occurrence. Such approach may give not accurate results what was shown by conducting analysis with weighted values which described probability of occurrence for selected failures. The last step was extending failure types which allowed to obtain information about the most probable failure which may occur during fluid flow. And finally allowed to identify causes and define preventing measures.

The matrix FMEA analysis seems to be very effective tool for improving quality and reliability of key components of systems which plays important role in engineering systems. However, it has to be supported by knowledge base of possible failures, malfunctioning or damages which can allow for prioritizing defects and failures.

**References**

[1] M.E. Stock, R.B. Stone, I.Y. Tumer, Going back in time to improve design: The elemental function-failure design method, in: ASME 2003 Int. Design Eng. Tech. Conf. and Computers and Information in Eng. Conf., September 2–6, 2003, Chicago, Illinois, USA, Vol.3b: 15th Int. Conf. Design Theory and Methodology, https://coi.org/10.1115/DETC2003/DTM-48638

[2] J. Fabis-Domagala, H. Momeni, M. Domagala, G. Filo, S. Bikass, P. Lempa, Matrix FMEA analysis of the flow control valve, QPI 1 (1) (2019) 590-595. https://doi.org/10.2478/cqpi-2019-0079

[3] X. Liu, H. Ji, W. Min, Z. Zheng, J. Wang, Erosion behavior and influence of solid particles in hydraulic spool valve without notches, Eng. Fail. Anal. 108 (2020) art. 104262. https://doi.org/10.1016/j.engfailanal.2019.104262

[4] M. Domagala, H. Momeni, J. Domagala-Fabis, G. Filo, M. Krawczyk, J. Rajda, Simulation of particle erosion in a hydraulic valve, Materials Research Proceedings 5 (2018) 17-24. https://doi.org/10.21741/9781945291814-4

[5] M. Domagala, H. Momeni, J. Domagala-Fabis, G. Filo, D. Kwiatkowski, Simulation of Cavitation Erosion in a Hydraulic Valve, Materials Research Proceedings 5 (2018) 1-6. https://doi.org/10.21741/9781945291814-1

[6] J. Fabis-Domagała, H. Momeni, M. Domagała, G. Filo, Matrix FMEA analysis as a preventive method for quality design of hydraulic components, CzOTO 1 (1) (2019) 684-691. https://doi.org/10.2478/czoto-2019-0087

[7] J. Fabis-Domagała, H. Momeni, G. Filo, M. Domagała, Instruments of identification of hydraulic components potential failures, MATEC Web Conf. 183 (2018) art. 03008. https://doi.org/10.1051/matecconf/201818303008

[8] A. Pacana, K. Czerwinska, R. Dwornicka, Analysis of non-compliance for the cast of the industrial robot basis, METAL 2019 28th Int. Conf. on Metallurgy and Materials (2019), Ostrava, Tanger 644-650. https://doi.org/10.37904/metal.2019.869

[9] E. Skrzypczak-Pietraszek, I. Kwiecien, A. Goldyn, J. Pietraszek, HPLC-DAD analysis of arbutin produced from hydroquinone in a biotransformation process in Origanum majorana L. shoot culture. Phytochemistry Letters 20 (2017) 443-448. https://doi.org/10.1016/j.phytol.2017.01.009

[10] E. Skrzypczak-Pietraszek, K. Piska, J. Pietraszek, Enhanced production of the pharmaceutically important polyphenolic compounds in Vitex agnus castus L. shoot cultures by precursor feeding strategy. Engineering in Life Sciences 18 (2018) 287-297. https://doi.org/10.1002/elsc.201800003

[11] Z. Ignaszak, P. Popielarski, T. Strek, Estimation of coupled thermo-physical and thermo-mechanical properties of porous thermolabile ceramic material using Hot Distortion Plus® test. Defect and Diffusion Forum 312-315 (2011) 764-769. https://doi.org/10.4028/www.scientific.net/DDF.312-315.764

[12] A. Tiziani, A. Molinari, J. Kazior, G. Straffelini, Effect of vacuum sintering on the mechanical-properties of copper-alloyed stainless-steel. Powder Metall. 22 (1990) 17-19.

[13] E. Radzyminska-Lenarcik, R. Ulewicz, M. Ulewicz, Zinc recovery from model and waste solutions using polymer inclusion membranes (PIMs) with 1-octyl-4-methylimidazole, Desalin. Water Treat. 102 (2018) 211-219. https://doi.org/10.5004/dwt.2018.21826

[14] D. Klimecka-Tatar, Electrochemical characteristics of titanium for dental implants in case of the electroless surface modification. Arch. Metall. Mater. 61 (2016) 923-26. https://doi.org/10.1515/amm-2016-0156

[15] J. Korzekwa, W. Skoneczny, G. Dercz, M. Bara, Wear mechanism of Al2O3/WS2 with PEEK/BG plastic. J. Tribol.-Trans. ASME 136 (2014) art. 011601. https://doi.org/10.1115/1.4024938

[16] S. Wojciechowski, D. Przestacki, T. Chwalczuk, The evaluation of surface integrity during machining of Inconel 718 with various laser assistance strategies. MATEC Web of Conf. 136 (2017) art. 01006. https://doi.org/10.1051/matecconf/201713601006

[17] N. Radek, A. Szczotok, A. Gadek-Moszczak, R. Dwornicka, J. Broncek, J. Pietraszek, The impact of laser processing parameters on the properties of electro-spark deposited coatings. Arch. Metall. Mater. 63 (2018) 809-816.

[18] T. Styrylska, J. Pietraszek, Numerical modeling of non-steady-state temperature-fields with supplementary data. ZAMM 72 (1992) T537-T539.

[19] Radek, N., Kurp, P., Pietraszek, J., Laser forming of steel tubes. Technical Transactions 116 (2019) 223-229. https://doi.org/10.4467/2353737XCT.19.015.10055

[20] J. Pietraszek, A. Gadek-Moszczak, The Smooth Bootstrap Approach to the Distribution of a Shape in the Ferritic Stainless Steel AISI 434L Powders. Solid State Phenomena 197 (2012) 162-167. https://doi.org/10.4028/www.scientific.net/SSP.197.162

[21] A. Gadek-Moszczak, J. Pietaszek, B. Jasiewicz, S. Sikorska, L. Wojnar, The Bootstrap Approach to the Comparison of Two Methods Applied to the Evaluation of the Growth Index in the Analysis of the Digital X-ray Image of a Bone Regenerate. New Trends in Comp. Collective Intell. 572 (2015) 127-136. https://doi.org/10.1007/978-3-319-10774-5_12

Terotechnology XI
Materials Research Proceedings **17** (2020) 16-22

Materials Research Forum LLC
https://doi.org/10.21741/9781644901038-3

# Improving the Non-Destructive Test by initiating the Quality Management Techniques on an Example of the Turbine Nozzle Outlet

SIWIEC Dominika[1,a*], DWORNICKA Renata[2,b] and PACANA Andrzej[3,c]

[1]Rzeszow University of Technology, Faculty of Mechanical Engineering and Aeronautics Rzeszów, Poland; orcid id: 0000-0002-6663-6621

[2]Cracow University of Technology, Faculty of Mechanical Engineering, Kraków, Poland, orcid: 0000-0002-2979-1614

[3]Rzeszow University of Technology, Faculty of Mechanical Engineering and Aeronautics Rzeszów, Poland, orcid id: 0000-0003-1121-6352

[a]d.siwiec@prz.edu.pl, [b]renata.dwornicka@mech.pk.edu.pl, [c]app@prz.edu.pl

**Keywords:** Mechanical Engineering, Turbine Nozzle Outlet, Quality Management, Non-Destructive Test, Fluorescent Method

**Abstract.** The NDT methods are an effective way to make a quality analysis of the product, mainly in the case of the aviation industry. However, although these studies effectively identify unconformities, they do not indicate the source of their occurrence. The aim was to analyze the quality of the product (turbine nozzle outlet) using the fluorescent method and identify the root of the unconformity by using the quality management technique sequence. The Ishikawa diagram and the 5Why method were the techniques used in the study and the turbine nozzle outlet made from 410 steel was its subject. The product using the fluorescent method was analyzed, after which the unconformity was identified (porosity cluster). By using the Ishikawa diagram, the potential causes were identified, from which two main causes were selected (molding sand and errors during production). The 5Why method was used, by means of which the root cause was identified - it was a faulty material form a supplier. On the example of the fluorescent method, it was shown that using the non-destructive test (NDT) in a sequential way with quality management techniques (Ishikawa diagram and 5Why method) allows for making a complex quality analysis of the product and identifying the root of eventual unconformity. The proposed technique sequence (fluorescent method, Ishikawa diagram and 5Why method) can be applied to analyze other products and other unconformities in production and service enterprises.

**Introduction**

Creating the quality of products required by the customer initiates the need for comprehensive qualitative analyses [1, 2]. It is very important in the case of the production and aviation industry, in which using the non-destructive test (NDT) is the main way to identify unconformities. These tests allow for identifying the unconformity without destroying the product [3]. One of the NDT methods is the penetration method, which includes the fluorescent method (FPI). This method applies mainly to aviation products [4]. Also, with the FPI method the tests were performed on the analyzed product (turbine nozzle outlet). A literature review of the selected literature positions shows that the fluorescent method was used in testing the quality of a product surface [5, 6] and detecting nonconformities that are the basis for further analysis [7]. Its efficiency was improved by using additional machines, substances [8] and was compared with other methods [9]. It was concluded that the area of NDT practice with quality management

Terotechnology XI                                                    Materials Research Forum LLC
Materials Research Proceedings **17** (2020) 16-22                https://doi.org/10.21741/9781644901038-3

techniques, in order to identify the root of pointed unconformity is not sufficiently discussed. Therefore, it was important to show that using the NDT method in a sequential way (on the example of the fluorescent method) with quality management techniques (Ishikawa diagram and 5Why method) allows for making a complex quality analysis of a product and identifying the root of eventual unconformity. Improvement of the NDT process is very important in the case of developing enterprises because, for these enterprises, it is not enough only to identify the unconformity, but they also need to identify their root. The proposition of the improvement of the NDT process on the fluorescent method in an enterprise localized in south-eastern Poland was made. In the enterprise, the non-destructive test was used, i.e. magnetic-powder and fluorescent method. After identifying the unconformity, the root cause of creation was not analyzed. In order to increase the effectiveness of the product analysis, application of the NDT with the Ishikawa diagram and the 5Why was proposed. The aim was to analyze the quality of the product (turbine nozzle outlet) using the fluorescent method and identify the root of the unconformity by using the quality management technique sequence. These techniques were the Ishikawa diagram and the 5Why method. By means of the fluorescent method, the unconformity (porosity cluster) was identified. By means of the Ishikawa diagram, the potential causes were identified, from which two main causes were selected, i.e. molding sand and errors during production. In the last step of the analysis, the 5Why method was used, by means of which the root cause was identified (it was the faulty material from the supplier). It was shown that using the NDT method in a sequential way (on the example of the fluorescent method) with quality management techniques (Ishikawa diagram and 5Why method) allows for making a complex quality analysis of the product and identifying the root of eventual unconformity.

In general, combining NDT methods with root cause analysis can be a very useful procedure also in other industrial areas, including mechanics [10, 11], hydraulics of working machines [12], heat flows [13], as well as diagnostics of modified surfaces [14, 15], including plasma-sprayed coatings [16] or electro-spark deposition ESD combined with laser machining [17, 18]. This approach may also be useful in the biotechnology industry [19, 20], where the cause-effect relationships of the processes are very complicated due to numerous feedbacks and the occurring circular processes. The inference scheme resulting from this approach should be taken into account in related data analyzes, e.g. stereological [21, 22] and statistical [23, 24].

## Material

The turbine nozzle outlet, used in the aviation industry, was the subject of the research. The choice of object for the analysis was determined by the type of unconformity that was identified on the product by the fluorescence method (cluster of porosity). In the enterprise, a large number of these types of unconformity (porosity cluster) existed. That is why it was decided that it is preferred to make the analysis of this type of problem. The turbine nozzle outlet was made from 410 steel. It is a martensitic chrome stainless steel resistant to corrosion [25, 26]. Martensitic steels are used to manufacture structural elements [25, 27]. They are characterized by excellent mechanical properties and good corrosion resistance. Selected properties of 410 steel are shown in the subject of the literature [25, 27, 28].

## Method

Product testing (turbine nozzle outlet) was carried out using the fluorescence method (FPI), due to the requirements of the customer who commissioned the product analysis using this method and the material from which the product was made. The fluorescent method allows for identifying the unconformity, in turn, the technique sequence i.e. the Ishikawa diagram and the 5Why method allow for pointing the root of unconformity. Therefore, it is effective to make the quality analysis of the product (identify the unconformity and root of the unconformity) by the connection of the NDT methods with quality management techniques [29-31]. The fluorescent method is one of penetrant tests. In FPI, a penetrant has colorant, whose indicators can be identified by ultraviolet radiation. To make this possible, it is necessary to darken the test stand [29, 32]. The methodology of conducting fluorescence research was presented in the literature [29]. By using the FPI method, the unconformity was identified (porosity cluster), and in order to identify the root of this unconformity the sequence of the Ishikawa diagram and 5Why method were used. The Ishikawa diagram (fishbone diagram) is a technique, by means of which it is possible to analyze the problem, and next point the potential causes and select the main causes [2, 33, 34]. For the analysis of the problem with the cluster of porosity on the turbine nozzle outlet, the problem was noted in the main part of the Ishikawa diagram. The main categories of Ishikawa (5M+E), which applied to this problem, were selected for its analysis. They were the following: man, method, material, machine, management and environment [1, 29-31]. To each of the categories, the intermediate causes were noted, from which the main causes were selected. Next, the analysis of the problem using the 5Why method was made. The 5Why (Why-Why method) is an effective way to identify the root cause of the problem. In the first step, the problem (porosity cluster) and the main causes of the problem were pointed. Next, the „Why?" question was made in a sequential way. The end of analysis was at the moment when the answer pointed that it was possible to make the improvement action, adequate to the problem [1, 2, 29, 31, 35]. After identifying the source of the problem, improvement actions were proposed.

**Results**
The quality of the turbine nozzle outlet was analyzed using a fluorescent method. After the analysis, the unconformity was identified (porosity cluster), which is shown in Fig. 1.

*Fig. 1. The cluster of porosity on the turbine nozzle outlet.*

The problem (porosity cluster on the turbine nozzle outlet) was analyzed using the Ishikawa diagram, which is shown in Fig. 2. From the potential causes shown in the criteria (5M+E), two causes were selected, i.e. molding sand and errors during production. In order to identify the root of the problem, the next analysis was made using the 5Why method shown in Fig. 3. It was concluded that the root cause of the porosity cluster on the turbine nozzle outlet was the faulty material from the supplier. The improvement actions, which were made in order to eliminate or minimalize the cause of the problem (porosity cluster) are to inform the customer about the root cause.

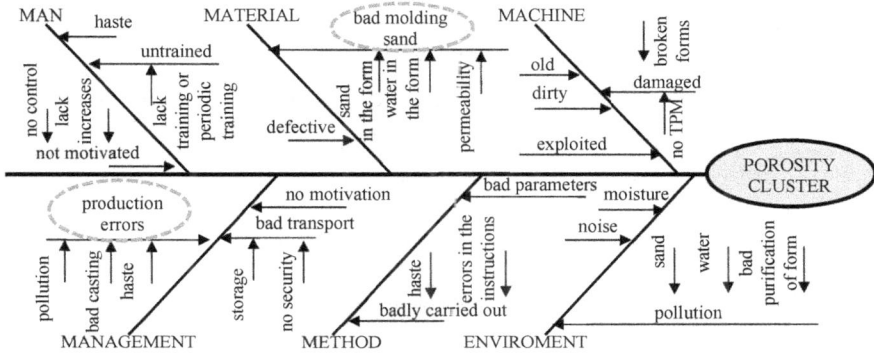

Fig. 2. The Ishikawa diagram for the porosity cluster on the turibne nozzle outlet.

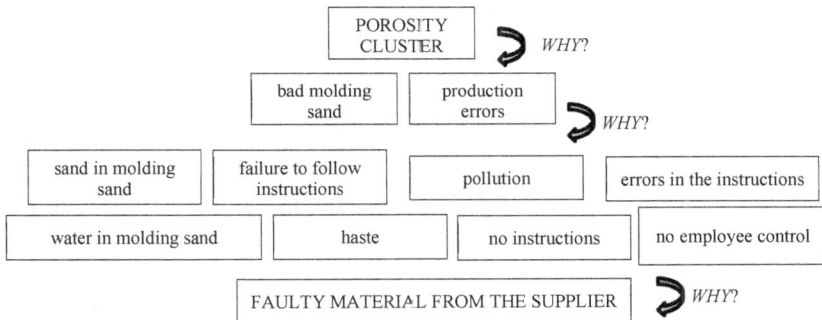

Fig. 3. The 5Why method for the porosity cluster on the turbine nozzle outlet.

## Summary

The NDT methods are an effective way to make a quality analysis of the product, mainly in the case of the aviation industry. However, although these studies effectively identify unconformities, they do not indicate the source of their occurrence. Therefore, it is purposeful to improve the non-destructive testing process by initiating quality management techniques. This was done in a production and service enterprise located in south-eastern Poland. The aim was to analyze the quality of the product (turbine nozzle outlet) using the fluorescent method and identify the root of the unconformity by using the quality management technique sequence. These techniques were the Ishikawa diagram and the 5Why method which, when used in a sequential way, allow for making an effective analysis whose results show the root cause of the problem. A turbine nozzle outlet made from 410 steel was the subject of the study. The problem (porosity cluster) was identified using the FPI method. By using the Ishikawa diagram, the potential causes of the problem were identified, i.e. bad molding sand and production errors. By

Terotechnology XI
Materials Research Proceedings **17** (2020) 16-22

Materials Research Forum LLC
https://doi.org/10.21741/9781644901038-3

using the 5Why method the root cause of the problem was identified – faulty material from the supplier. The improvement actions, which were made in order to eliminate or minimalize the cause of the problem (porosity cluster) are to inform the customer about the root cause. It was shown that using the NDT in a sequential way (on the example of the fluorescent method) with quality management techniques (Ishikawa diagram and 5Why method) allows for making a complex quality analysis of the product and identifying the root of eventual unconformity. The proposed technique sequence (fluorescent method, Ishikawa diagram and 5Why method) can be applied to analyze other products and other unconformities in production and service enterprises.

## References

[1] R. Ulewicz, Quality control system in production of the castings from spheroid cast iron, Metalurgija 42(1) (2003) 61-63.

[2] A. Pacana, L. Bednárová, I. Liberko etal., Effect of selected production factors of the stretch film on its extensibility, Przemysl Chemiczny 93(7) (2014) 1139-1140

[3] J. Zheng, W. F. Xie, M. Viens et al., Design of an advanced automatic inspection system for aircraft parts based on fluorescent penetrant inspection analysis, INSIGHT 57(1) (2015) 18-34. https://doi.org/10.1784/insi.2014.57.1.18

[4] L. J. H. Brasche, R. Lopez, D. Eisenmann, Characterization of developer application methods used in fluorescent penetrant inspection, Review of Progress in Quantitative Nondestructive Evaluation. Vols. 25a and 25b, AIP Conference Proceedings 820 (2006) 598-605. https://doi.org/10.1063/1.2184582

[5] H. Fischer, F. Karaca, R. Marx, Detection of microscopic cracks in dental ceramic materials by fluorescent penetrant method, Journal of Biomedical Materials Research 61(1) (2002) 153-158. https://doi.org/10.1002/jbm.10148

[6] N. J. Shipway, P. Huthwaite, M. J. S. Lowe et al., Performance Based Modifications of Random Forest to Perform Automated Defect Detection for Fluorescent Penetrant Inspection, Journal of Nondestructive Evaluation 38(2) (2019) 1-11. https://doi.org/10.1007/s10921-019-0574-9

[7] J. Zheng, W. F. Xie, M. Viens, M. et al., Design of an advanced automatic inspection system for aircraft parts based on fluorescent penetrant inspection analysis, INSIGHT 57(1) (2015) 18-34. https://doi.org/10.1784/insi.2014.57.1.18

[8] K. Daneshvar, B. Dogan, Application of quantum dots as a fluorescent-penetrant for weld crack detection, Materials at High Temperatures 27(3) (2010) 179-182. https://doi.org/10.3184/096034010X12813744660988

[9] Y. Guo, F. R. Ruhge, Comparison of detection capability for acoustic thermography, visual inspection and fluorescent penetrant inspection on gas turbine components, Review of Progress in Quantitative Nondestructive Evaluation, (28A and 28B), AIP Conference Proceedings, 1096 (2009) 1848. https://doi.org/10.1063/1.3114183

[10] G. Filo, E. Lisowski, M. Domagala, J. Fabis-Domagala, H. Momeni, Modelling of pressure pulse generator with the use of a flow control valve and a fuzzy logic controller. MSM 2018 14th Int. Conf. Mechatronic Systems and Materials, AIP Conference Proceedings, vol. 2029, art. 020015-1. https://doi.org/10.1063/1.5066477

Terotechnology XI                                                              Materials Research Forum LLC
Materials Research Proceedings **17** (2020) 16-22                          https://doi.org/10.21741/9781644901038-3

[11] P. Krawiec, M. Grzelka, J. Kroczak, G. Domek, A. Kolodziej, A proposal of measurement methodology and assessment of manufacturing methods of nontypical cog belt pulleys. Measurement 132 (2019) 182-190. https://doi.org/10.1016/j.measurement.2018.09.039

[12] M. Domagala, H. Momeni, J. Domagala-Fabis, G. Filo, M. Krawczyk, J. Rajda, Simulation of particle erosion in a hydraulic valve. Materials Research Proceedings 5 (2018) 17-24. https://doi.org/10.21741/9781945291814-4

[13] L. J. Orman, Boiling heat transfer on meshed surfaces of different aperture. AIP Conf. Proc. 1608 (2014) 169-172. https://doi.org/10.1063/1.4892728

[14] D. Przestacki, M. Kuklinski, A. Bartkowska, Influence of laser heat treatment on microstructure and properties of surface layer of Waspaloy aimed for laser-assisted machining. Int. J. Adv. Manuf. Technol. 93 (2017) 3111-3123. https://doi.org/10.1007/s00170-017-0775-2

[15] S. Wojciechowski, D. Przestacki, T. Chwalczuk, The evaluation of surface integrity during machining of Inconel 718 with various laser assistance strategies. MATEC Web of Conf. 136 (2017) art. 01006. https://doi.org/10.1051/matecconf/201713601006

[16] W. Zorawski, R. Chatys, N. Radek, J. Borowiecka-Jamrozek, Plasma-sprayed composite coatings with reduced friction coefficient. Surf. Coat. Technol. 202 (2008) 4578-4582. https://doi.org/10.1016/j.surfcoat.2008.04.025

[17] N. Radek, K. Bartkowiak, Laser treatment of electro-spark coatings deposited in the carbon steel substrate with using nanostructured WC-Cu electrodes. Physics Procedia. 39 (2012) 295-301. https://doi.org/10.1016/j.phpro.2012.10.041

[18] I. Pliszka, N. Radek, A. Gadek-Moszczak, P. Fabian, O. Paraska, Surface improvement by WC-Cu electro-spark coatings with laser modification. Materials Research Proceedings 5 (2018) 237-242. https://doi.org/10.5604/01.3001.0010.5906

[19] E. Skrzypczak-Pietraszek, A. Hensel, Polysaccharides from Melittis melissophyllum L. herb and callus. Pharmazie 55 (2000) 768-771.

[20] E. Skrzypczak-Pietraszek, A. Urbanska, P. Zmudzki, J. Pietraszek, Elicitation with methyl jasmonate combined with cultivation in the Plantform™ temporary immersion bioreactor highly increases the accumulation of selected centellosides and phenolics in Centella asiatica (L.) Urban shoot culture. Engineering in Life Sciences. 19 (2019) 931-943. https://doi.org/10.1002/elsc.201900051

[21] A. Szczotok, D. Karpisz, Application of two non-commercial programmes to image processing and extraction of selected features occurring in material microstructure. METAL 2019: 28th Int. Conf. on Metallurgy and Materials, Ostrava, TANGER, 1721-1725. https://doi.org/10.37904/metal.2019.971

[22] L. Wojnar, A. Gadek-Moszczak, J. Pietraszek, On the role of histomorphometric (stereological) microstructure parameters in the prediction of vertebrae compression strength. Image Analysis and Stereology 33 (2019) 63-73. https://doi.org/10.5566/ias.2028

[23] J. Pietraszek, E. Skrzypczak-Pietraszek, The optimization of the technological process with the fuzzy regression. Adv. Mater. Res-Switz. 874 (2014) 151-155. https://doi.org/10.4028/www.scientific.net/AMR.874.151

[24] J. Pietraszek, A. Gadek-Moszczak, T. Torunski, Modeling of Errors Counting System for PCB Soldered in the Wave Soldering Technology. Advanced Materials Research 874 (2014) 139-143. https://doi.org/10.4028/www.scientific.net/AMR.874.139

[25] D. C. Moreira., H. C. Furtado, J. S. Buarque, et al., Failure analysis of AISI 410 stainless-steel piston rod in spillway floodgate, Engineering Failure Analysis 97 (2019) 506-517. https://doi.org/10.1016/j.engfailanal.2019.01.035

[26] F. Hejripour,D. K. Aidun, Consumable selection for arc welding between Stainless Steel 410 and Inconel 718, Journal of Materials Processing Technology 245 (2017) 287-299. https://doi.org/10.1016/j.jmatprotec.2017.02.013

[27] M. Moradi, H. Arabi., S. J. Nasab et al., A comparative study of laser surface hardening of AISI 410 and 420 martensitic stainless steels by using diode laser, Optics and Laser Technology 111 (2019) 347-357. https://doi.org/10.1016/j.optlastec.2018.10.013

[28] T. S. Oliveira, E. S Silva, S. F Rodrigues et al., Softening Mechanisms of the AISI 410 Martensitic Stainless Steel Under Hot Torsion Simulation, Materials Research-Ibero-American Journal of Materials 20(2) (2017) 395-406. https://doi.org/10.1590/1980-5373-mr-2016-0795

[29] A. Pacana, D. Siwiec, L. Bednárová, Analysis of the incompatibility of the product with fluorescent method, Metalurgija 58 3-4 (2019) 337-340.

[30] D. Siwiec, L. Bednarova, A. Pacana., M. Zawada, M. Rusko, Decision support in the selection of fluorescent penetrants for industrial non-destructive testing, Przemysl Chemiczny 98(10) (2019) 1594-1596.

[31] A. Pacana, A. Radon-Cholewa, J. Pacana J. et al., The study of stickiness of packaging film by Shainin method, Przemysl Chemiczny 94(8) (2015) 1334-1336.

[32] http://www.e-spawalnik.pl/?wybor-metody-badannieniszczacych,168 (access: 11.10.2019)

[33] D. Malindzak, et al., An effective model for the quality of logistics and improvement of environmental protection in a cement plant, Przemysl Chemiczny 96(9) (2019) 1958-1962.

[34] R. Wolniak, Application methods for analysis car accident in industry on the example of power, Support Systems in Production Engineering 6(4) (2017) 34-40.

[35] Radek, N., Kurp, P., Pietraszek, J., Laser forming of steel tubes. Technical Transactions 116 (2019) 223-229. https://doi.org/10.4467/2353737XCT.19.015.10055

Terotechnology XI
Materials Research Proceedings 17 (2020) 23-30

Materials Research Forum LLC
https://doi.org/10.21741/9781644901038-4

# The Conditions for Application of Foundry Simulation Codes to Predict Casting Quality

POPIELARSKI Paweł[1, a *]

[1]Poznan University of Technology, 3 Piotrowo street 60-965 Poznan, Poland

[a]pawel.popielarski@put.poznan.pl

**Keywords:** Casting, Simulation Codes, Casting Defects, Validation, Thermal Properties

**Abstract.** Casting processes are widely used to produce metal components wherein the cast iron castings represent more than 70% of the world production of castings. Designing a new casting technology requires incurring large costs associated with the preparation of instrumentation necessary to perform casting moulds. Therefore, the simulation codes currently applied in the foundry industry are used primarily to optimize the casting quality, quality mainly connected with the defects location such as shrinkage porosity. In this case, it is very important for the simulation code user to master the phase of pre-processing, which is the best possible, corresponding to the actual casting-mould system, formulation of the model which along with the relevant differential equations also includes defined certain conditions (geometric conditions, the physical parameters of casting-mould, initial and boundary conditions). The lack of as complete as possible identification of these values, used in modeling dependencies, is the cause of limitation of the development and scope of models describing casting solidification - which sometimes translates into a foundry's negative attitude to the usefulness of the simulation codes, because of incorrect predictions on casting quality. Correct model installation and the use of a database corresponding to the model are the development condition of the simulation code in the foundry practice. The paper describes the utilitarian aspects connected with these problems.

## Introduction

Casting of metals and alloys is a method of production, called simply casts, which gives wide possibilities of their geometric shape and simultaneously the impact on their local functional features, impossible to achieve in this area (on the sections of the walls) with other materials processing technology [1-3]. Conscious control of a structure, and thus functional features of castings, in particular for control of mechanical properties [4-6], is made possible by a synergistic linkage of theoretical and practical knowledge of metallurgical processes and different casting techniques in the context of application in practice.

In the classical approach for technology design, manufacturing of castings of good (assumed by the designer / user) quality, with the lowest possible amount of defects even below the threshold of, the so called, permitted defects in accordance with the usual and currently applied procedure, requires specimen castings to check (by means of control tests, including non-destructive testing on castings, their fragments, cast-on test bars) the effectiveness of this technology. Making specimen castings in the case of the assumed correction of casting technology increases the cost of production preparation and greatly affects the time for obtaining the first castings for the sale, which is of particular importance in relation to the prototype castings, and especially single castings. Therefore, a rational procedure of optimization of design of casting technology has involved the use of computer systems to support this process, also known in foundry as simulation codes (hard modelling). Technology design using simulation

Terotechnology XI                                                Materials Research Forum LLC
Materials Research Proceedings **17** (2020) 23-30        https://doi.org/10.21741/9781644901038-4

codes is increasingly supported by soft modelling, i.e. the area of mathematical modelling based on empirical data set using artificial intelligence algorithms (Data Mining) [7, 8].

**The place of simulation codes in the foundry industry**

With the development of microelectronics and the increase in computing power in the 90s of the twentieth century, the mechanical engineering industry gradually introduced CAD / CAM systems in the field of machining and CAD / CAE programs to support design activities by means of simulation. Founding with aided design and manufacturing of castings was without doubt a pioneer in the area of simulation.

In the area of founding, systems based on geometric solid models used in foundries inevitably created conditions for the use of full virtualization of the casting process by technologists. A further significant increase in computing power favored the creation and proposing within the so-called up-grades, new modules, extensions of CAE systems, which gradually used algorithms based on new models of physical phenomena processes (the so-called hard modelling described by differential equations and soft modelling, using empirical formulas), which enabled getting a visual image of more and more phenomena that accompany casting. More and more advanced systems were used, such as CAE (simulation codes) ProCAST, QuickCAST, MagmaSOFT, NovaFlow&Solid, Simtec, Calcosoft, CastCAE, Vulcan and others, including increasingly more complex algorithms, numerical solutions using FDM (Finite Difference) and / or FEM (finite element). There was a gradual transition from the classical "paper" casting technology design to the more and more often implemented design technology with computer-aided casting which became a standard in foundry.

Simulation codes currently used in the foundry industry are primarily used to predict casting quality, quality mainly connected with the location of defects such as shrinkage (voids of shrinkage origin). Prediction of zones exposed to other casting defects (i.e. erosion of mould, the presence of non-metallic inclusions, zones exposed to "hot tears", to penetration of the mould by the liquid alloy) takes place on the basis of models - empirical formulas (called soft) or indirectly the user's knowledge and analysis of the results of simulation, for example, speed field of metal stream in the mould cavity or the time-temperature image of cast and mould interaction. These activities are the basis for decisions concerning the selection of optimal casting technology, with the expectation of obtaining the final, acceptable version of the concept, taking into account the criterion of the best relationship of quality / price of the casting. This made it possible to eliminate the classical method of *trial and error* commonly used over the years in the design process of casting technology, completed only through intuition of engineers and experimental tests [9]. While technologists with a simulation system and with the results of calculations using this tool, take decision on the basis of their assessment and based on the above, they suggest the next version after the change of technology / casting design or approve designed technology [9]. The simulation code is assumed to enable the verification of the (first or subsequent) technology concept and it is therefore necessary to have confidence in the quality of predictions analyzed in the framework of the so-called post-processing. The importance of professionally implemented experimental validation which increases the probability of forecasting accuracy and accumulation of knowledge on the subject should be reiterated at this point. The relations of validation and virtualization are an endless challenge and, to a large extent, they jointly decide about the success of the application code as a supportive tool.

The situation is different in the case of the thermo-physical parameters of cast alloy and mould and boundary conditions, where significant errors may appear, often resulting from the lack of material characteristics (thermo-physical) which are adequate to reality and

Terotechnology XI                                                   Materials Research Forum LLC
Materials Research Proceedings **17** (2020) 23-30            https://doi.org/10.21741/9781644901038-4

simplifications used in the model, which, as already stated, critically affects the accuracy of a forecasting simulation. Deviations in the accuracy of forecasts are most commonly caused by material data mismatches in the conditions of production in the foundry. The problem of the sensitivity of the model to the errors of coefficients has been developed, among others, in [10-12]. The creators of casting simulation codes, as a principle (customary) do not take responsibility for the quality of the parameters contained in the database and at the same time they suggest modifications of material data provided there and supplementing the database with material parameters and the boundary conditions corresponding to the material and assortment conditions of casting in the particular foundry.

**The problem of efficient defining and introducing certain conditions in the issues of modelling phenomena in the casting mould system**
The simulation model contains a system of differential equations for solutions with the FDM finite difference method or FEM finite elements method and must also include certain properly defined conditions (geometric conditions, the physical parameters of a casting-mould system, initial conditions and boundary conditions). The credibility of simulation results depends on the credibility of these certain conditions.

It should be remembered that a number of responsible decisions related to defining and introducing certain conditions must be taken between the stage of defining the basis and assumptions for modeling and obtaining credible results (awareness of the influence of certain conditions on the margin of error of conclusion). Synergy of knowledge is consistently required, its usefulness is significant at any stage, including inter alia, the stage of creating and supplementing the material data base (the physical parameters of the casting-mould system) [9, 13, 14].

The effective use of a simulation system requires identification and knowledge of physical parameters (thermal) of the casting-mould system which are adapted to reality. The lack of as complete as possible identification of these values, used in modeling dependencies, is the cause of limitation of the development and scope of models describing casting solidification [1, 9, 15, 16], which sometimes translates into a foundry's negative attitude to the usefulness of the simulation codes, because of incorrect predictions on casting quality.

The sets of coefficients in handbooks and manuals are, for users of simulation codes, the first source of acquisition of thermo-physical data formally needed for simulation calculations [17-20]. While singing them, users should note that the coefficients as the research results presented in these sources were determined in conditions that were often not corresponding to the conditions of the actual time-temperature casting process (not going into details as to the type of alloy and forms). So one should pay special attention to the usefulness of data used for the calculation and simulation methods and look for their completing and verification methods.

Validation of a model is a condition absolutely required and necessary in view of the diversity of materials used for mould manufacturing. This validation is based on the effective adjustment of model components with parameters of processes occurring in reality and connected with the tests of different variants, which allows for the margin of error to be determined. This validation determines the usefulness of applying a simulation system for solving models and for optimizing casting technology. Validation of material data determined in this way would necessitate a simultaneous validation of the whole model (comparison of the results of simulation and of real casting, Fig. 1) [21]. In the case of thermal casting processes, such a procedure may be called a numerical identification of thermo-physical data. In the procedure of simplifying a physical

Terotechnology XI
Materials Research Proceedings **17** (2020) 23-30

Materials Research Forum LLC
https://doi.org/10.21741/9781644901038-4

model, it is important to take temperature–time conditions, simplifications of geometrical model and mesh design into account.

In conclusion, in order to assess the quality of available thermo-physical databases, the user must take into account their future purpose, and therefore consider the following issues:

- the source and the conditions of determining thermo-physical coefficients, such as accuracy of their determination (if possible), a comparison with the time and temperature range characteristic for specific conditions of the casting process,
- criteria and selection of the best (according to the user's and / or consultant's of a foundry knowledge) data set to solve the problem,
- the sensitivity of simulation results (uncertainty) - assessed in the static and dynamic manner - to the dispersion of coefficients (resulting from errors), which is identifiable by the user by means of simulation tests [10, 11].

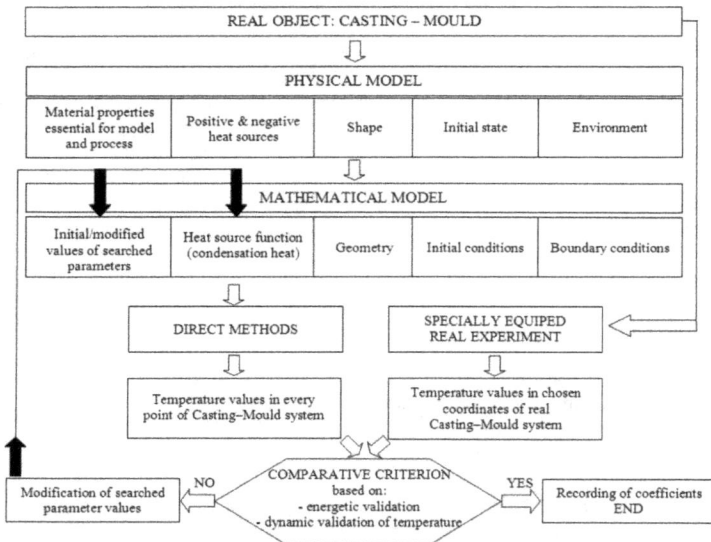

*Fig. 1. Scheme of energetic/dynamic validation with the comparison criterion [21]*

The problem of missing or unreliable coefficients in databases, essential for modeling processes, concerns not only the hard models (based on differential equations of elementary processes such as fluid and heat flows, diffusion, mechanical impact - stress), but also occurs in the case of the use of soft models (empirical equations including modeling of nucleation and grain growth, supply and the formation of shrinkage defects, local structures and their mechanical properties, expansion movements) [6, 22-25].

Initial conditions, which are one of the two remaining certain conditions, are relatively easier to identify and define, especially when the virtualization of a casting process includes the stage of pouring (it is enough to provide the temperature of metal stream entering the gating system). The last condition - the boundary condition, and actually boundary conditions - as they should be related to each pair of contacting elements of the casting-mould system - should be considered in terms of the significance of the effect of thermal resistance on the control surfaces (the problem of shrinkage slot variable in time) on weakening the heat flux in relation to the thermal resistance of the layers adjacent to the two surfaces in contact and resulting from the material thermos-physical parameters. Thus, for example, a correctly estimated value of thermal resistance is much more important and necessary to determine for the casting-metal mould contact than the casting-sand mould contact.

## Casting process simulation

Modeling and simulation of complex phenomena during crystallization of alloys in the simulation codes are based on mathematical and physical models, which in different ways reflect the complexity of physical and chemical processes and their coupling. Simplifications they are subjected to are not in a sufficiently comprehensive manner disclosed to users in help windows and in manuals of codes. While developing the project of experimental validation of casting simulation codes, it must be assumed that the scope of these simplifications is assigned unambiguously to the code that is used. And, that this also applies to "soft" models within the compliance of virtual and actual result of simulation calculations, most commonly the predictions of porosity of shrinkage origin.

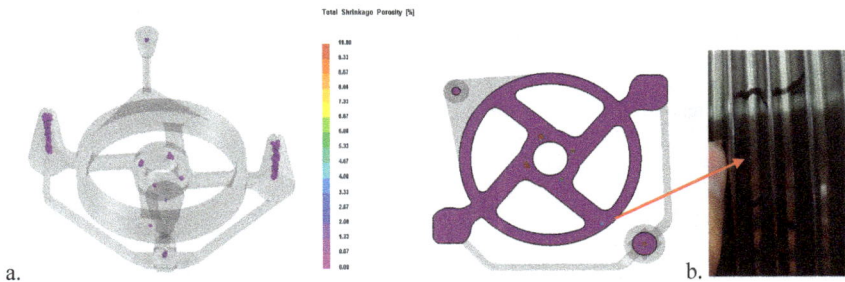

Fig. 2. Shrinkage porosity: a) results of the simulation – Procast,
b) shrinkage porosity visible on the real casting surface after machining

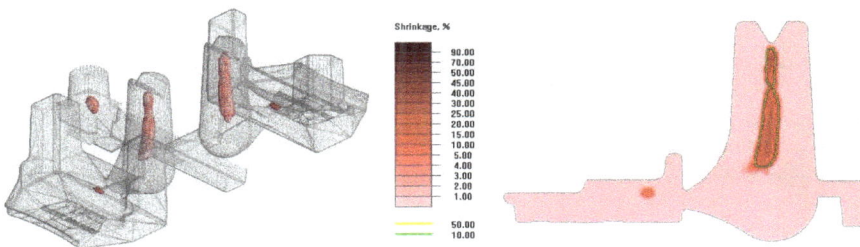

Fig. 3. Results of the simulation (shrinkage porosity) - NovaFlow&Solid

Thus, the basic expectation formulated by users (especially beginners) to process a virtualization system, as to confirm the effectiveness of pre-developed technology, is to check its correctness in terms of the exclusion of defects of shrinkage origin. Prediction of these defects and comparing them with the acceptance criterion formulated in the conditions of acceptance by the customer, is to admit the casting for the production (by initially developed  and then corrected version of technology).

The thermo-physical parameters determined on the basis of the described validation procedure were used for a simulation of casting process. The results of the simulation using the NovaFlow&Solid and Procast codes (forecasting of shrinkage porosity) were compared with the real casting. The results of simulation are presented in the Fig. 2 and Fig. 3.

**Conclusions**

In conclusion, the proper use of the casting simulation code requires knowledge, a proper understanding and recognition of procedures to identify the parameters of a modeled thermal phenomena and appropriate approach to their validation.

Only effective validation activities determine the actual usefulness of a simulation code to optimize the concept of casting technology. The first validation step should always involve the adjustment of thermo-physical properties and boundary conditions in the database of a simulation code to the actual conditions of the casting-mold system, to perform experiments based on thermal analysis of the real-time duration of the process.

The issue of corrective modifications of material databases of each specific code of simulation introduced to the foundry still remains one of the most important conditions for the full use of casting simulation codes.

The results concerning the applicability of the software and relevant material models obtained during the research should also be of interest to other technological areas in which there are thermomechanical issues, e.g. the production of protective coatings by ESD and laser machining [26-28], nanocomposite materials [29], hydraulics of heavy working machines [30] and related methods for the analysis of experimental results [31, 32].

**References**

[1]  B. Ravi, Computer-aided Casting Design and Simulation, STTP, V.N.I.T. Nagpur, July 21, 2009.

[2]  M. Kujawski, P.Krawiec, Analysis of Generation Capabilities of Noncircular Cog belt Pulleys on the Example of a Gear with an Elliptical Pitch Line. J. Manuf. Sci. Eng.-Trans. of the ASME 133(5) (2011) 051006. https://doi.org/10.1115/1.4004866

[3]  P. Krawiec, A. Marlewski, Profile design of noncircular belt pulleys, J. Theor. Appl. Mech. 54 (2) (2016) 561-570. https://doi.org/10.15632/jtam-pl.54.2.561

[4]  K. Gawdzińska, K. Bryll, D. Nagolska, Influence of heat treatment on abrasive wear resistance of silumin matrix composite castings, Arch. Metall. Mater. 61(1) (2016) 177-182. https://doi.org/10.1515/amm-2016-0031

[5]  T. Chwalczuk, M. Wiciak, A. Felusiak, P. Kieruj, An Investigation of Tool Performance in Interrupted Turning of Inconel 718, MATEC Web of Conf. 237 (2018) art. 02008. https://doi.org/10.1051/matecconf/201823702008

[6]  T. Chwalczuk, D. Przestacki, P. Szablewski, A. Felusiak, Microstructure characterisation of Inconel 718 after laser assisted turning, MATEC Web of Conf. 188 (2018) art. 02004. https://doi.org/10.1051/matecconf/201818802004

[7]  Z. Ignaszak, R. Sika, M. Rogalewicz, Contribution to the assessment of the data acquisition effectiveness in the aspect of gas porosity defects prediction in ductile cast iron castings, Arch. Foundry Eng. 18 (2018) 35-40.

[8]  R. Sika, J. Hajkowski, Synergy of modeling processes in the area of soft and hard modeling, MATEC Web of Conf. 121 (2017) art. 04009. https://doi.org/10.1051/matecconf/201712104009

[9]  Z. Ignaszak, J. Hajkowski, P. Popielarski, Examples of new models applied in selected simulation systems with respect to database, Arch. Foundry Eng. 13 (2013) 45-50. https://doi.org/10.2478/afe-2013-0009

[10] X.L. Yang, P.D. Lee, R.F. Brooks, and R. Wunderlich, The Sensitivity of Investment Casting Simulations to the Accuracy of Thermophysical Property Values, TMS Ed., Superalloys, The Minerals, Metals & Materials Society, 2004. https://doi.org/10.7449/2004/Superalloys_2004_951_958

[11] Z. Ignaszak, P. Popielarski, J. Hajkowski, Sensitivity of models applied in chosen simulation systems with respect to database quality for resolving the casting problems, Defect and Diffusion Forum 336 (2013) 135-146. https://doi.org/10.4028/www.scientific.net/DDF.336.135

[12] P. Twardowski, M. Wiciak-Pikuła, Prediction of Tool Wear Using Artificial Neural Networks during Turning of Hardened Steel, Materials 12(19), 3091-1 - 3091-15 (2019). https://doi.org/10.3390/ma12193091

[13] J.-M. Drezet, M. Rappaz, G.-U. Grun, M. Gremaud, Determination of Thermophysical Properties and Boundary Conditions of Direct Chill–Cast Aluminum Alloys Using Inverse Methods, Metallurgical and Materials Trans. A 31 (2000) 1627-1634. https://doi.org/10.1007/s11661-000-0172-5

[14] E. Majchrzak, B. Mochnacki, Identification of Thermal Properties of the System Casting-Mould, Materials Science Forum 539-543 (2007) 2491-2496. https://doi.org/10.4028/www.scientific.net/MSF.539-543.2491

[15] B. Ravi, Casting Simulation – Best Practices, Transactions of 58th IFC, Ahmedabad (2010).

[16] M. Jolly, Casting simulation: How well do reality and virtual casting match? State of art. review, Int. J. Cast Metals Res. 14 (2002) 303-313. https://doi.org/10.1080/13640461.2002.11819448

[17] D.R. Lide, Handbook of chemistry and physics, Boca Raton, CRC Press, 2000.

[18] A. Goldsmith, T. Waterma, H.J. Hirschbaum, Handbook of thermo-physical properties of solid materials, ARMOUR Research Foundation, 1961.

[19] K. Gawdzińska, L. Chybowski, A. Bejger, S. Krile, Determination of technological parameters of saturated composites based on SiC by means of a model liquid, Metalurgija 55 (2016) 659-662.

[20] M. Ratajczak, M. Ptak, L. Chybowski, K. Gawdzińska, R. Będziński, Material and Structural Modeling Aspects of Brain Tissue Deformation under Dynamic Loads. Materials, 12 (2019) art. 271. https://doi.org/10.3390/ma12020271

[21] P. Popielarski, Z. Ignaszak, Effective modelling of phenomena in over-moisture zone existing in porous sand mould subjected to thermal shock, in: Drying and Energy Technologies,

Springer International Publishing, Switzerland, 2016, 181-206. https://doi.org/10.1007/978-3-319-19767-8_10

[22] Z. Ignaszak, P. Popielarski, J. Hajkowski, J.B. Prunier , Problem of Acceptability of Internal Porosity in Semi-Finished Cast Product as New Trend "Tolerance of Damage" Present in Modern Design Office, Defect and Diffusion Forum 326-328 (2012) 612-619. https://doi.org/10.4028/www.scientific.net/DDF.326-328.612

[23] A. Bartkowska, A. Pertek, M. Jankowiak, K. Jóźwiak, Borided layers modyfied by chromium and laser treatment on C45 steel, Arch Metall Mater. 57 (2012) 211-214. https://doi.org/10.2478/v10172-012-0012-9

[24] S. Wojciechowski, D. Przestacki, T. Chwalczuk, The evaluation of surface integrity during machining of inconel 718 with various laser assistance strategies, MATEC Web of Conf. 136 (2017) art. 01006. https://doi.org/10.1051/matecconf/201713601006

[25] D. Przestacki, T. Chwalczuk, The analysis of surface topography during turning of Waspaloy with the application of response surface method, MATEC Web of Conf. 136 (2017) art. 02006. https://doi.org/10.1051/matecconf/201713602006

[26] M. Kukliński, A. Bartkowska, D. Przestacki, Investigation of laser heat treated Monel 400,  MATEC Web of Conf. 219 (2018) art. 02005. https://doi.org/10.1051/matecconf/201821902005

[27] R. Dwornicka, N. Radek, M. Krawczyk, P. Osocha, J. Pobedza, The laser textured surfaces of the silicon carbide analyzed with the bootstrapped tribology model. METAL 2017 26th Int. Conf. on Metallurgy and Materials (2017), Ostrava, Tanger 1252-1257.

[28] N. Radek, A. Szczotok, A. Gadek-Moszczak, R. Dwornicka, J. Broncek, J. Pietraszek, The impact of laser processing parameters on the properties of electro-spark deposited coatings. Arch. Metall. Mater. 63 (2018) 809-816.

[29] E. Piesowicz, I. Irska, K. Bryll, K. Gawdzinska, M. Bratychak, Poly(butylene terephthalate/carbon nanotubes nanocomposites part ii. Structure and properties. Polimery 61 (2016) 24-30. https://doi.org/10.14314/polimery.2016.024

[30] J. Krawczyk, A. Sobczyk, J. Stryczek, P. Walczak, Tests of new methods of manufacturing elements for water hydraulics. Materials Research Proceedings 5 (2018) 200-205.

[31] J. Pietraszek, Fuzzy Regression Compared to Classical Experimental Design in the Case of Flywheel Assembly. In: Rutkowski L., Korytkowski M., Scherer R., Tadeusiewicz R., Zadeh L.A., Zurada J.M. (eds) Artificial Intelligence and Soft Computing ICAISC 2012. Lecture Notes in Computer Science, vol 7267. Berlin, Heidelberg: Springer, 2012, 310-317. https://doi.org/10.1007/978-3-642-29347-4_36

[32] L. Wojnar, A. Gadek-Moszczak, J. Pietraszek, On the role of histomorphometric (stereological) microstructure parameters in the prediction of vertebrae compression strength. Image Analysis and Stereology 38 (2019) 63-73. https://doi.org/10.5566/ias.2028

Terotechnology XI
Materials Research Proceedings **17** (2020) 31-35

Materials Research Forum LLC
https://doi.org/10.21741/9781644901038-5

# The Ex Ante Risk Assessment in the Project in Interval Analysis Description

KOZIEN Ewa[1, a *] and KOZIEN Marek S.[2,b]

[1]Cracow University of Economics, College of Economics, Finance and Law, Institute of Economics, Department of Organization Development, ul. Rakowicka 27, 30-155 Cracow, Poland

[2]Cracow University of Technology, Faculty of Mechanical Engineering, Institute of Applied Mechanics, Al. Jana Pawla II 37, 31-864 Cracow, Poland

[a]koziene@uek.krakow.pl, [b]kozien@mech.pk.edu.pl

**Keywords:** Interval Analysis, Risk Assessment, Project

**Abstract.** Ongoing enterprise projects often have a unique and unrepeatable character. Referring to future solutions, their results may be different from their assumptions and aims. The uncertainty and its description is usually the main problem in quantitative risk estimation goals. Mathematical methods for taking into account the impact of uncertainty are based on the following attempts: deterministic with elements of statistical analysis, stochastic, using fuzzy logic and using interval computations. In risk analysis, the following areas of risks are usually taken into account: project milieu, client and contract, suppliers, maturity of the organization, project characteristics and project team. The aim of the paper is the *ex ante* risk assessment of the project based on six partial risks specified in the scope of project milieu (client, terms and conditions of the contract, suppliers, maturity of an organization, scope of the project, project team) with application of the interval analysis arithmetic for description of uncertainty.

## Introduction

The analysis of risk for projects is not easy to realize in a quantitative form. One of the interesting and useful forms of the quantitative ex ante assessment was proposed by K. Bradley [1]. Application and generalization of the Bradley's method using the risk-list method for risk-assessment in the project was done by the authors in [2,3].

The uncertainty existing in realization of projects is difficult to include in quantitative risk estimation. Mathematical models used for its description are based on the following four attempts: deterministic with elements of statistical analysis, stochastic, using fuzzy logic or using interval computations. Statistical analysis and interpretation was applied by the author for the identification of stage phase growth [4] or for the non-parametric assessment of uncertainty in the analysis of airfoil blade traces [5]. The application of fuzzy logic attempt for the description of uncertainty for the *ex ante* risk assessment in the project was discussed by the authors in [6]. There are combined statistical-fuzzy logic attempts, as it is presented in [7].

The theoretical background of interval arithmetic was formalised in the 60s of the XX century. In this theory, the real value is identified with individually defined range on the real axis. The basic arithmetic operations such as summation, subtraction, multiplication and division can be defined. Moreover, the elementary functions can be defined and used in the interval analysis description [8]. The authors applied the interval arithmetic to choose the values of a dynamical damper [9] or to identify the stage phase in companies [10].

Terotechnology XI                                                                      Materials Research Forum LLC
Materials Research Proceedings **17** (2020) 31-35                    https://doi.org/10.21741/9781644901038-5

In the article, a comparison is made of the results obtained by the application of the interval analysis with the results obtained by the application of statistic and the fuzzy logic attempts for the *ex ante* risk assessment in the project.

Such an analysis may be very useful in many industrial and research fields, in which there is a high uncertainty of activities and the associated high level risk of failure, e.g. heat flow [11], hydraulics of heavy-duty machines [12, 13], surface layer enhancement [14], designing of materials [15, 16] and biomaterials [17]. It should be particularly advantageous in the case of such risky activities as biotechnology [18, 19] or with such fuzzy and unreliable data as in any case where the human factor is involved [20, 21].

**Total risks and partial risks in the project**
Analyses of risk were conducted in many scientific disciplines. In 1964, D.B. Hertz introduced the notion of risk in the context of analysis of uncertainty concerning capital investments [22]. It is worth noting that in 1921, F. Knight pays attention to identify the two notions, namely uncertainty and risk. In his fundamental work entitled *Uncertainty and Profit,* F. Knight separated the *sensu stricto* unmeasurable uncertainty from measurable uncertainty, namely the risk [23, 24].

Risk identification in a project is a dynamic process and requires determining the time of identification. In a project, a risk assessment may be performed:
- *ex ante*, which involves the anticipation of probability and effects of risk occurrence in a project. *Ex ante* risk assessment conducted at an initial stage of project preparation bears impact on a decision to take up implementation of a project,
- *on-line*, refers to ongoing monitoring or partial risks assessed in an initial phase of project implementation as well as identification of new ones, which occurred during its implementation,
- *ex post*, which concerns the final risk assessment connected with closing a project.

The *ex ante* assessment of risk in a project gives the possibility of real management of partial risks in an initial phase of preparation, as well as during the implementation of a project.

To take into account the external proximal project milieu, the following six areas are quantitatively rating: partial risk connected with a client, terms and conditions of a contract, suppliers, as it was proposed by K. Bradley [1] and generalized and applied by the author [2, 6].

**Interval analysis**
Interval analysis based on interval arithmetic is a mathematical theory whose methods give possibility to desribe uncertainty in the value of a generalised number. The uncertainity is taken into account by defining the lower limit $a_{min}$ and upper limit $a_{max}$ of the value *[a]* defined in form (1). Hence, the real (point) value has interval arithmetic interpretation as the point range *[a]=[a,a]*.

$$[a] = [a_{min}, a_{max}] = \{a \in R : a_{min} \leq a \leq a_{max}\} \tag{1}$$

The distance between interval elements can be defined in form (2). In this case, a set of all the real compact ranges of the real set with metrics (2) builds a complete metric space.

$$q([a],[b]) = \max\{|a_{min} - b_{min}|, |a_{max} - b_{max}|\} \tag{2}$$

The arithmetic operations can be defined for the elements of an interval set [8]. Especially operations of summation, used in the analysis, are defined in form (3).

$$[a]+[b]=[a_{min}+b_{min}, a_{max}+b_{max}]$$  (3)

**Comparable analysis**

Application of the above mentioned Bradley's concept related to the six areas [1] is a useful method of ex ante risk analysis in the project. The formulated criteria are related to the following six areas: client, terms and conditions of the contract, suppliers, maturity of the organization, scope of the project and project team. Bradley orders the approval of applicable weight for each of the criteria $w_i$, at the same time proposing the scope of variability of value of these weights. In such a formulation, the value of total project risk coefficient $RP$ is designated according to formula (4), where $r_i$ are values describing the partial risks ($N = 1,...,6$). They are taking values from the range [0,4].

$$RP = \frac{\sum_{i=1}^{N} w_i r_i}{\sum_{i=1}^{N} w_i}$$  (4)

The obtained value of total project risk coefficient $RP$ is a basis for classification of a project as a project with: (Bradley, 2003):

- low risk – for the value of resulting coefficient below 2.0;
- moderate risk - for the value of resulting coefficient from the division [2.0, 2.2];
- high risk - for the value of resulting coefficient from the division [2.2, 2.6];
- very high risk - for the value of resulting coefficient above 2.6.

*Table 1. Partial and total risk interval assessment*

| No | Area of risk estimation | $w_i$ | $r_i$ |
|----|-------------------------|-------|-------|
| 1 | Client | 0.075 | [2,3] |
| 2 | Terms and conditions of the contract | 0.080 | [1,2] |
| 3 | Suppliers | 0.130 | [2.4] |
| 4 | Maturity of the organization | 0.100 | [2,3] |
| 5 | Scope of the project | 0.375 | [1,3] |
| 6 | Project team | 0.240 | [1,3] |
| | **Total project risk estimation coefficient RP** | - | **[1.31,3.05]** |

The analysis was performed following the acceptance of values of partial risks given in Table 1, which are equivalent to those analyzed in the articles [2, 6].

The performed analysis which uses interval analysis allows for the designation of complete project risk assessment, such as the average value of partial risks, to the mid value of the range equal to range of the value of 2.18, which in the descriptive interpretation means moderate risk. The same linguistic description of risk is applied in the risk-list method and by application of the fuzzy logic attempt.

**Summary**

Quantitative *ex* ante risk assessment may be a very useful tool for managers during realization of a project, especially a technical one. A difficult and important problem in estimation is the way of taking into account uncertainty in the obtained values of quantitative score of some criterion. Among a few attempts, application of the interval arithmetic is a convenient solution.

Results of analyses show the same quantity and descriptive (moderate risk) results of risk assessment after the application of a risk-list method with statistical attempt (2.12) [2], the fuzzy login one (2.13) [6] and the interval one (2.18). The fact of sensitivity results of the analyses on the length of the assumed ranges for partial risk estimation parameters should be kept in mind during the application of interval description.

**References**

[1] K. Bradley, The basic properties of the PRINCE2™ methodology, Warszawa, Centrum Rozwiazan Menadzerskich, 2003.

[2] E. Kozien, Using the risk-list method for risk-assessment in the project, in: M. Cingula, D. Rhein, M. Machrafi (eds.), Economic and Social Development (Book of Proceedings), 31st International Scientific Conference on Economic and Social Development, Varazdin Development and Entrepreneurship Agency, Varazdin, 2018, 152-158.

[3] E. Kozien, M.S. Kozien, Ex-ante risk estimation in the project, System Safety: Human-Technical Facility-Environment 1 (1) (2019) 708-715. https://doi.org/10.2478/czoto-2019-0090

[4] E. Kozien, Identification of stage phase growth in the checklist method using different statistical parameters, in: Proc. 20th Int. Sci. Conf. Economic and Social Development, Varazdin, Varazdin Development and Entrepreneurship Agency, 2017, 538-545.

[5] J. Pietraszek, A. Szczotok, M. Kolomycki, N. Radek, E. Kozien, Non-parametric assessment of the uncertainty in the analysis of the airfoil blade traces, in: METAL 2017: 26th Int. Conf. on Metallurgy and Materials, Tanger Ltd., Ostrava, 2017, 1412-1418.

[6] E. Kozien, M.S. Kozien, Using the fuzzy logic description for the ex ante risk assessment in the project, in: Proc. 35th Int. Sci. Conf. on Economic and Social Development, Varazdin, Varazdin Development and Entrepreneurship Agency, 2018, 224-231.

[7] J. Pietraszek, A. Sobczyk, E. Skrzypczak-Pietraszek, M. Kolomycki, The fuzzy interpretation of the statistical test for irregular data, Technical Transactions 113 (14) (2016) 119-126.

[8] G. Alefeld, J. Herzberger, Introduction to interval computations, New York, Academic Press, 1983.

[9] M.S. Kozien, D. Smolarski, Choosing of the values of dynamic damper with application of interval arithmetic, Technical Transactions 114 (2) (2011) 69-74.

[10] E. Kozien, M.S. Kozien, Interval analysis as a method of measurement of uncertainity in the check-list method applied to identification of stage phase in companies, in: Proc. 26th Int. Sci. Conf. on Economic and Social Development, Varazdin, Varazdin Development and Entrepreneurship Agency, 2017, 210-215.

[11]L. J. Orman, Boiling heat transfer on single phosphor bronze and copper mesh microstructures. EPJ Web of Conf. 67 (2014) art. 02087. https://doi.org/10.1051/epjconf/20146702087

[12]M. Domagala, H. Momeni, J. Domagala-Fab.s, G. Filo, D. Kwiatkowski, Simulation of cavitation erosion in a hydraulic valve. Materials Research Proceedings 5 (2018) 1-6. https://doi.org/10.21741/9781945291814-1

[13]M. Domagala, H. Momeni, J. Domagala-Fabis, G. Filo, M. Krawczyk, J. Rajda, Simulation of particle erosion in a hydraulic valve. Materials Research Proceedings 5 (2018) 17-24. https://doi.org/10.21741/9781945291814-4

[14]D. Przestacki, M. Kuklinski, A. Bartkowska, Influence of laser heat treatment on microstructure and properties of surface layer of Waspaloy aimed for laser-assisted machining. Int. J. Adv. Manuf. Technol. 93 (2017) 3111-3123. https://doi.org/10.1007/s00170-017-0775-2

[15]R. Skulski, P. Wawrzala, J. Korzekwa, M. Szymonik, The electrical conductivity of PMN-PT ceramics. Arch. Metall. Mater. 54 (2009) 935-941.

[16]R. Ulewicz, P. Szataniak, F. Novy, Fatigue properties of wear resistant martensitic steel. METAL 2014: 23rd Int. Conf. on Metallurgy and Materials. Ostrava, TANGER (2014) 784-789.

[17]D. Klimecka-Tatar, Electrochemical characteristics of titanium for dental implants in case of the electroless surface modification. Arch. Metall. Mater. 61 (2016) 923-26. https://doi.org/10.1515/amm-2016-0156

[18]E. Skrzypczak-Pietraszek, J. Pietraszek, Phenolic acids in in vitro cultures of Exacum affine Balf. f. Acta Biol. Crac. Ser. Bot. 51 (2009) 62-62.

[19]E. Skrzypczak-Pietraszek, I. Kwiecien, A. Goldyn, J. Pietraszek, HPLC-DAD analysis of arbutin produced from hydroquinone in a biotransformation process in Origanum majorana L. shoot culture. Phytochemistry Letters 20 (2017) 443-448. https://doi.org/10.1016/j.phytol.2017.01.009

[20]A. Gadek-Moszczak, J. Pietaszek, B. Jasiewicz, S. Sikorska, L. Wojnar, The Bootstrap Approach to the Comparison of Two Methods Applied to the Evaluation of the Growth Index in the Analysis of the Digital X-ray Image of a Bone Regenerate. New Trends in Comp. Collective Intell. 572 (2015) 127-136. https://doi.org/10.1007/978-3-319-10774-5_12

[21]A. Pacana, K. Czerwinska, R. Dwornicka, Analysis of non-compliance for the cast of the industrial robot basis, METAL 2019 28th Int. Conf. on Metallurgy and Materials (2019), Ostrava, Tanger 644-650. https://doi.org/10.37904/metal.2019.869

[22]S.B. Hertz, Risk analysis in capital investment, Harvard Business Review 58 (5) (1979) 169-171.

[23]F. Knight, Risk, Uncertainty and profit, New York, Dover Publication Inc., 2006.

[24]A. De Meyer, Ch. Loch, M. Pich, Managing project uncertainty. From variation to chaos, MIT Sloan Management Review 43 (2) (2002) 60-67.

Terotechnology XI
Materials Research Proceedings **17** (2020) 36-42

Materials Research Forum LLC
https://doi.org/10.21741/9781644901038-6

# Fluid-Structure Interaction Simulation of Flow Control Valve

DOMAGALA Mariusz[1, a *], MOMENI Hassan[2,b], DOMAGALA-FABIS Joanna[1,c],

SAEED Bikass[2,d], FILO Grzegorz[1,e] and AMZIN Shokri[2,b]

[1]Cracow University of Technology, Institute of Applied Informatics, Al. Jana Pawla II 37 31-841 Krakow, Poland

[2]Western Norway University of Applied Sciences, Department of Mechanical and Marine Engineering, Postboks 7030, 5020 Bergen, Norway

[a]domagala@mech.pk.edu.pl, [b]hassan.momeni@hvl.no, [c]fabis@mech.pk.edu.pl, [d]saeed.bikass@hvl.no, [e]filo@mech.pk.edu.pl, [f]shokri.amzin@hvl.no

**Keywords:** FSI Simulation, CFD Simulation, Flow Control Valve

**Abstract.** Flow control valves are commonly used in fluid power systems. Controlled by proportional solenoid allows to control flow rate irrespective of pressure on inlet or outlet of the valve. Simulation of such valve is complex task due to the usage of throttle and compensating valves inside one housing. This work presents an attempt of using CFD simulation with implemented Fluid Structure Interaction for simulation of flow control valve under changeable working conditions.

## Introduction

Flow control valves which main function is maintaining constant flow rate during system operation irrespective of possible pressure changes are commonly used in fluid power systems. Constant flow rate is achieved by combining hydraulic components (valves) with proportional controllers. Presented in this work flow control valve consists of throttle valve adjusted by electronically controlled solenoid and compensation valve. Regardless of using proportional control system research on ensuring adequate quality and reliability of such valves is required. Quality of control depends on knowledge of phenomena which occurs inside valve during fluid flow. The key issue are flow forces generated on valve components. Research on them are conducted for decades and theirs origin has been explained by Lee and Blackburn [1]. Presented theory based on momentum conservation theory and defines flow forces as a function of flow rate and valve opening. However, this not allows to describe flow forces with acceptable quality today. New possibilities brought numerical methods in flow simulations (CFD) [2]. From the early beginning such tools have found application in modelling flow inside valves and particularly calculation of flow forces. Borghi at al. [3] have investigated flow forces and jet angles in spool control valve. Lisowski at al. [4] have made an experimental tests and simulations for solenoid operated control valve. Along with CFD simulations Lugowski [5] have conducted experimental research on pressure distribution inside spool valves. Despite huge achievements in simulation tools recent research have been conducted on fixed position of valves components, while valve components (spools or poppets) depends directly on flow conditions and in consequences on forces values. Latest implementation of possibilities in simulation of fluid and structure interaction into CFD tools opens new era in simulation of hydraulic valves.

This works is an attempt of implementation of Fluid-Structure Interaction (FSI) for flow control valve. Numerical simulations have been conducted with general purpose CFD code:

Terotechnology XI
Materials Research Proceedings **17** (2020) 36-42

Materials Research Forum LLC
https://doi.org/10.21741/9781644901038-6

ANSYS CFX. Created model of flow control valve has been tested for variable flow conditions, which may appear during normal operation.

The experience gained during the simulation and the developed guidelines will be useful for analogous thermomechanical calculations of flows in biotechnological equipment [6, 7], impact modelling of steel [8], pressing and sintering powders [9] and laser treatment of ESD coatings [10-12]. They can also have a significant impact on the practical use of the thermodynamic adjustment calculus [13] and estimation of the uncertainty of the obtained results [14, 15]. Some optimization techniques may also be useful in general optimization problems [16-18].

**Flow control valve**

Flow control valve presented in Fig.1 is solenoid controlled proportional vale which consists of throttle valve with spool (3) and compensation valve with spool (2). Required flow rate is maintained by setting position of throttle valve (2) by the usage of solenoid. When pressure on port A (inlet) or port B (outlet) changes than required flow rate is adjusted by proper position of compensation valve with spool (2).

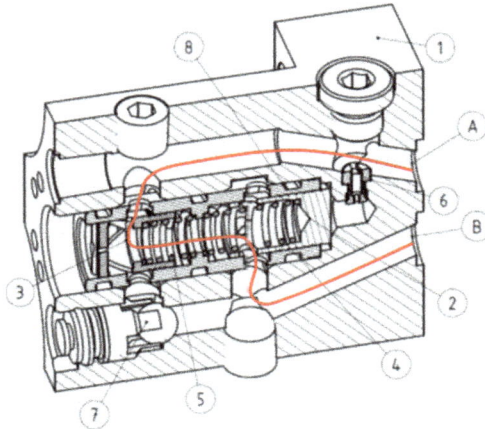

*Fig. 1. Flow control valve: 1-valve body, 2,3-spools, 4,5-springs, 6-nozzle, 7-non return valve, red line indicates flow direction (from port A to port B), 8 sleeve.*

Both spools (2,3) are inserted into sleeve (8) with springs which make common chamber for both valves. Working fluid from supply line flows into the vale through port A and through throttle spool (3) to common chamber and flowing out through spool (2) to port B. During fluid flow position of spool (2) depends on pressure on both ports A and B. As position of spool (3) is controlled by solenoid, position of spool (2) is determined by forces caused by fluid flow and spring (4). Proper design of this spool is crucial for obtaining adequate quality control of flow rate while value of hydrodynamic forces generated on spool (2) plays major role.

According to force balance (Fig.2), the motion equation for spool (2) can be expressed as:

$$m_e \ddot{x} = F_{hs1} - F_{hs2} - F_s - F_{hd} - F_t \qquad (1)$$

$$F_{hs1} = p_1 A \tag{2}$$

$$F_{hs2} = p_2 A \tag{3}$$

$$F_s = F_0 + kx \tag{4}$$

$$F_t = \mu \pi d l \frac{\dot{x}}{c_r \sqrt{\left(1 - \left(\frac{e}{c_r}\right)^2\right)}} \tag{5}$$

$$F_{hd} = \rho Q v \cos(\alpha) \tag{6}$$

where:

$m_e$ – effective moving mass (spool, spring and moving fluid), $F_{hs1}$, $F_{hs2}$ – hydrostatic forces, $F_s$ – spring force, $F_{hd}$ – hydrodynamic force, $F_t$ – friction force, $p_1, p_2$ – pressure, $A$ – spool area, $\mu$ – fluid viscosity, $c_r$ – radial clearance, $e$ – eccentricity, $l$ – spool length, $d$ – spool diameter,

$k$ – spring rate, $\rho$ – fluid density, $Q$ – fluid flow rate, $\alpha$ – jest angle.

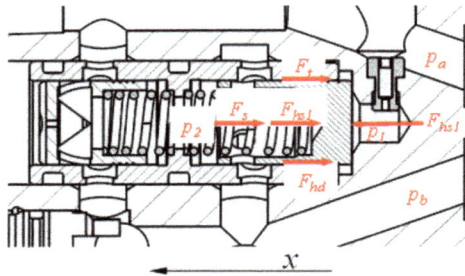

*Fig. 2. Force balance on spool 2.*

The last two terms of Eq. (1) seems to be the most problematic. As the friction forces might be neglected for steady state conditions the hydrodynamic forces are still the issue. Due to that CFD tools have to be used, in which flow forces are calculated as an integral of stress tensor over surface:

$$F_{hd} = \int \tau_n n dS \tag{7}$$

where:

$\tau_n$ – stress tensor, n – normal vector.

**CFD simulation**

CFD simulations have been performed in two stages. First the full model of the valve have been used. It has confirmed that velocity distribution in both throttle and compensating value is not uniform in all valves holes. Selected results of path lines colored by fluid velocity have been presented on Fig. 3.

Terotechnology XI
Materials Research Proceedings **17** (2020) 36-42

Materials Research Forum LLC
https://doi.org/10.21741/9781644901038-6

Most of research have estimated forces for fixed position of valve components. Lisowski and Filo [4] have evaluated jet angles in CFD simulations and they have implemented them in Simulink model. Other approach was using deforming mesh and FSI simulation. Beune et al. [19] have used mesh deformation to simulate high pressure safety valve in 2D model. Menendez-Blanco et al. [20] have used 2D and 3D model with various techniques of mesh deformation to simulate diaphragm pump and check valves.

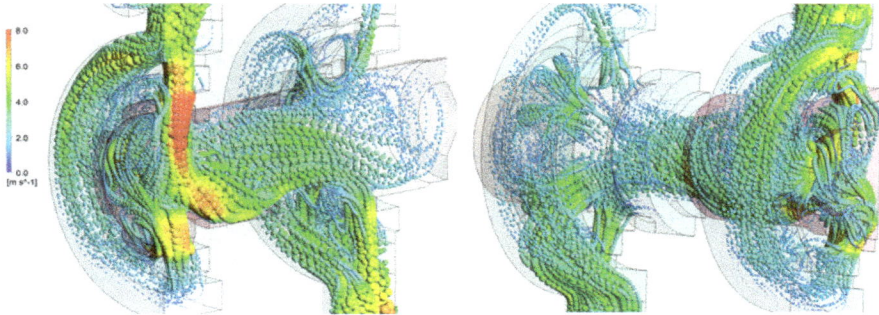

*Fig. 3. Path lines colored by fluid velocity.*

There are few approaches of performing FSI simulation and enabling motion of rigid bodies. Grid may be stretched and expanded, but as previous works [21-23] have shown problems may happen with distorted cells. Others methods like layering, sliding mesh or overset mesh [24, 25] are also available. After analysis of deforming mesh capabilities a sliding mesh has been used. CFD model of flow control valve has been split into two parts. Model of compensating valve has been used in FSI simulations, remaining part of the model has been used to define input data for FSI simulations. The way the model has been split is presented on Fig. 4.

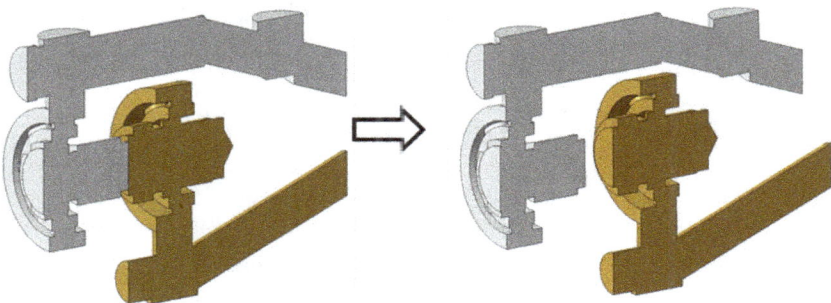

*Fig. 4. Splitting of CFD model.*

Grid between spool chamber and sleeve have been connected by interface which allows for mass transfer between domains. Grid interface is presented on Fig. 5.

Terotechnology XI                                                                      Materials Research Forum LLC
Materials Research Proceedings **17** (2020) 36-42                    https://doi.org/10.21741/9781644901038-6

Interaction between flowing fluid and model of spool (2) has been enabled by applying Eq. (1) to spool model by using Newmark scheme [9], which is trapezoid rules with second order accurate scheme:

$$x_{n+1} = x_n + \Delta t \dot{x}_n + \frac{\Delta t^2}{4}(\ddot{x}_n + \ddot{x}_{n+1})$$ (8)

where: index "$n$" is a step number and $\Delta t$ is time step, $x$ – spool displacement.

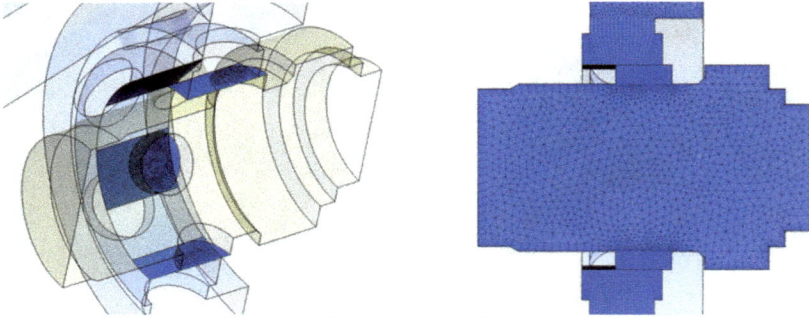

Fig. 5. Grid interface.

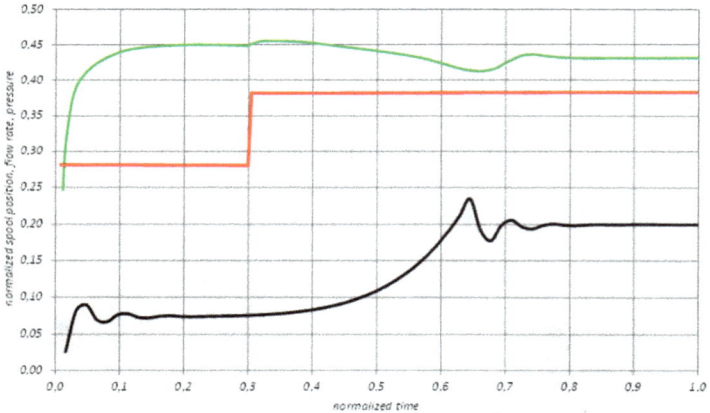

Fig. 6. Spool displacement: black, pressure at outlet: red, flow rate: green.

Numerical simulations have been conducted for case when pressure on outlet has changed from initial value to new value with a step function. At the first stage (to 0.2 of normalized time) spool is oscillating. After achieving steady position pressure on outlet have increased (at value of 0.3 of normalized time). Spool has been moved making flow gap smaller to maintain constant flow rate. Spool displacement and pressure on valve outlet and flow rate value are presented on Fig. 6 as a normalized values.

Terotechnology XI                                                        Materials Research Forum LLC
Materials Research Proceedings 17 (2020) 36-42                    https://doi.org/10.21741/9781644901038-6

## Conclusions

CFD FSI simulation have been used to simulate operation of flow control valve. A sliding mesh with grid interface has been used to for interaction between flowing fluid and spool of compensating valve. Numerical simulation for case study have been performed in Ansys CFX a general purpose CFD code. Thanks to created model and direct implementation of motion equation for spool it was possible to simulate response of the spool for changes of outlet pressure. Presented in the work approach seems to be an effective method of investigating dynamic phenomena which occurs during fluid flow in hydraulic valves. It is more time consuming process than simulations with fixed spool position but brings new quality in modeling of valve dynamics.

## References

[1] S.-Y. Lee, J. Blackburn, Contributions to Hydraulic Control: 1 Steady-State Axial Forces on ControlValve Pistons, Transactions of the ASME 74 (1952) 1005–1011.

[2] T.J.R. Hughes, The Finite Element Method, Englewood Cliffs, N.J., Prentice-Hall, 1987.

[3] M. Borghi, M. Milani, R. Paoluzzi, Stationary axial flow force analysis on compensated spool valves, International Journal of Fluid Power 1 (2000) 17-25. https://doi.org/10.1080/14399776.2000.10781079

[4] E. Lisowski, W. Czyzycki, J. Rajda, Three cimensional CFD analysis and experimental test of flow force acting on the spool of solenoid operated directional control valve, Energy Conversion and Management 70 (2013) 200-229. https://doi.org/10.1016/j.enconman.2013.02.016

[5] E. Lisowski, G. Filo, Analysis of a proportional control valve flow coefficient with the usage of a CFD method, Flow Measurement and Instrumentation 53 (2017), 269-278. https://doi.org/10.1016/j.flowmeasinst.2016.12.009

[6] E. Skrzypczak-Pietraszek, K. Reiss, P. Zmudzki, J. Pietraszek, J. Enhanced accumulation of harpagide and 8-O-acetyl-harpagide in Melittis melissophyllum L. agitated shoot cultures analyzed by UPLC-MS/MS. PLOS One. 13 (2018) art. e0202556. https://doi.org/10.1371/journal.pone.0202556

[7] E. Skrzypczak-Pietraszek, K. Piska, J. Pietraszek, Enhanced production of the pharmaceutically important polyphenolic compounds in Vitex agnus castus L. shoot cultures by precursor feeding strategy. Engineering in Life Sciences 18 (2018) 287-297. https://doi.org/10.1002/elsc.201800003

[8] M. Mazur, K. Mikova, Impact resistance of high strength steels. Materials Today-Proceedings 3 (2016) 1060-1063. https://doi.org/10.1016/j.matpr.2016.03.048

[9] T. Pieczonka, J. Kazior, A. Szewczyk-Nykiel, M. Hebda, M. Nykiel, Effect of atmosphere on sintering of Alumix 431D powder. Powder Metall. 55 (2012) 354-360. https://doi.org/10.1179/1743290112Y.0000000015

[10] S. Wojciechowski, D. Przestacki, T. Chwalczuk, The evaluation of surface integrity during machining of Inconel 718 with various laser assistance strategies. MATEC Web of Conf. 136 (2017) art. 01006. https://doi.org/10.1051/matecconf/201713601006

[11] N. Radek, M. Scendo, I. Pliszka, O. Paraska, Properties of Electro-Spark Deposited Coatings Modified Via Laser Beam. Powder Metall. Met. Ceram. 57 (2018) 316-324. https://doi.org/10.1007/s11106-018-9984-y

[12] N. Radek, A. Szczotok, A. Gadek-Moszczak, R. Dwornicka, J. Broncek, J. Pietraszek, The impact of laser processing parameters on the properties of electro-spark deposited coatings. Arch. Metall. Mater. 63 (2018) 809-816.

[13] T. Styrylska, J. Pietraszek, Numerical modeling of non-steady-state temperature-fields with supplementary data. ZAMM 72 (1992) T537-T539.

[14] Z. Ignaszak, P. Popielarski, J. Hajkowski, Sensitivity of models applied in selected simulation systems with respect to database quality for resolving of casting problems. Defect and Diffusion Forum 336 (2013) 135-146. https://doi.org/10.4028/www.scientific.net/DDF.336.135

[15] A. Gadek-Moszczak, J. Pietraszek, B. Jasiewicz, S. Sikorska, L. Wojnar, The Bootstrap Approach to the Comparison of Two Methods Applied to the Evaluation of the Growth Index in the Analysis of the Digital X-ray Image of a Bone Regenerate. New Trends in Comp. Collective Intell. 572 (2015) 127-136. https://doi.org/10.1007/978-3-319-10774-5_12

[16] J. Pietraszek, A. Gadek-Moszczak, T. Torunski, Modeling of Errors Counting System for PCB Soldered in the Wave Soldering Technology. Advanced Materials Research 874 (2014) 139-143. https://doi.org/10.4028/www.scientific.net/AMR.874.139

[17] D. Malindzak, A. Pacana, H. Pacaiova, An effective model for the quality of logistics and improvement of environmental protection in a cement plant. Przem. Chem. 96 (2017) 1958-1962.

[18] A. Pacana, K. Czerwinska, R. Dwornicka, Analysis of non-compliance for the cast of the industrial robot basis, METAL 2019 28th Int. Conf. on Metallurgy and Materials (2019), Ostrava, Tanger 644-650. https://doi.org/10.37904/metal.2019.869

[19] A. Beune, J.G.M. Kuerten, J.P.C. van Heumen, CFD analysis with fluid-structure interaction simulation of opening high-pressure safety valve, Computers & Fluids 64 (2012) 108-116. https://doi.org/10.1016/j.compfluid.2012.05.010

[20] A. Menendez-Blanco, J.M. Fernandez Oro, A. Meana-Fernandez, Unsteady three-dimensional modeling of the Fluid–Structure Interaction in the check valves of diaphragm volumetric pumps, Journal of Fluids and Structures 90 (2019) 432–449. https://doi.org/10.1016/j.jfluidstructs.2019.07.008

[21] M. Domagala, Metodyka modelowania zaworów maksymalnych bezpośredniego działania, Ph.D. thesis, Krakow, Politechnika Krakowska, 2007.

[22] M. Domagala, H. Momeni, J. Domagala-Fabis, G. Filo, M. Krawczyk, J. Rajda, Simulation of particle erosion in a hydraulic valve, Materials Research Proceedings 5 (2018) 17-24. https://doi.org/10.21741/9781945291814-4

[23] M. Domagala, H. Momeni, J. Domagala-Fabis, G. Filo, D. Kwiatkowski, Simulation of Cavitation Erosion in a Hydraulic Valve, Materials Research Proceedings 5 (2018) 1-6. https://doi.org/10.21741/9781945291814-1

[24] Radek, N., Kurp, P., Pietraszek, J., Laser forming of steel tubes. Technical Transactions 116 (2019) 223-229. https://doi.org/10.4467/2353737XCT.19.015.10055

[25] M. Domagala, H. Momeni, J. Fabis-Domagała, G. Filo, P. Lempa, Simulations of Safety Vales for Fluid Power Systems, System Safety: Human-Technical Facility-Environment 1(1) (2019) 670-677. https://doi.org/10.2478/czoto-2019-0085

Terotechnology XI
Materials Research Proceedings 17 (2020) 43-49

Materials Research Forum LLC
https://doi.org/10.21741/9781644901038-7

# The Concept of a Modular Integrated Test Stand for Testing Hydraulic Drive and Control Systems

GUZOWSKI Artur[1,a*] and SOBCZYK Andrzej[1,b]

[1]Cracow University of Technology, Al. Jana Pawla II 37, 31-864, Cracow, Poland

[a]artur.guzowski@mech.pk.edu.pl, [b]andrzej.sobczyk@mech.pk.edu.pl

**Keywords:** Hydraulic Power and Control Systems Testing, Modular Construction, Integrated Test Stand, Control and Measurement System

**Abstract.** The paper presents proposals for the construction of a stand for testing elements of hydraulic drive systems. The presented solution is an excellent example which can be used to design and build such a system for companies or scientific centers that provides testing services for hydraulic pumps and motors. The article shows the demand of companies in Poland for the construction of test stands, necessary in the technical assessment of new and used elements, used in stationary stands and mobile working machines or vehicles. This approach clearly shows how to increase company's competitiveness and implement a new service. The proposed high power station (up to 350 kW) allows for testing of elements in a wide range of volumetric flow rate (up to 500 dm3 / min) and working under high pressure (up to 50 MPa). The new modular solution enables any integrated configuration of the station, depending on the company's needs.

## Introduction

Hydraulic drive elements play one of the key roles in the development and position of domestic and foreign industry. For many years, many industries have used numerous components or comprehensive systems that create the manufacturing technology of a given company, including innovative projects leading to a competitive product. Today, hydraulic drive is widely used, among others in the aviation, mining and machine industries. The transfer of energy using liquids as a working medium, in particular in terms of high forces and moments, is still unmatched and many machines would not be able to function without its use. In each country, including Poland, you can identify companies that manufacture components or comprehensive hydraulic systems, operate stationary and mobile machines and devices, as well as service or regenerate parts. Especially at the production and service stages, it is necessary to have workstations and tools for the correct assessment of the properties and functionality of components, e.g. Bosch-Rexroth, Danfoss, Sun Test Systems, Parker Hannifin or MH Hydraulics. The correct selection of pumps, engines and other elements for hydraulic systems, especially for responsible hydrostatic drive systems and vehicle suspension systems [1], requires a good knowledge of the characteristics of these elements. Such characteristics are plotted on the basis of experimental data of these machines. Depending on the type of desired characteristics, a specialized test stand is configured [2]. A hydraulic motor without a speed reduction gearing can be used to directly drive the running wheels of a low-speed working machine or vehicle. The necessary condition, however, is stable motor operation at very low speed. Therefore, such a motor must have very small leaks in the timing slots. Another desirable feature of hydrostatic drive system components is their low weight. Among the known designs of pumps and hydraulic motors, the lowest weight-to-power ratio is characterized by pumps and satellite motors [3]. In addition, an important feature of mobile machines is the low level of emitted sound [4]. At the same time, manufacturers and research labs are looking for new solutions and applications of hydraulics, including using water

Terotechnology XI                                                      Materials Research Forum LLC
Materials Research Proceedings **17** (2020) 43-49           https://doi.org/10.21741/9781644901038-7

as a working medium. In this case, pumps, motors and valves made of traditional steel can be coated with special coatings [5, 6], which will allow, among others, to replace more expensive and more difficult to process stainless steel. Regardless of the working medium, hydraulic systems based on numerous components must be tested for exposure to low and high operating temperatures in accordance with regulations [7]. Based on the above, it can be concluded that it is necessary to build multi-functional stands for testing hydraulic components or systems and verification of simulation tests. The complexity of tests and, in particular, a wide range of power required shapes the design of such solutions that allow for a flexible approach to testing. The implementation of such tasks is facilitated by modular solutions that can be integrated and tailored to your needs.

The proposed test stand is of interest to many industries, not only directly related to the production of power hydraulics [8], but also pressure components in heat transport [9] or chemically aggressive compounds in biotech [10, 11]. It could also be useful for testing protective coatings [12, 13], even in corrosive conditions [14]. The availability of such a research stand can significantly increase the efficiency of management and industrial logistics [15-17], as well as the analytical methods used [18-20] and pipe shaping [21].

**Assumptions**
When designing the station, it was assumed that the system would be based on an electric motor with a power of 350 [kW] and speed regulated by the inverter within the range of 800 ÷ 2300 [rpm]. The electric motor will supply two independent systems with two hydraulic units with a maximum flow rate of 500 [l / min] and an operating pressure of 50 [MPa], each with a mechanical load system. Control with the use of Load Sensing (LS) systems will be implemented by two triple sets of proportional control signals with maximum operating pressures of 4 [MPa] and 37 [MPa]. In addition, the station will allow for the testing of pumps and motors with variable flow direction and the measurement of such parameters as pressure, temperature, flow rate, rotational speed and load torque. The concept of the layout - the modular station for testing hydraulic pumps and motors consists of the following modules: 1.pump power supply module, 2. motor module (load), 3.circuit connection module, 4. LS control module.

The modules should be connected by means of additional hydraulic pipes and/or hoses taking into account the required pressure values and volumetric flow rate, which are set by the producer of the tested element. During testing, the appropriate duty cycle should be adopted. Control and monitoring of work parameters are carried out using a control and measuring system, the construction and principle of operation of which is omitted.

**Pump module (power supply)**
The power module (Fig. 1) performs the function of a pump drive with the elements necessary to connect the hydraulic suction lines with the connections p1T, p2T, p3T and leak lines p1R, p2R, p3R (if required). The module allows you to connect on one drive shaft a maximum of three sectional pumps set with a maximum power of 350 [kW] and a rotational speed in the range from 800 to 2300 [rpm], controlled by an inverter. The electric motor (8) acting as a drive allows for transferring a nominal driving torque of 2300 [Nm]. In addition, the module contains elements for controlling and measuring the pump operating parameters, such as pressure on the suction lines p1T, p2T, p3T; rotational speed e.g. and torque Mp (3); volumetric flow rate on the leakage lines Q1R, Q2R, Q3R (4 and 5); pressure p1 and oil temperature T1 and volumetric flow rate Q1 to 120 [l / min] (6). For these pumps, there is also the use of an overflow valve up to 50 [MPa] (9) and a relief for the system by means of an adjustable throttle valve (12). The above elements should also be used in tests of supplementary pumps in installations with variable flow pumps.

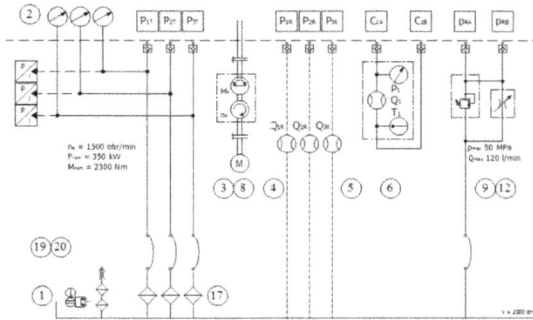

*Fig. 1. Schema of the stand system power module.*

**Module of the hydrostatic motor load**

The load module (Fig.2) performs the function of continuing the hydraulic line from the supply module and its load with maximum pressure, thanks to the use of the HM mechanical brake (13).

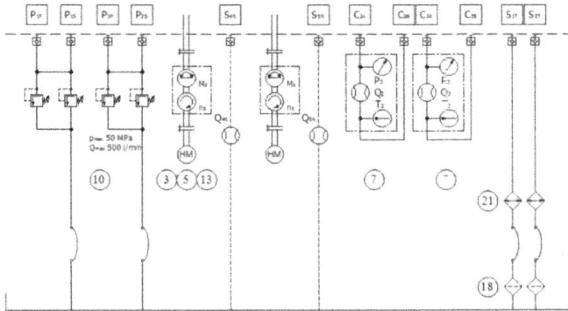

*Fig. 2. Diagram of the load system.*

In addition, the module allows for the use of elements necessary to connect hydraulic filtered (18) drain lines with the use of S1T, S2T connections and internal leaks with the use of S4R, S5R connections and leaks lines S4R, S5R (if required). The stand allows you to connect two motors at the same time, while the total power of both systems may not exceed the input power of 350 [kW]. In addition, the module contains elements for controlling and measuring motor operating parameters, such as rotational speed ns and torque Ms (3); volumetric flow rate on the leakage lines Q4R, Q5R; pressure p2, p3, oil temperature T2, T3 and volumetric flow rate on supply lines Q2, Q3 up to 500 [l / min] (7). For the used motors, it is also possible to use adjustable overflow valves up to the pressure of 50 [MPa] (10).

**Closed circuit connection module**

In the case of testing units operating in a closed system, both pumps and motors, the elements shown in Fig. 3 are necessary for use.

The non-return valve block (11) mounted on the plate (14) is used to connect the supplementary pump, and the non-return valve block (11) mounted on the plate (16) protects the motor against overload due to braking. The overflow valve block (10) mounted on the plate (15) protects the supply pump connected to the supply module. All components can operate at a pressure of 50 [MPa] and a maximum flow of 500 [l / min].

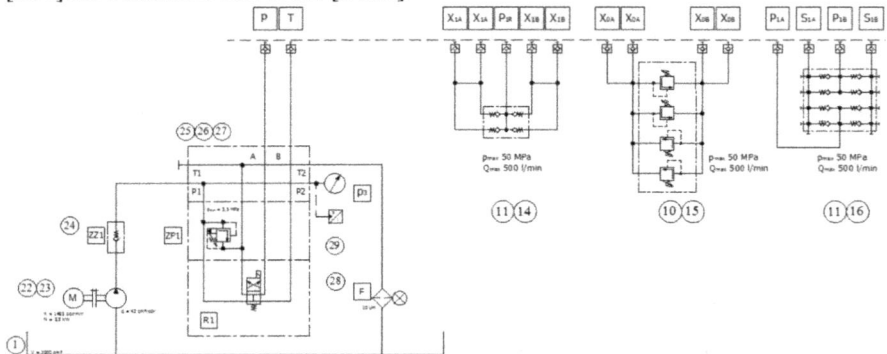

*Fig. 3. Elements of additional connections for the closed system.*

Hydraulic power pack based on a gear pump with constant unit volume of 42 [cm3] (23) and an electric motor with the power of 5.5 [kW] (22) allows for generating a control signal with a maximum pressure of 3.5 [MPa] for elements that require its use.

**Load Sensing (LS) control module**
The LS control system (Fig. 4.) plays the role of an additional system enabling the control signal to be supplied to pumps and motors with variable operating parameters, whose main parameters are the following: volume flow rate Qmax (10 [l / min]); 3 control signals pmax1 (37 [MPa]); 3 control signals pmax2 (4 [MPa]); supply pressure p (20 [MPa]).

The proportional valves 4/3 from R3 to R8 used allow for the adoption of control in accordance with the required duty cycle or load of the components tested. Pressure sensors pLS1 ÷ pLS6 allow for reading the current setting or save it in the control and measurement system.

**Summary**
The article presents a design of building a modular/integrated test stand for pumps and hydraulic motors with high performance parameters - pressure 50 [MPa] and volume flow rate 500 [l / min]. The proposed hydraulic stand is characterized by a modular construction, which allows for its free configuration, depending on the user's needs regarding the size of the tested elements and the method or algorithm of their control.

The presented hydraulic diagrams show a detailed structure, the analysis of which allows further development of the concept of connecting pumps and motors working in both directions. A set of valve blocks enables the connection of open and closed systems, and the multi-section Load Sensing control module allows for comprehensive simulation of work cycles at variable load settings. Numerous measurement signals of the main hydraulic and mechanical parameters allow for the creation of a history of tests and analysis of results, e.g. determining the efficiency of pumps or engines or the technical condition of the tested component. Extensive connection sets and a 350 [kW] electric motor allow for testing a set of several hydraulic components at the same time. This work is an excellent benchmark for designing or building similar hydraulic

46

stations. The company that implements the proposed solution will use niches on the market in the aspect of technical tests of high-pressure power hydraulics.

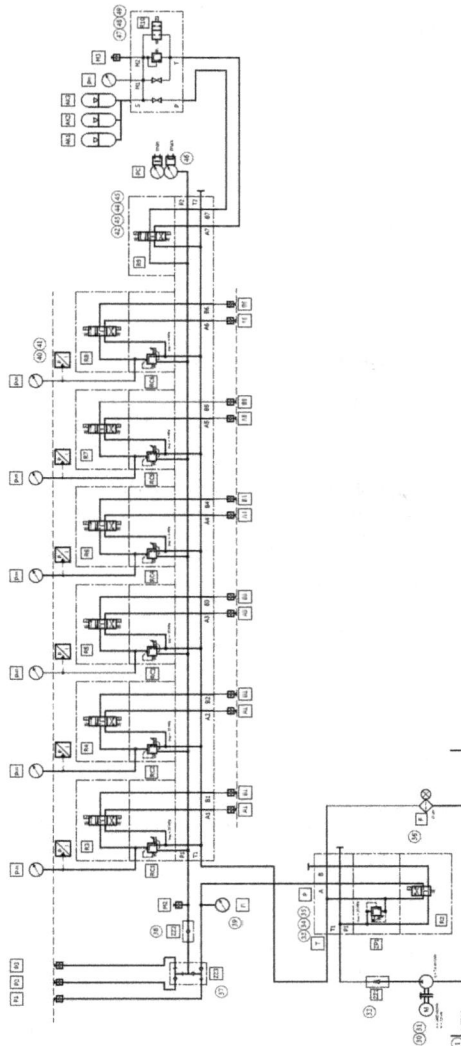

Fig.4. Schema of the LS control module.

## References

[1] A. Guzowski, A. Sobczyk, Reconstruction of hydrostatic drive and control system dedicated for small mobile platform, Proc. 8th Fluid Power Net International PhD Symposium, ASME (2014). https://doi.org/10.1115/FPNI2014-7862

[2] P. Śliwiński, The influence of water and mineral oil on mechanical losses in the displacement pump for offshore and marine applications. Polish Maritime Research 25 (2018) 178-188. https://doi.org/10.2478/pomr-2018-0040

[3] P. Osiński, W. Huss, P. Bury, K. Kiec, Badania mocy cieplnej w pompie zębatej 3PZ4, Napędy i Sterowanie 3 (2018) 110-114.

[4] P. Śliwiński, Influence of water and mineral oil on the leaks in satellite motor commutation unit clearances, Polish Maritime Research 24 (2017) 58-67. https://doi.org/10.1515/pomr-2017-0090

[5] J. Pobędza, A. Sobczyk, Properties of high pressure water hydraulic components with modern coatings, Advanced Materials Research 849 (2014) 100-107. https://doi.org/10.4028/www.scientific.net/AMR.849.100

[6] J. Krawczyk, A. Sobczyk, J. Stryczek, P. Walczak, Tests of new methods of manufacturing elements for water hydraulics, Materials Research Proceedings 5 (2018) 200-205. https://doi.org/10.21741/9781945291814-35

[7] P. Kucybała, A. Gawlik, J. Pobędza, P. Walczak, The usage of a thermoclimatic chamber for technoclimatic tests of special vehicles and mobile machines, Journal of KONES Powertrain and Transport 25 (2018) 199-206.

[8] M. Domagala, H. Momeni, J. Domagala-Fabis, G. Filo, M. Krawczyk, J. Rajda, Simulation of particle erosion in a hydraulic valve. Materials Research Proceedings 5 (2018) 17-24. https://doi.org/10.21741/9781945291814-4

[9] L. Dabek, A. Kapjor, L.J. Orman, Boiling heat transfer augmentation on surfaces covered with phosphor bronze meshes. MATEC Web of Conf. 168 (2018) art. 07001. https://doi.org/10.1051/matecconf/201816807001

[10] E. Skrzypczak-Pietraszek, J. Pietraszek, Phenolic acids in in vitro cultures of Exacum affine Balf. f. Acta Biol. Crac. Ser. Bot. 51 (2009) 62-62.

[11] E. Skrzypczak-Pietraszek, I. Kwiecien, A. Goldyn, J. Pietraszek, HPLC-DAD analysis of arbutin produced from hydroquinone in a biotransformation process in Origanum majorana L. shoot culture. Phytochemistry Letters 20 (2017) 443-448. https://doi.org/10.1016/j.phytol.2017.01.009

[12] R. Dwornicka, N. Radek, M. Krawczyk, P. Osocha, J. Pobedza, The laser textured surfaces of the silicon carbide analyzed with the bootstrapped tribology model. METAL 2017 26th Int. Conf. on Metallurgy and Materials (2017), Ostrava, Tanger 1252-1257.

[13] S. Wojciechowski, D. Przestacki, T. Chwalczuk, The evaluation of surface integrity during machining of Inconel 718 with various laser assistance strategies. MATEC Web of Conf. 136 (2017) art. 01006. https://doi.org/10.1051/matecconf/201713601006

Terotechnology XI                                                     Materials Research Forum LLC
Materials Research Proceedings **17** (2020) 43-49          https://doi.org/10.21741/9781644901038-7

[14] D. Klimecka-Tatar, Electrochemical characteristics of titanium for dental implants in case of the electroless surface modification. Arch. Metall. Mater. 61 (2016) 923-26. https://doi.org/10.1515/amm-2016-0156

[15] P. Krawiec, A. Marlewski, Spline description of non-typical gears for belt transmissions. J. Theor. Appl. Mech. 49 (2011) 355-367.

[16] D. Malindzak, A. Pacana, H. Pacaiova, An effective model for the quality of logistics and improvement of environmental protection in a cement plant. Przem. Chem. 96 (2017) 1958-1962.

[17] A. Pacana, K. Czerwinska, R. Dwornicka, Analysis of non-compliance for the cast of the industrial robot basis, METAL 2019 28$^{th}$ Int. Conf. on Metallurgy and Materials (2019), Ostrava, Tanger 644-650. https://doi.org/10.37904/metal.2019.869

[18] J. Pietraszek, Response surface methodology at irregular grids based on Voronoi scheme with neural network approximator. In: Rutkowski L., Kacprzyk J. (eds) Neural Networks and Soft Computing. Advances in Soft Computing, vol 19. Physica, Heidelberg: 2003, 250-255. https://doi.org/10.1007/978-3-7908-1902-1_35

[19] I. Dominik, J. Kwasniewski, K. Lalik, R. Dwornicka, Preliminary signal filtering in self-excited acoustical system for stress change measurement. CCC 2013 32$^{nd}$ Chinese Control Conf. (2013) 7505-7509.

[20] J. Pietraszek, A. Gadek-Moszczak, T. Torunski, Modeling of Errors Counting System for PCB Soldered in the Wave Soldering Technology. Advanced Materials Research 874 (2014) 139-143. https://doi.org/10.4028/www.scientific.net/AMR.874.139

[21] Radek, N., Kurp, P., Pietraszek, J., Laser forming of steel tubes. Technical Transactions 116 (2019) 223-229. https://doi.org/10.4467/2353737XCT.19.015.10055

Terotechnology XI
Materials Research Proceedings 17 (2020) 50-56

Materials Research Forum LLC
https://doi.org/10.21741/9781644901038-8

# Analysis of the Quality of Bearing Housing and Improving the Method of Identifying the Root of Incompatibility

SIWIEC Dominika[1,a] and PACANA Andrzej [2,b]

[1]Rzeszow University of Technology, The Faculty of Mechanical Engineering and Aeronautics
Poland

[2]Rzeszow University of Technology, The Faculty of Mechanical Engineering and Aeronautics
Poland

[a]d.siwiec@prz.edu.pl, [b]app@prz.edu.pl

**Keywords:** Mechanical Engineering, Bearing Housing, Non-Destructive Test, Quality Management, Fluorescent Method

**Abstract.** In improving the quality of a product, it is important to use actions which allow for identifying problems and their root, which is possible by using adequate techniques. In industrial practice the non-destructive tests are used, because they are effective in the product quality assessment. By their use, it is possible to identify the incompatibility of a product, but not the root of its occurrence. Therefore, it is effective to integrate the non-destructive test with other methods that allow for it. The aim of the study was to identify the root of linear indications detected by the FPI method on a bearing housing using the sequence of quality management techniques. An enterprise localized in the Podkarpacie had a problem with not identified root of incompatibility. In the enterprise, linear indications were often identified on different types of products, and the root of this problem was not known. To solve the problem and identify the root of linear indications, a sequence of selected techniques was used. These methods included the fluorescent method inspection (FPI), Ishikawa diagram and 5Why method. A bearing housing made from 410 steel, on which linear indications were often identified was the subject of the research. By using the Ishikawa diagram, the potential and main causes were identified (production errors and bad storage). By the 5Why method the root of the linear indications was identified - it was faulty material from the supplier. It was shown that integrating the sequence of quality management techniques with the FPI method is effective to identify the root causes of a product. These sequences can be used to quality analyze other types of products and also to identify the root of other types of incompatibilities.

## Introduction

Quality management techniques have applications in improving the process of quality analysis of a product. A complex analysis of the problem by using the quality management techniques is possible when the selected techniques are used in a sequential way [1, 2]. It is possible by using, for example, the Ishikawa diagram with the 5Why method [3, 4]. By the Ishikawa diagram, the potential causes of a problem can be pointed and next the main causes of the problem can be selected [5, 6]. But for a complex quality analysis, the most important is to identify the root of the problem. So, in the next steps, it is effective to use the 5Why method, which allows for it. This strategy in using these quality management techniques to identify the root of the problem is known and often used [7]. But it is very useful to show that this simple and effective strategy can be used like an integrating technique with some non-destructive test. It is a way to improve the process of the quality analysis of a product when it is possible to identify the incompatibility but its root is not known [4, 8]. The integration quality of management techniques (Ishikawa diagram and 5Why method) was proposed in an enterprise localized in Podkarpacie.

The aim of the study was to identify the root of linear indications detected by the FPI method on a bearing housing using a sequence of quality management techniques. In the enterprise, the linear indications were often identified on different types of products, and the root of this problem was not known. So, it was important to propose this sequence to analyze the quality of the bearing housing and improve the method of identifying the root of incompatibility.

This approach, presented in this article, may also be used in those types of industrial activities where the identification of the sources of non-compliance is of particular importance, e.g. biotechnology [9, 10], fatigue failures in mechanics [11, 12], hydraulics in heavy-duty machines [13], protective coatings [14, 15] deployed by electro-spark deposition [16, 17] and enhanced by laser machining [18]. It may also be used to detect errors in research analyzes involving largely subjective human judgment, e.g. nonparametric models [19] or image analysis [20, 21].

**The subject of research and material**

A bearing housing, which was made from 410 steel was the subject of the research. The bearing housing allows for simple exploration and change, so by using the bearing housing it is possible to fix a product in the machine well [8]. The 410 steel is nickel alloy which is heat and corrosion-resistant. Selected properties of 410 steel are shown in the subject of the literature [22-24]. The choice of the subject of research depended on often identified types of incompatibilities (linear indications).

**Method**

In order to identify the root of the incompatibility on the bearing housing, a sequence of selected techniques was used. These techniques included the fluorescent method (FPI), Ishikawa diagram and 5Why method. The choice of these techniques was conditioned on their effectiveness in identifying the incompatibility on the product in the first part of the method (fluorescent method) and, in the second part of the method, the potential causes of the incompatibility (Ishikawa diagram) and its root (5Why method) [25-27]. Also, the choice of the fluorescent method was dependent on the individual needs of the customer who ordered product inspection and the type of product material.

In the first part of the method, the fluorescent method was applied to check the quality of the product (bearing housing). The fluorescent method is one of penetrant tests. In this method, the penetrant has a colorant, so the indicators can be identified by ultraviolet radiation. To make analysis using the FPI, it is necessary to darken the test stand [28-30]. The method of FPI has been presented in the literature [31]. After applying the FPI, the incompatibility on the bearing housing was identified, and it was linear indications. In order to identify the causes of the linear indications, in the second part of the method, the Ishikawa diagram was prepared. The Ishikawa diagram allows for the analysis of a problem using the main categories of the causes of a problem, i.e.: man, material, method, machine, management and environment [3, 5]. And these categories were used to analyze the problem with linear indications on the bearing housing because they were adequate to this problem. Potential causes were noted to each of the categories, from which the main causes were selected (these were two main causes of the linear indications). These main causes were used in the next analysis – the 5Why method. The 5 Why method was applied to identify the root of incompatibility. In this method, the analysis of the root of the problem began from the problem and the main causes of the problem, which were identified by the previous methods (FPI and Ishikawa diagram). In the 5Why method, the

Materials Research Forum LLC
https://doi.org/10.21741/9781644901038-8

"Why?" question was asked until the root of the problem was identified [3, 5]. After identifying the problem, improving actions were proposed.

**Results**

After the analysis of the quality of the bearing housing using the fluorescent method, the incompatibilities were identified. These incompatibilities were linear indications which are shown in Figure 1.

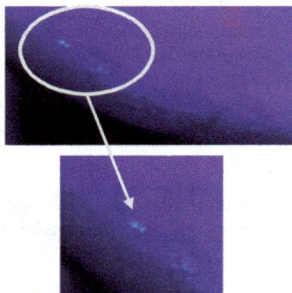

*Fig. 1. The linear indications on the bearing housing.*

In the next part of the method, in order to identify the potential and main causes of the linear indications on the bearing housing, the Ishikawa diagram was prepared, which is shown in Fig.2.

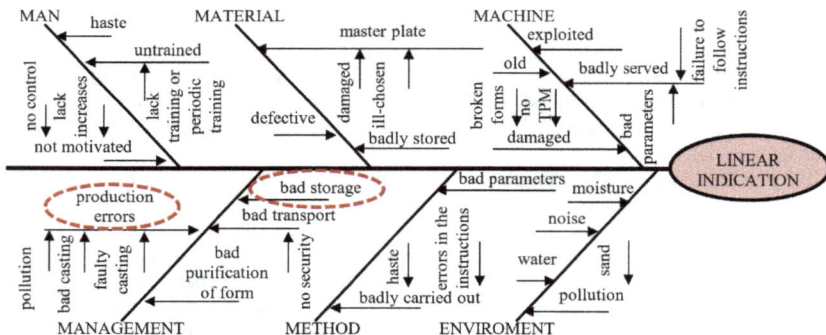

*Fig. 2. The Ishikawa diagram for the linear indications on the bearing housing.*

The main causes which were selected were production errors and bad storage, and these causes were used in the next part of the analysis with the 5Why method, which is shown in Figure 3.

After the analysis of the problem with the 5Why method, the root causes of the problem were identified. The root cause of the linear indications on the bearing housing was the faulty material from the supplier. In order to eliminate or minimalize the problem with linear indications on the product, improvement actions were taken, which informed the supplier about the root of the problem.

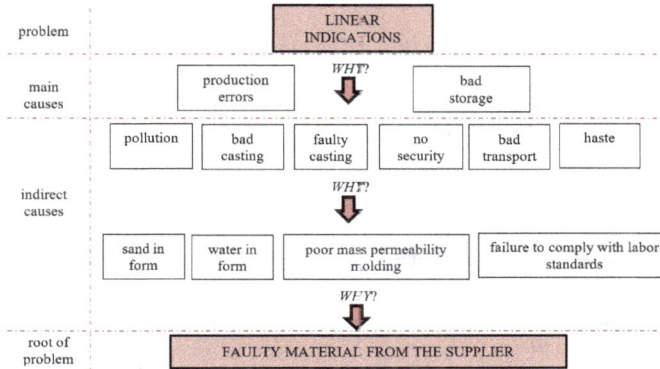

*Fig. 3. The 5Why method for the linear indications on the bearing housing.*

## Summary

Improving the quality of products is an important step for manufacturing companies. In order to make a complex analysis of a product, it is necessary to use adequate techniques that allow for it. The main part of the quality analysis of the product are non-destructive tests, which allow for identifying the problem on the product without destroying it. But it is not effective when the root of the problem is not known. So, to identify the root of the problem it is necessary to use other techniques, i. e. Ishikawa diagram and 5Why method. The use of these techniques was proposed in an enterprise localized in Podkarpacie. In the enterprise, non-destructive tests were applied to different types of products. It was noted that the problem with linear indications on products is often repeated. In order to identify the root of the linear indications, a sequence of the techniques (FPI, Ishikawa diagram and 5Why method) was proposed to use. The aim of the study was to identify the root of linear indications detected by the FPI method on a bearing housing using a sequence of quality management techniques. In the first part of the method, the FPI was made, after which the problem on the bearing housing was identified – linear indications. In order to identify the root of the linear indications, the sequence of selected techniques was used. In the second part of the method, using the Ishikawa diagram the potential causes of the problem were identified from which two main causes were pointed. The main causes were production errors and bad storage, and these main causes were used in the next analysis of the problem (in 5Why method). Therefore, in order to identify the root of the linear indications, the 5Why method was applied. After the analysis of the incompatibility (linear indications) on the product (bearing housing) with the 5why method, it was concluded that root causes were the faulty material from the supplier. In order to eliminate or minimalize the problem with linear indications on the product, improvement actions were taken, which informed the supplier about the root of the problem. It was shown that integrated sequence of quality management techniques with the FPI method is effective to identify the root causes of a product. These sequences can be used to quality analyze other types of products and also to identify the root of other types of incompatibilities.

**References**
[1]  M. Nowicka-Skowron, R. Ulewicz, Quality management in logistics processes in metal branch, METAL 2015 24th Int. Conf. on Metallurgy and Materials, Ostrava, Tanger, 2015, 1707-1712.

[2]  E. Tillová, M. Chalupová, L. Kuchariková, Quality control of cylinder head casting, Production Engineering Archives 14 (2017), 3-6. https://doi.org/10.30657/pea.2017.14.01

[3]  A. Pacana, D. Siwiec, L. Bednárová, Analysis of the incompatibility of the product with fluorescent method, Metalurgija 58 (2019) 337-340.

[4]  D. Siwiec, L. Bednarova, A. Pacana., M. Zawada, M. Rusko, Decision support in the selection of fluorescent penetrants for industrial non-destructive testing, Przemysł Chemiczny 98 (10) (2019) 1594-1596.

[5] R. Ulewicz, Quality Control System in Production of the Castings from Spheroid Cast Iron, Metalurgija 42 (1) (2003) 61-63.

[6]  R. Wolniak, Application methods for analysis car accident in industry on the example of power, Support Systems in Production Engineering 6 (4) (2017) 34-40.

[7]  A. Pacana, A. Radon-Cholewa, J. Pacana, et al., The study of stickiness of packaging film by Shainin method, Przemysl Chemiczny 94 (8) (2015) 1334-1336.

[8]  T. Ma, Z. Sun, Q. Chen, Study on crack features in images of fluorescent magnetic particle inspection for railway wheelsets. Insight: Non-Destructive Testing and Condition Monitoring 60 (9) (2018) 519-524. https://doi.org/10.1784/insi.2018.60.9.519

[9]  L. Skrzypczak, E. Skrzypczak-Pietraszek, E. Lamer-Zarawska, B. Hojden, Micropropagation of Oenothera-Biennis L. and an assay of fatty-acids. Acta Soc. Bot. Pol. 63 (1994) 173-177. https://doi.org/10.5586/asbp.1994.023

[10] E. Skrzypczak-Pietraszek, K. Piska, J. Pietraszek, Enhanced production of the pharmaceutically important polyphenolic compounds in Vitex agnus castus L. shoot cultures by precursor feeding strategy. Engineering in Life Sciences 18 (2018) 287-297. https://doi.org/10.1002/elsc.201800003

[11] R. Ulewicz, P. Szataniak, F. Novy, Fatigue properties of wear resistant martensitic steel. METAL 2014: 23rd Int. Conf. on Metallurgy and Materials. Ostrava, TANGER (2014) 784-789. https://doi.org/10.12693/APhysPolA.125.A-164

[12] E. Augustyn, M.S. Kozien, A study on possibility to apply piezoelectric actuators for active reduction of torsional beams vibrations. Acta Phys. Pol. A 125 (2014) A164-A168.

[13] M. Domagala, H. Momeni, J. Domagala-Fabis, G. Filo, M. Krawczyk, J. Rajda, Simulation of particle erosion in a hydraulic valve. Materials Research Proceedings 5 (2018) 17-24. https://doi.org/10.21741/9781945291814-4

[14] W. Zorawski, R. Chatys, N. Radek, J. Borowiecka-Jamrozek, Plasma-sprayed composite coatings with reduced friction coefficient. Surf. Coat. Technol. 202 (2008) 4578-4582. https://doi.org/10.1016/j.surfcoat.2008.04.026

Terotechnology XI
Materials Research Proceedings **17** (2020) 50-56

Materials Research Forum LLC
https://doi.org/10.21741/9781644901038-8

[15] D. Klimecka-Tatar, S. Borkowski, P. Sygut, The kinetics of Ti-1Al-1Mn alloy thermal oxidation and characteristic of oxide layer. Arch. Metall. Mater. 60 (2015) 735-38. https://doi.org/10.1515/amm-2015-0199

[16] R. Dwornicka, N. Radek, M. Krawczyk, P. Osocha, J. Pobedza, The laser textured surfaces of the silicon carbide analyzed with the bootstrapped tribology model. METAL 2017 26th Int. Conf. on Metallurgy and Materials (2017), Ostrava, Tanger 1252-1257.

[17] N. Radek, A. Szczotok, A. Gadek-Moszczak, R. Dwornicka, J. Broncek, J. Pietraszek, The impact of laser processing parameters on the properties of electro-spark deposited coatings. Arch. Metall. Mater. 63 (2018) 809-816.

[18] N. Radek, K. Bartkowiak, Laser treatment of electro-spark coatings deposited in the carbon steel substrate with using nanostructured WC-Cu electrodes. Physics Procedia. 39 (2012) 295-301. https://doi.org/10.1016/j.phpro.2012.10.041

[19] J. Pietraszek, Response surface methodology at irregular grids based on Voronoi scheme with neural network approximator. In: Rutkowski L., Kacprzyk J. (eds) Neural Networks and Soft Computing. Advances in Soft Computing, vol 19. Physica, Heidelberg: 2003, 250-255. https://doi.org/10.1007/978-3-7908-1902-1_35

[20] A. Gadek-Moszczak, History of stereology. Image Anal. Stereol. 36 (2017) 151-152. https://doi.org/10.5566/ias.1867

[21] L. Wojnar, A. Gadek-Moszczak, J. Pietraszek, On the role of histomorphometric (stereological) microstructure parameters in the prediction of vertebrae compression strength. Image Analysis and Stereology 38 (2019) 63-73. https://doi.org/10.5566/ias.2028

[22] D. C. Moreira., H. C. Furtado, J. S. Buarque, et al., Failure analysis of AISI 410 stainless-steel piston rod in spillway floodgate, Engineering Failure Analysis 97 (2019) 506-517. https://doi.org/10.1016/j.engfailanal.2019.01.035

[23] M. Moradi, H. Arabi., S. J. Nasab et al., A comparative study of laser surface hardening of AISI 410 and 420 martensitic stainless steels by using diode laser, Optics and Laser Technology 111 (2019) 347-357. https://doi.org/10.1016/j.optlastec.2018.10.013

[24] T.S. Oliveira, E. S Silva, S. F Rodrigues et al., Softening Mechanisms of the AISI 410 Martensitic Stainless Steel Under Hot Torsion Simulation, Materials Research 20 (2) (2017) 395-406. https://doi.org/10.1590/1980-5373-mr-2016-0795

[25] J. Zheng, W. F. Xie, M. Viens et al., Design of an advanced automatic inspection system for aircraft parts based on fluorescent penetrant inspection analysis, Insight: Non-Destructive Testing and Condition Monitoring 57 (1) (2015) 18-24 and 34. https://doi.org/10.1784/insi.2014.57.1.18

[26] D. Siwiec, A. Pacana, The use of quality management techniques to analyse the cluster of porosities on the turbine outlet nozzle, Production Engineering Archives 24 (2019) 33-36. https://doi.org/10.30657/pea.2019.24.08

[27] E. Nedeliaková, V. Štefancová, M. P. Hranický, Implementation of six sigma methodology using DMAIC to achieve processes improvement in railway transport, Production Engineering Archives 23 (2019) 18-21. https://doi.org/10.30657/pea.2019.23.03

Terotechnology XI                                                                  Materials Research Forum LLC
Materials Research Proceedings **17** (2020) 50-56                  https://doi.org/10.21741/9781644901038-8

[28] R. Ulewicz, M. Nowicka-Skowron, Total quality management in the practice of Polish metallurgical enterprises, METAL 2017 26th Int. Conf. on Metallurgy and Materials, Ostrava, Tanger, 2017, 2338-2343.

[29] N.J. Shipway, P. Huthwaite, M.J.S. Lowe, T.J. Barden, Performance Based Modifications of Random Forest to Perform Automated Defect Detection for Fluorescent Penetrant Inspection, Journal of Nondestructive Evaluation 38 (2) (2019) art. 37. https://doi.org/10.1007/s10921-019-0574-9

[30] J. Zheng, W. F. Xie, M. Viens, M. et al., Design of an advanced automatic inspection system for aircraft parts based on fluorescent penetrant inspection analysis, Insight: Non-Destructive Testing and Condition Monitoring 57 (1) (2015) 18-34. https://doi.org/10.1784/insi.2014.57.1.18

[31]     Lovejoy D., Penetrant testing. A practical guide. New York, Chapman & Hall, 1991.

Terotechnology XI
Materials Research Proceedings 17 (2020) 57-64

Materials Research Forum LLC
https://doi.org/10.21741/9781644901038-9

# Applying a Traditional Casting Defect Classification to Categorize Casting Defects in Metal Matrix Composites with Saturated Reinforcement

GAWDZIŃSKA Katarzyna[1, a *], BRYLL Katarzyna[1,b], KOSTECKA Ewelina [1,c] and STOCHŁA Dorota [1,d]

[1]Maritime University of Szczecin, Faculty of Marine Engineering, Szczecin, Willowa 2 Street, 71-650 Szczecin, Poland

[a]k.gawdzinska@am.szczecin.pl, [b]k.bryll@am.szczecin.pl, [c]e.kostecka@am.szczecin.pl, [d]d.stochla@am.szczecin.pl

**Keywords:** Metal Matrix Composites, Saturated Composite, Casting Defects, Classification

**Abstract.** Traditional casting defects in metal-matrix composites with saturated reinforcement. This classification forms a casting defect group called "structure defects," while the remaining defect groups (shape defects and raw surface defects) under this new classification method include groups present in casting defects in traditionally cast materials. This group (structure defects) contains 5 subgroups, including both structural defects in traditionally cast materials, which correspond to structural defects in saturated composite castings, as well as defects specific to these castings. The proposed classification is still being refined.

## Introduction

Metal composites are increasingly replacing traditional construction materials used in aviation and in the construction of machinery and equipment. This is due to the ability to obtain virtually any desirable functional properties using a composite material, such as a high damping factor, high resistance to abrasion, high Young's modulus, low specific weight, and low coefficient of thermal expansion which was generally widely described in classic books [1, 2], and detailed issues in conference papers, earlier [3, 4] and recently [5-8].

A composite is defined [1, 2] as an ideal material with a perfect structure; however, real composite materials, especially cast composites, usually have imperfect structures because they contain various defects [4, 5, 9-12]. These defects arise because castings have specific structures that are affected by the manufacturing process sequence. Classifying such material irregularities makes it possible to:
- identify them precisely,
- determine why they form,
- determine which manufacturing stage causes their formation,
- promptly take countermeasures.

For metal composite castings, especially those produced by saturation, there is no such classification [13-16]. The classification of casting defects of traditional materials (cast iron, cast steel, and non-ferrous metal alloys) is insufficient and must be supplemented with specific defects for metal composites. This problem, noticed during the manufacturing of castings from saturated metal composites, was the reason for creating such a classification in this paper.

**Findings and discussion**

All metal castings have defects of various types and origins. We define a casting defect as a deviation in a material's structure and mechanical or physicochemical properties from its specifications [16]. Defects can be identified based on their features, which in turn leads to the creation of a *casting defect classification*. This classification is useful for:

1) transferring information in research work, during an educational process, or in a manufacturing process;
2) eliminating defective castings from further stages of a manufacturing process;
3) intervention aimed at removing the causes of defects from the manufacturing process.

As for the second case, the classification criteria of casting defects can be divided into three casting groups [14-19]:

a) good castings (without defects) or with acceptable defects,
b) castings with repairable defects,
c) castings with disqualifying defects.

- **Casting defect classification systems**

For castings made from traditional materials, there are standards, atlases or catalogues of defects [2, 14-21], which:

- enable unambiguous identification of defects;
- provide methods to detect them;
- determine why they form;
- suggest technological measures to prevent their formation.

Classification diagrams for casting defects are shown in Figure 1.

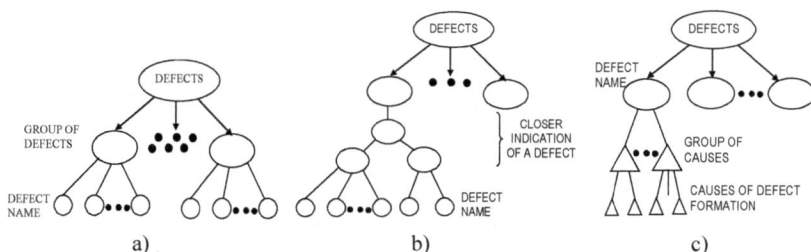

*Figure 1. Classification diagrams for casting defects in classic materials [17-21]
a – according to Polish standards; b – according to French standards; c – according to English and German systems.*

In Poland, a division described in [18-19] is used, with two distinguished levels (Figure 1a). At the upper level, 4 defect groups are identified; at the lower level, each group is assigned defects with specific features that are given names which help unambiguously identify them. In the French foundry industry, a multi-stage structure is used [17, 20-21] whose first level contains 7 groups:

- outward growth of a metal;
- external and internal bubbles;
- discontinuities in a casting;

– surface defects;
– incompleteness of a product;
– inaccurate dimensions or shape;
– structural inclusions or anomalies.

The lowest level also contains the names of individual defects; however, between this level and the definition of a defect group, there are two intermediate levels containing additional features of the group or subgroup (Figure 1b). In this way, each defect is assigned certain characteristics to make it easier to identify its causes and take preventive actions [17, 20-21].

In the English and German literature, defects are classified in a different manner [17, 20-21]. The principle of this classification is presented in Figure 1c. Defects are given names here, and they are assigned to cause groups and particular causes of defect formation. This division would be very convenient for identification, but some causes were defined somewhat inaccurately. Since several defects may have a common cause, this division may not always be used objectively.

**Classification of casting defects according to Polish Standards**
Polish classification of defects in metal castings is one of the few classifications covered by governmental standards. As presented in the review in the previous section, this classification is the simplest because of its two-stage arrangement. It also clearly divides defects in castings made of different materials. The features, wide availability, and familiarity of this classification in Poland make it necessary to refer to this classification while attempting to create another. Thus, any further classifications will be based on the standard Polish defect classification, making it necessary to elaborate on the Polish classification system. According to the Polish standard [19] and national studies [2, 13, 18-19], this classification contains 4 groups of defects, presented in Figure 2.

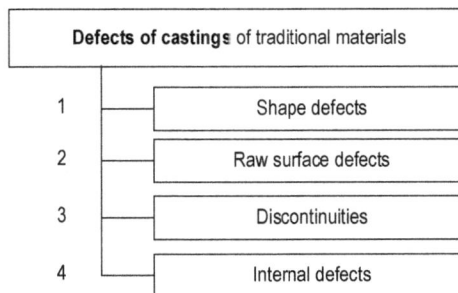

*Figure 2. Defects in castings of traditional materials*
*(e.g., cast steel, cast iron, non-ferrous alloys).*

The order of defect groups is consistent with the sequence of operations in a casting acceptance by a quality control department. Shape defects are observed first, followed by raw surface defects, and discontinuities. Internal defects are detected during non-destructive and destructive testing and machining of castings. Each of the four groups is assigned certain defects, which are then marked with "W", and the type of material in which they occur is indicated in Table 1. The Polish Standard (PN-85/H-83105) also indicates the causes of defect formation. In

Materials Research Forum LLC

https://doi.org/10.21741/9781644901038-9

other studies [2, 3, 5, **8-18**, 22-27], which properly extend this standard, methods of detection, description, and repair of defects were also found.

*Table 1 Classification of defects in castings of traditional materials [19]*

| Defect name | Marking | Occurrence: |
|---|---|---|
| Group 1 – Shape defects | | |
| Mechanical damage | W-101 | all alloys |
| Misrun | W-102 | all alloys |
| Knob | W-103 | all alloys |
| Flash | W-104 | all alloys |
| Mismatch (shift) | W-105 | all alloys |
| Swelling | W-106 | all alloys |
| Warping | W-107 | all alloys |
| Group 2 – Raw surface defects | | |
| Roughness | W-201 | all alloys |
| External bubble | W-202 | all alloys |
| Pitted skin | W-203 | cast steel |
| Pock-marking | W-204 | all alloys |
| Pinholes | W-205 | all alloys |
| Shrinkage depression | W-206 | all alloys |
| Cold lap | W-207 | all alloys |
| Sand buckle | W-208 | all alloys |
| Rat tails | W-209 | all alloys |
| Sand holes | W-210 | all alloys |
| Crush | W-211 | all alloys |
| Contamination | W-212 | all alloys |
| Scale | W-213 | malleable cast-iron |
| Galling | W-214 | non-ferrous metals |
| Partial melting (during annealing) | W-215 | malleable cast-iron |
| Elephant skin | W-216 | spheroidal graphite iron |
| Sweat | W-217 | non-ferrous metals |
| Flowers | W-218 | non-ferrous metals |
| Metal penetration | W-219 | all alloys |
| Veins | W-220 | all alloys |
| Burning-on (of sand) | W-221 | all alloys |
| Sand holes | W-222 | all alloys |
| Oxidation | W-223 | non-ferrous metals |
| Peel | W-224 | malleable cast-iron |
| Group 3 – Discontinuities | | |
| Hot cracks | W-301 | all alloys |
| Cold cracks | W-302 | all alloys |
| Shrinkagcrack | W-303 | all alloys |
| Annealing cracks | W-304 | malleable cast-iron |
| Transgranular cracks | W-305 | cast steel, non-ferrous metals |

| Group 4 – Internal defects | | |
|---|---|---|
| Gas bubble | W-401 | all alloys |
| Porosity | W-402 | all alloys |
| Shrinkage cavity | W-403 | all alloys |
| Microporosity | W-404 | all alloys |
| Slag inclusion | W-405 | all alloys |
| Sand drops | W-406 | all alloys |
| Cold shots | W-407 | all alloys |
| Foreign metal | W-408 | all alloys |
| Segregation | W-409 | non-ferrous metals |
| Coarse-grained structure | W-410 | non-ferrous metals |
| Hard spots | W-411 | cast iron |
| Gray spots | W-412 | malleable cast-iron |
| White fracture | W-413 | malleable cast-iron |
| Bright fracture | W-414 | malleable cast-iron |
| Bright border | W-415 | malleable cast-iron |
| Heterogeneity | W-416 | all alloys |

**Summary**
The literature concerning defects in metal matrix saturated composite castings is scarce, and a description of the quality of such castings requires unambiguous defect classification. In the technical literature, there is no such classification – only attempts – to which the authors have contributed [10, 28-29]. In publications on composite castings, these irregularities (defects) are often defined imprecisely [1, 5, 17, 22], and the descriptions often include several similar defects occurring at different stages of the manufacturing process, possibly due to various non-interrelated reasons. The only common feature of these defects is often their shape, size, form, etc.

The first two groups of defects and part of the third and fourth groups in castings made of conventional materials (Fig.2 and Table 1) and in composite castings, including saturated composites, are consistent with the previously presented defect identification methods. Metal saturated composite castings have a characteristic structure that results from the presence of reinforcement, most often fibres, located in the metal matrix and permanently connected with it.

The matrix and reinforcement can contact in locations where defects not present in conventional castings may appear. Some can be classified as a lack of continuity, e.g., a discontinuity at phase boundaries, while others are classified as internal defects, e.g., pores or defects in the matrix structure. Other defects are not found in traditional classifications and mainly include reinforcement and matrix defects or a combination of these components; however, they may also include pores if the matrix incompletely saturates the reinforcement. Hence, there is a need to develop a classification for specific defects of metal matrix composite castings with saturated reinforcement. Therefore, it was decided to develop a defect classification system which was an intermediate system between the Polish and French systems (Figure 1), dividing defects into groups, while shape defects and raw surface defects would correspond to the Polish classification. Due to the specific structures of composites, continuity defects, internal defects, and defect characteristics for saturated composites fell under a single group, which was divided into subgroups. However, making this division requires a detailed analysis of the manufacturing process of saturated metal composite castings, which should be performed concerning a possible defect formation during various stages of this process. "Structure defects" would be the best-suited name for that group. The general structure of this classification is shown in Fig. 3.

The proposed classification, including both ontology and taxonomy, may be interesting not only in the material [30] area, but also in the mechanical [31-33] and technological [34, 35] area.

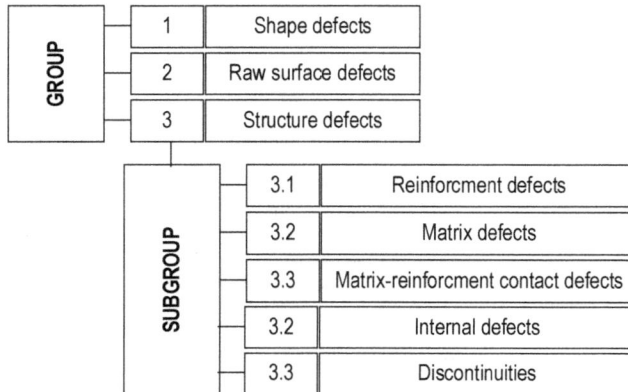

*Fig. 3. Proposed classification of defects in saturated metal composite castings.*

**References**

[1] T.W.Clune, P.J. Withers, An Introduction of Metal-Matrix Composites, Cambridge University Press 1993. https://doi.org/10.1017/CBO9780511623080

[2] J. Sobczak , Metalowe materiały kompozytowe, Instytut Odlewnictwa 1996.

[3] Z. Ignaszak, P.Popielarski, Sensitivity Tests of Simulation Models Used in Chosen Calculation Codes on Uncertainty of Thermo-Mechanical Parameters during Virtual Mechanical Stress Estimation for Ferrous Alloy Castings, Defect and Diffusion Forum 312-315 (2011) 758-763. https://doi.org/10.4028/www.scientific.net/DDF.312-315.758

[4] B. Anasori, E. Caspi, and M. Barsoum, Fabrication and Mechanical Properties of Pressureless Melt Infiltrated Magnesium Alloy Composites Reinforced with TiC and Ti2AlC Particles, Mater. Sci. Eng. 618 (2014) 511-522. https://doi.org/10.1016/j.msea.2014.09.039

[5] K. Gawdzińska, D. Nagolska, P. Szymański, Determination of duration and sequence of vacuum pressure saturation in infiltrated MMC castings, Arch. Foundry Eng. 18 (2018) 23-28

[6] M. Wiciak, T. Chwalczuk, A. Felusiak, Experimental investigation and performance analysis of ceramic inserts in laser assisted turning of Waspaloy, MATEC Web of Conf. 237 (2018) art. 01003. https://doi.org/10.1051/matecconf/201823701003

[7] T. Chwalczuk, M. Wiciak, A. Felusiak, P. Kieruj, An Investigation of Tool Performance in Interrupted Turning of Inconel 718, MATEC Web of Conf. 237 (2018) art. 02008. https://doi.org/10.1051/matecconf/201823702008

[8] Hajkowski J., Popielarski P., Sika R. (2018) Prediction of HPDC casting roperties made of AlSi₉Cu₃ alloy. In: Hamrol A., Ciszak O., Legutkc S., Jurczyk M. (eds) Advances in Manufacturing. Lecture Notes in Mechanical Engineering. Springer, Cham. https://doi.org/10.1007/978-3-319-68619-6_59

[9] D. Przestacki, T. Chwalczuk, The analysis of surface topography during turning of Waspaloy with the application of response surface method, MATEC Web of Conf. 136 (2017) art. 02006. https://doi.org/10.1051/matecconf/201713602006

[10] K. Gawdzińska, L.Chybowski, W. Przetakiewicz, Study of Thermal Properties of Cast Metal-Ceramic Composite Foams. Arch. Foundry Eng. 17 (2017) 47-50. https://doi.org/10.1515/afe-2017-0129

[11] A. Felusiak, T. Chwalczuk, M. Wiciak, Surface Roughness Characterization of Inconel 718 after Laser Assisted Turning, MATEC Web of Conf. 237 (2018) art. 01004. https://doi.org/10.1051/matecconf/201823701004

[12] Z. Ignaszak, P. Popielarski, J. Hajkowski, Sensitivity of Models Applied in Selected Simulation Systems with Respect to Database Quality for Resolving of Casting Problems, Defect and Diffusion Forum 336 (2013) 135-146. https://doi.org/10.4028/www.scientific.net/DDF.336.135

[13] P. Twardowski, M. Wiciak-Pikuła, Prediction of Tool Wear Using Artificial Neural Networks during Turning of Hardened Steel, Materials 12 (2019), art. 3091. https://doi.org/10.3390/ma12193091

[14] P.Szymański, M. Borowiak, Evaluation of castings surface quality made in 3D printed sand moulds using 3DP technology, Lecture Notes in Mechanical Engineering, Advances in Manufacturing II, Vol. 4 - Mechanical Engineering, pp 201-212. https://doi.org/10.1007/978-3-030-16943-5_18

[15] A. Alagarsamy. Defect Analysis Procedure and Case History. Ductile Iron News 2004.

[16] American Foundry Society.(2015). Analysis of Casting Defects. American Foundry Society, pp. 117-120.

[17] S. Kluska-Nawarecka, Metody komputerowe wspomagania diagnostyki wad odlewów, Instytut Odlewnictwa, Kraków 1999.

[18] Z. Falęcki, Analiza wad odlewów, Wyd. AGH, Kraków 1997.

[19] Polska norma PN-85/H-83105. Odlewy. Podział i terminologia wad.

[20] International Atlas of Casting Perfect, Internat. Comit. of Found Tech. Associat. American Foundrymen's Society 1986.

[21] International atlas of foundry defects. International Committee of Foundry Technical Associations. Committee of Metallurgy and Foundry Properties. English Edition 1974.

[22] T. Chwalczuk, D. Przestacki, P. Szablewski, A. Felusiak, Microstructure characterisation of Inconel 718 after laser assisted turning, MATEC Web of Conf. 188 (2018) art. 02004. https://doi.org/10.1051/matecconf/201818802004

[23] S. Wojciechowski, P. Twardowski, T. Chwalczuk, Surface roughness analysis after machining of direct laser deposited tungsten carbide, Journal of Physics: Conference Series 483 (2014) art. 012018. https://doi.org/10.1088/1742-6596/483/1/012018

[24] M. Kawalec, D. Przestacki, K. Bartkowiak, Jankowiak, M., Laser assisted machining of aluminium composite reinforced by SiC particle. ICALEO 2008 27th Int. Congress on Applications of Lasers and Electro-Optics (2008) 895-900. https://doi.org/10.2351/1.5061278

[25] S. Wojciechowski, D. Przestacki, T. Chwalczuk, The evaluation of surface integrity during machining of inconel 718 with various laser assistance strategies, MATEC Web of Conf. 136 (2017) art. 01006. https://doi.org/10.1051/matecconf/201713601006

[26] A. Bartkowska, A. Pertek, M. Popławski, D. Przestacki, A. Miklaszewski, Effect of laser modification of B-Ni complex layer on wear resistance and microhardness Optics and Laser Tech. 72 (2015) 116-124. https://doi.org/10.1016/j.optlastec.2015.03.024

[27] M. Kukliński, A. Bartkowska, D. Przestacki, Investigation of laser heat treated Monel 400, MATEC Web of Conf. 219 (2018) art. 02005. https://doi.org/10.1051/matecconf/201821902005

[28] K. Gawdzińska, Quality features of metal matrix composite castings, Arch. Metall. Mater. 58 (2013) 659-662. https://doi.org/10.2478/amm-2013-0051

[29] Gawdzińska K., Bryll K., Nagolska D., Influence of heat treatment on abrasive wear resistance of silumin matrix composite castings, Arch. Metall. Mater. 61 (2016) 177-182. https://doi.org/10.1515/amm-2016-0031

[30] L. Mosinska, K. Fabisiak, K. Paprocki, M. Kowalska, P. Popielarski, M. Szybowicz, A. Stasiak, Diamond as a transducer material for the production of biosensors. Przem. Chem. 92 (2013) 919-923.

[31] P. Krawiec, A. Marlewski, Profile design of noncircular belt pulleys. J. Theor. Appl. Mech. 54 (2016) 561-570. https://doi.org/10.15632/jtam-pl.54.2.561

[32] P. Krawiec, G. Domek, L. Wargula, K. Walus, J. Adamiec, The application of the optical system atos ii for rapid prototyping methods of non-classical models of cogbelt pulleys. MATEC Web of Conf. 157 (2018) art. 01010. https://doi.org/10.1051/matecconf/201815701010

[33] P. Krawiec, M. Grzelka, J. Kroczak, G. Domek, A. Kolodziej, A proposal of measurement methodology and assessment of manufacturing methods of nontypical cog belt pulleys. Measurement 132 (2019) 182-190. https://doi.org/10.1016/j.measurement.2018.09.039

[34] A. Bartkowska, M. Kuklinski, P. Kieruj, The influence of laser heat treatment on the geometric structure of the surface and condition of the surface layer and selected properties of Waspaloy. MATEC Web of Conf. 121 (2017) art. UNSP 03006. https://doi.org/10.1051/matecconf/201712103006

[35] P. Kieruj, M. Kuklinski, Tool life of diamond inserts after laser assisted turning of cemented carbides. MATEC Web of Conf. 121 (2017) art. UNSP 03011. https://doi.org/10.1051/matecconf/201712103011

Terotechnology XI
Materials Research Proceedings 17 (2020) 65-71

Materials Research Forum LLC
https://doi.org/10.21741/9781644901038-10

# Logistic Controlling Processes and Quality Issues in a Cast Iron Foundry

ULEWICZ Robert[1, a], MAZUR Magdalena[1, b] *, KNOP Krzysztof[1, c]
and DWORNICKA Renata[2, d]

[1]Czestochowa University of Technology, J.H. Dabrowskiego 69, 42-201 Czestochowa, Poland

[2] Krakow University of Technology, Warszawska 24, 31-155 Krakow, Poland

[a] robert.ulewicz@pcz.pl, [b] magdalena.mazur@pcz.pl, [c] krzysztof.knop@pcz.pl,
[d] renata.dwornicka@mech.pk.edu.pl

**Keywords:** Iron Foundry, Quality, Process, Logistics, Controlling, Work Ergonomics

**Abstract.** The article shows issues related to the creation and implementation of added value through logistics and processing in a cast iron foundry. The possibilities of improving the casting production systems through the assessment of their labor consumption and the analysis of manufacturing prime costs were discussed. In order to increase the efficiency of iron foundry management, quality control was introduced. The comparative method was used in the studies presented here. Two areas of organization management were compared: the quality management area and the financial management area. The purpose of this comparison is to find common features when supervising tasks in both zones.

## Introduction

The activity of a person managing the process comes down to transforming resources into a desired effect in such a way as to meet the requirements, while the control should be understood as the activity aimed at ensuring the desired course of the process [1]. The final effect of the company's work may take various forms, and one of the basic effects of starting a production activity are generally understood products. High quality of products can be achieved, among others by building a measurement and control system in a manner enabling the generation of information related to the company's needs or the needs of the system monitoring the company's situation [2, 3].

For most companies, high-quality delivery is a key area that determines the production process. In the presented case, quality is the rhythm of deliveries as well as the quantity and technical parameters in accordance with the orders. On the other hand, sales is an area that determines the company's market success. In this perspective, purchasing and selling represent two extreme activities of the company. The first one determines the commencement of production within a strictly defined deadline, provided that the appropriate input material is obtained. The second enables an organized co-operation with recipients. In both cases, it is necessary to ensure efficient control of logistics processes, elimination of interruptions, delays in order fulfillment, etc. Optimizing the involvement of stock and capital usually leads to lowering running costs and increasing their fluctuation, especially when the manufactured product is an industrial product, e.g. a cast mold, where a significant part of the total cost (about 50%) is related to the material.

### The quality of the supply chain

The quality of the supply chain must be considered from the very beginning. In the case of iron foundries, the logistic chain of supplier-foundry-customer may be of a different nature due to the

specificity of production [4]. Here you can deal with long-term contracts covering both serial production and complex individual castings. In both cases, however, key elements can be identified:
– product quality depends on the quality of raw materials, materials and semi-finished products,
– the quality of information flow and the level of customer service.
In the conditions of fierce competition on the market of foundry products, apart from good quality castings, customer service is also important, depending on logistic processes in the area of customer service. Only a product that meets customer expectations and customer service alone is able to provide the company with a competitive advantage. Many companies, in order to improve their competitiveness, create meta-logistics chains, which is one of the key issues in the uniform understanding of quality.

The problem is quite serious because, in common terminology, a 'defect' triggers procedures, control system check-ups, possibly misunderstandings, downtime, deadline failures and thus generates financial losses. In solving this problem, the management system leaned on ISO 9001 is very helpful, in which the authors introduced a trial approach with the definition of an internal customer and an external customer [5].

The use of comprehensive management includes essential elements of TQM, and the goal is to ensure high quality parameters of raw materials, materials, semi-finished products and final products. If the quality level of these products is high, the logistics system will also achieve its goals: cost reduction, inventory reduction, storage space reduction.

**Controlling of logistics processes**
Controlling is the only logical and concise response to the processes taking place in the company and its environment. Material flows as well as production and management processes play an important role [6, 7]. The controlling system compares the standard defined by the schedule and plan with the actual state, taking into account the economic dimension of the existing derivations. A comparative assessment of the actual and the assumed state may bring two results:
a) first – when the actual state meets the limitations of the standard (no corrective actions) and
b) second – when the actual condition does not meet the limits of the standard (corrective actions should be taken) [8].
The controlling department, using the available information, analyzes to what extent the production status depends on the stock levels, human resources, development research, available technologies and the market situation. Fig. 1 shows controlling in the analyzed company against the background of key areas.

In order to be able to speak of universal controlling, it is first necessary to present four basic procedures that will have to be carried out regardless of where controlling is introduced. These treatments include:
a) procedure "0" – building the controlling base. This procedure covers both the reporting part and the control. In particular, the first one produces market technical and organizational indicators, as well as financial reporting related to the total effect basic items, capital flows and income statement related to the basic total effect items, capital flows and profit and loss ratios.
b) procedure "1" – organizational and information security of the controlling system in the company's organizational structures. This procedure concerns determining the scope of responsibility for creating a base of individual controlling services, keeping records and actual monitoring of the size, as well as calculating and analyzing gagers size deviations: indicators

of the technical and organizational market plan, indicators, financial reporting, control indicators.

c) procedure "2" – determination of the level of acceptable deviations of indicators used in the controlling system. This procedure concerns the determination of acceptable deviations of the baseline indicators. It shows a variant approach to creating an acceptable limit method.

d) procedure "3" – corrective actions. This procedure pertains to deciding on an acceptable gager size. When the amount of deviation from the base is measured, at least the cause of this phenomenon must be determined within a legal interval.

*Figure 1. The main links of the controlling system*
*influencing the logistic processes of purchasing and sales*

When designing controlling procedures, the logistics system may use the following gagers:
a) level of the customer's service
   - state of material provisions,
   - state of provisions of production in progress,
   - state of provisions of final goods,
   - state of the physical flow of materials,
   - state of computer processes,
b) market position
   - market share in basic assortments,
   - market share in all assortments,
   - changes in the market share in general,
   - valuable participation of own logistic processes in demand for logistics in general;
c) remunerativeness
   - level of expenses of logistics,
   - participation of expenses of logistics in expenses of producing,
   - level of producing expenses,
   - level of sale expenses,
   - income in general,
   - remunerativeness of logistics,
   - remunerativeness of producing.

Terotechnology XI                                                    Materials Research Forum LLC
Materials Research Proceedings **17** (2020) 65-71                    https://doi.org/10.21741/9781644901038-10

**Controlling of the quality**

The research used the methodology described by Polak [9], and the techniques of document examination, interview and participant observation were used to collect the data. The following documents were analyzed: processes, supplementary order cards, audit reports, proxy quality plans, quality reports, reports on quality council sessions and supervisory board meetings, and financial reporting. The following action plan was established:

- defining aspects and goals,
- planning the implementation of activities,
- implementation of activities,
- control of the implementation of activities,
- analysis of the final results and summary.

In the first stage of work, the goals of the organization were defined in terms of quality, environment, work safety and finances. Risky spots were created in the implementation of the enterprise in relation to standards, regulations regarding the rights and expectations of the client. Properly identified aspects define the goals of the organization and related processes. Properly identified aspects define the goals of the organization, as well as processes related to them. Usually, these are separate goals concerning quality, environment, work safety and finances, which simplifies the delegation of responsibility and controlling. It is not recommended to combine goals in all spheres of management. When planning quality tasks for the entire organization, the proxy determines the main priorities: periodic reviews, audits, corrective tasks. The final stage of activities results in an annual report containing breakdowns and existing threats. The report should include an analysis of management effectiveness and measures to improve the quality of management. A similar plan of action also applies to the sphere of financial management. There are two boards in the area of quality and finance: the quality board oversees the achievement of universal goals, and the supervisory board oversees the achievement of results-oriented goals. There is a great similarity in controlling, operating in the sphere of quality management and in the financial sphere, which creates the possibility of quick integration of these spheres. Table 1 presents a comparison of the sphere of quality and the sphere of finance.

*Table 1. Comparison of financial sphere and qualities.*

| Element | Defining of element | Quality management zone | Finance management zone |
|---|---|---|---|
| **Responsibility for controlling process** | Initiator | Management proxy of quality system | Finance controller |
| | References center | Quality board | Chairmen board |
| | Recipient | External auditor | Accountant |
| **Tools** | | Audit, inspection, periodical survey. | Stock-taking, inventory, survey |
| **Documents related to controlling process** | Initiating | Quality plan | Finance plan |
| | Periodical survey | Periodical report | Finance report |
| | Finalizing | Quality state report | Company balance |
| | Improving | Business plan | |

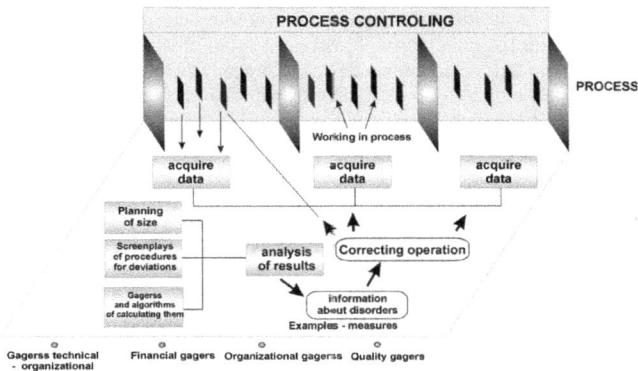

*Figure 2. The main relations of the controlling system affecting the logistics processes of purchases and sales*

The collection and analysis of all kinds of information is the basis of the controlling unit's operation. The type of information flow between the various departments of the company is of great importance for the method of production control. The flow of information in the organization affects, among other things, the state of stock and its relationship with the current production and purchases. Thanks to the available information, the controlling department analyzed the dependence of the production status on the stock level, human resources, development research, available technologies and the market situation. The level of coordination of the activities of the marketing department, production department and development department turned out to be very important. The coordinator here is the controlling unit. Controlling plays a significant role in quality assurance and optimizes the product control process.

The functioning of a company in a trial system with the use of controlling requires the introduction of absolute information discipline. Persons participating in these processes should inform about the degree of implementation of tasks on an ongoing basis, so that process managers have up-to-date information about emerging disorders. An example may be Fig. 2, where data is recorded as part of the activities carried out, which in turn creates the necessary basic information resource, for example, for the effective operation of controlling.

## Summary

Controlling in logistics and quality, properly implemented in the company, manifests itself in two ways. On the one hand, it helps and enables the coordination of logistics and management processes with quality, with a special emphasis on planning and quality control of supply chains. Through the quality of production, it also works to increase the competitiveness of the offered products and increase the company's revenues. On the other hand, controlling is forcing processes to constantly manage the expenses incurred in the company and looking for an opportunity to reduce them. Therefore, controlling, by direct impact on expenses and indirectly on income, increases the profitability of the company.

The implementation of the company's process management system makes it possible to organize the company in terms of the processes carried out. The controlling system helps in filling that part of the management element, which has also been subordinated to the processes. A team that is built with the company's processes in mind must use simple tools that are best suited to its processes to ensure compliance in the company's "internal markets". The currently identified centers of responsibility in the casting production process mean that the final goods can be delivered at the right time, to the right place and possibly produced at a sufficiently low cost. The controlling system must adapt to changing internal customers and the conditions of the operations performed. Process controlling is intended to help in the complete shaping, control and evaluation of these processes. Further work is underway to find the appropriate explanatory factors describing the effects of the implementation of the processes and to capture their indicators appropriate to the form or measured quantity.

Logistics controlling presented in this article may be useful in various industrial areas, including hydraulics [10, 11] and heavy working machines [12, 13], corrosion protection [14], as well as production of machine parts [15, 16] and improvement of their properties [17]. It may also have a direct impact on the management of similar industrial processes [18], and require appropriate modifications in analytical research methods [19] even in biotechnology [20], especially those that use statistical methods [21, 22].

**References**

[1] R. Wolniak, Operation manager and its role in the enterprise, Production Engineering Archives 24 (2019) 1-4. https://doi.org/10.30657/pea.2019.24.01

[2] A. Pacana, K. Czerwińska, Analysis management instruments of the causes of control panel inconsistencies in the gravitational casting process by means of quality, Production Engineering Archives 25 (2019) 12-16. https://doi.org/10.30657/pea.2019.25.03

[3] B. Jereb, J. Rosak-Szyrocka, Quality of investments in logistics, Production Engineering Archives 2 (2014) 6-8. https://doi.org/10.30657/pea.2014.02.02

[4] M. Nowicka-Skowron, R. Ulewicz, Quality management in logistics processes in metal branch, METAL 2015 24th Int. Conf. on Metallurgy and Materials, Ostrava, Tanger, 2015, 1707-1712.

[5] R. Ulewicz (eds.), Narzędzia jakości w praktyce. Poradnik dla biznesu, Czestochowa, SMJiP, 2018.

[6] P. Nyhuis, H.P. Wiendahl, Fundamentals of production logistic – theory, tools and applications. Berlin, Springer, 2009. https://doi.org/10.1007/978-3-540-34211-3

[7] H. Kerzner, Project management best practices: achieving global excellence. New York, Wiley, 2010.

[8] S. Kukla, Costs analysis of iron casts manufacturing. Arch. Foundry Eng. 12 (2012), 45-48. https://doi.org/10.2478/v10266-012-0034-9

[9] A. S. Polak. Quality controlling – trends. Problemy Jakosci 2 (2004), 18-21.

[10] P. Walczak, A. Sobczyk, Simulation of water hydraulic control system of Francis turbine. Proc. 8th FPNI Ph.D Symposium on Fluid Power, 2014, art. V001T04A001. https://doi.org/10.1115/FPNI2014-7814

[11] M. Domagala, H. Momeni, J. Domagala-Fabis, G. Filo, D. Kwiatkowski, Simulation of cavitation erosion in a hydraulic valve. Materials Research Proceedings 5 (2018) 1-6. https://doi.org/10.21741/9781945291814-1

[12] M. Domagala, H. Momeni, J. Domagala-Fabis, G. Filo, M. Krawczyk, J. Rajda, Simulation of particle erosion in a hydraulic valve. Materials Research Proceedings 5 (2018) 17-24. https://doi.org/10.21741/9781945291814-4

[13] P. Krawiec, A. Marlewski, Spline description of non-typical gears for belt transmissions. J. Theor. Appl. Mech. 49 (2011) 355-367.

[14] T. Lipinski, D. Karpisz, Corrosion rate of 1.4152 stainless steel in a hot nitrate acid. METAL 2019: 28th Int. Conf. on Metallurgy and Materials, Ostrava, TANGER, 2019, 1086-1091. https://doi.org/10.37904/metal.2019.911

[15] T. Pieczonka, J. Kazior, A. Szewczyk-Nykiel, M. Hebda, M. Nykiel, Effect of atmosphere on sintering of Alumix 431D powder. Powder Metall. 55 (2012) 354-360. https://doi.org/10.1179/1743290112Y.0000000015

[16] J. Pietraszek, A. Gadek-Moszczak, The Smooth Bootstrap Approach to the Distribution of a Shape in the Ferritic Stainless Steel AISI 434L Powders. Solid State Phenomena 197 (2012) 162-167. https://doi.org/10.4028/www.scientific.net/SSP.197.162

[17] D. Przestacki, R. Majchrowski, L. Marciniak-Podsadna, Experimental research of surface roughness and surface texture after laser cladding. App. Surf. Sci. 388 (2016) 420-423. https://doi.org/10.1016/j.apsusc.2015.12.093

[18] A. Pacana, L. Bednarova, I. Liberko, A. Wozny, Effect of selected production factors of the stretch film on its extensibility. Przem. Chem. 93 (2014) 1139-1140.

[19] L. Wojnar, A. Gadek-Moszczak, J. Pietraszek, On the role of histomorphometric (stereological) microstructure parameters in the prediction of vertebrae compression strength. Image Analysis and Stereology 38 (2019) 63-73. https://doi.org/10.5566/ias.2028

[20] E. Skrzypczak-Pietraszek, A. Urbanska, P. Zmudzki, J. Pietraszek, Elicitation with methyl jasmonate combined with cultivation in the Plantform™ temporary immersion bioreactor highly increases the accumulation of selected centellosides and phenolics in Centella asiatica (L.) Urban shoot culture. Engineering in Life Sciences. 19 (2019) 931-943. https://doi.org/10.1002/elsc.201900051

[21] T. Styrylska, J. Pietraszek, Numerical modeling of non-steady-state temperature-fields with supplementary data. ZAMM 72 (1992) T537-T539.

[22] J. Pietraszek, E. Skrzypczak-Pietraszek, The optimization of the technological process with the fuzzy regression. Adv. Mater. Res-Switz. 874 (2014) 151-155. https://doi.org/10.4028/www.scientific.net/AMR.874.151

Terotechnology XI
Materials Research Proceedings 17 (2020) 72-78

Materials Research Forum LLC
https://doi.org/10.21741/9781644901038-11

# Assessment of the Technological Position of a Selected Enterprise in the Metallurgical Industry

KLIMECKA-TATAR Dorota[1, a *] and INGALDI Manuela[1, b]

[1] Department of Production Engineering and Safety, Faculty of Management, Czestochowa University of Technology, Al. Armii Krajowej 19b, 42-218 Czestochowa, Poland

[a] d.klimecka-tatar@pcz.pl, [b] manuela.ingaldi@pcz.pl

**Keywords:** Technological Portfolio, Technology Assessment, Metallurgical Industry, Matrix 3x3

**Abstract.** A dynamic production market, in particular for the metallurgical industry, strongly influences the level of production efficiency, product quality, and mainly a company's position on the market and its competitiveness. This paper presents the results of research on the assessment of the technological position of a selected average company from the metallurgical industry in relation to development strategies. The analysis of a company's technological position has been made using the 3x3 matrix and Parker rating scale. The 3x3 matrix helped to determine an enterprise's technological position as well as factors that affect it. As follows from the presented research, an enterprise is located as "ordinary, average" both in terms of its technological possibilities and its position on the market. However, it was also possible to indicate the factors to which the company should pay special attention in order to strengthen their importance.

**Introduction**

According to various definitions, technology is the process of producing the necessary products and services, implemented in a hierarchical production system. It is important that all elements and connections are identified and have their order [1]. Available theoretical and practical knowledge [2] also play an important role for the implementation of technology. What is worth noticing, the functioning of enterprises in the 21st century is based on the constant pursuit of improvement and constant development [3]. The changing and dynamic external environment of each enterprise enforces the need to strive for improvement by using increasingly newer and innovative technologies [4,5]. Therefore, the scientific and technical resources (machines, devices, human capital, knowledge, experience, etc.) are becoming a very important competitive factor on the market [6–9]. The technological level of an enterprise determines the development of the organization, creates the possibility of strategy building, affects the efficiency and effectiveness of the enterprise, as well as determines the production volume, quality and unit costs [10,11]. The indicated competitiveness factors become particularly important when technology assessment is the basis for implementation or commercialization [12].

The assessment of technologies and potential technological possibilities is extremely important, because it allows the company to introduce changes aimed at production optimization, improvement of the products quality and increase of production capacity [1,12]. The results of technology assessment are also the basis for the implementation of modernization and reorganization of production processes. All activities related to the improvement of the technological potential of the company contribute to an increase in competitive advantage. Company's technological competitiveness depends primarily on the value of "applied research" and the competence of "development" teams, linking technology with the company's core

Terotechnology XI                                                    Materials Research Forum LLC
Materials Research Proceedings **17** (2020) 72-78              https://doi.org/10.21741/9781644901038-11

business, time advantage over competition and financing potential [13]. The goal of a technology portfolio assessment is to maximize the expected profit with demand and technological uncertainty. Due to research on the assessment of a company's technological position, it is possible to increase its efficiency [14–16]. This is mainly to contribute to the risk assessment in two scenarios. Scenario 1: a company does not develop new technology because there is sufficient currently used - the risk that other companies in this industry will largely dominate the market due to the changing nature of industry (Industry 4.0). Scenario 2: a company will introduce new technology without proper preparation - the risk of a significant weakening of the position of an enterprise caused by the diversion of power to implement the technology for which it was not prepared [17].

## Methodology

The 3x3 matrix, which determines the relationship between technological possibilities of the enterprise with its position on the market, has been used in the paper. This matrix was described in the previous works [18–20]. In the original matrix [20], the X-axis is represented by the technological possibilities and the Y-axis by the position on the market.

The 3x3 matrix consists of 9 fields, which correspond to the respective technological position of the companies and help to identify the future action in terms of technology. Therefore it can be used as an element defining the strategy of a company. The objective of each company is the field marked with number 1, that is "Focus on the revealed chance". This is a field where both elements get high evaluations. So it can characterize a company as the one with a very good position on the market (competitive) and with high technological possibilities.

In the paper, also the 9-point Parker scale has been used. In the original version its scale is following: 1-3 means weak, 4-6 average, 7-9 strong influence [20]. But in the paper both positive and negative factors have been used so authors decided to change the interpretation of the scale for the following: 1-3 negative influence of the factor, 4-6 neutral (there is a factor but it has minimal influence or does not have it), 7-9 positive with regard to technological possibilities and position on the market. It allows for indicating the technological position of companies.

The research has been conducted in a chosen metallurgical enterprise. It is the enterprise operating on the market for about 50 years, producing various types of steel profiles for the European and global market. The first stage of the research determined the factors that well described the functioning of the enterprise. All these factors were divided into two groups: those that determine the technological capabilities of the enterprise and those that determine its position on the market. Those factors that did not match any of the groups were omitted. All these factors were evaluated on a scale of 1-9. 1 means negative influence, 5 means neutral influence, however the factor occurs, while 9 means positive influence. Then the average values for both groups of factors were calculated. These averages placed on the 3x3 matrix helped to determine the enterprise's technological position and determine the factors which affect it.

## Results

The summary of factors which decide about technological possibilities of the research enterprise and factors deciding about its position on the market are presented in Table 1, while the 3x3 matrix and its technological position is presented in Figure 1.

From the analysis whose results are presented in Figure 1, it can be concluded that the research enterprise is located in the middle field of the 3x3 matrix, e.g. "Search for occasions". This means that the enterprise is located as "ordinary, average" both in terms of its technological possibilities and its position on the market.

*Table 1. List of factors for 3x3 matrix [own study]*

| Factors deciding about technological possibilities | | Evaluation |
|---|---|---|
| 1 | Modernity of the production line | 6 |
| 2 | New technology of production | 5 |
| 3 | Time of changeovers | 6 |
| 4 | Synchronization between production operations | 7 |
| 5 | Repeatability of process | 8 |
| 6 | Use of SMED | 7 |
| 7 | 5S methodology | 6 |
| 8 | Storage area | 5 |
| 9 | Modern means of transport | 5 |
| 10 | Flexibility of assortment | 3 |
| 11 | Quality of the products | 7 |
| 12 | Staff experience | 5 |
| 13 | A large number of young, qualified people on the labor market | 2 |
| 14 | Fluctuating prices of raw materials | 4 |
| | **Average** | **5.43** |
| Factors deciding about position in market | | Evaluation |
| 1 | Regular customers | 6 |
| 2 | Opinion of the customers | 7 |
| 3 | New markets | 4 |
| 4 | Market demand | 5 |
| 5 | Product warranty | 8 |
| 6 | Advertisement | 2 |
| 7 | Competitive enterprises on the market | 2 |
| 8 | New competitors on the market | 3 |
| 9 | Substitutes on the market | 3 |
| 10 | Chinese products | 1 |
| 11 | Offers from foreign enterprises | 6 |
| 12 | Image of the enterprise | 7 |
| 13 | Low prices on the market | 4 |
| | **Average** | **4.46** |

There are factors that should be emphasized because they have good influence on technological position of the research of the enterprise. When it comes to technological possibilities, the enterprise is able to repeat its processes (5) and obtains good synchronization between production operations (4). Both factors make the enterprise manufacture products of high quality (11). In the production process, the enterprise uses SMED (6), which has some influence on short production cycle. From the point of view of the position on the market, the enterprise gives a reasonable product warranty (5). This is of the factors that can have an influence on good opinion of the customers about the enterprise (2) and its image (12).

Materials Research Proceedings **17** (2020) 72-78 https://doi.org/10.21741/9781644901038-11

| POSITION ON THE MARKET | | | |
|---|---|---|---|
| **S** | Buy the ready technology | Develop your technological potential | Focus on the revealed chance |
| **A** | Keep in the background | Search for occasions | Improve the marketing |
| **W** | Keep in the background | Discover the incidental market | Search for partners |
| | **W** | **A** | **S** |

**TECHNOLOGICAL POSSIBILITIES**

*Fig. 1. Technological position of the research enterprise [own study]*

## Summary

It results from the 3x3 matrix that the enterprise should think how to improve both variables. Between factors that affect technological possibilities, there are those independent of the enterprise, e.g. a large number of young, qualified people on the labor market (13) and fluctuating prices of raw materials (14). In the operating region there are not too many schools and universities that teach students about metallurgical industry, this kind of study is not popular at all among young people, who prefer easier subjects. Prices depend on their producers and situation on the market of materials. But the enterprise can have small impact on the flexibility of assortment (3). Maybe managers of the enterprise should think about introducing some more types of steel profiles in the assortment of the enterprise. The biggest problem when it comes to the position of the enterprise on the markets are Chinese products (10). This is a problem for enterprises from different industries. In the metallurgical industry it is especially bothersome, because enterprises from this industry do not have many possibilities to change their production profile and assortment. For the research enterprise there are many competitive enterprises on the market (7), not only those from China. Thus, it is difficult to acquire new customers and to have a big share of the market. There is also a problem with advertisement (6), but at the same time it is difficult to advertise any steel products and to conduct reasonable marketing campaign in order to convince customers in this way. Maybe it is a good idea to ask any professional marketing organization for help.

The method presented in this article can be useful in various areas, both industrial and related research, such as improving the surface of parts by ESD [21, 22] and testing the properties of the obtained surface layers [23, 24], including specific image analysis methods [25]. It may be also useful in mechanics [26, 27] and related quality management analysis [28] as well as in so high risk business as biotechnology [29, 30]. In heat transfer problems [31, 32], it should be also fruitful, especially with support from fuzzy approach [33].

## References

[1] M. Szary, K. Knop, Evaluation of technology and technological capabilities of the company from the metal industry, Archives of Engineering Knowledge 3 (2018) 31–34.

[2]  H. Mao, S. Liu, J. Zhang, Z. Deng, Information technology resource, knowledge management capability, and competitive advantage: The moderating role of resource commitment, International Journal of Information Management 36 (2016) 1062–1074. https://doi.org/10.1016/j.ijinfomgt.2016.07.001

[3]  S. Şener, E. Sarıdoğan, The Effects of Science-Technology-Innovation on Competitiveness and Economic Growth, Procedia – Social and Behavioral Sciences 24 (2011) 815–828. https://doi.org/10.1016/j.sbspro.2011.09.127

[4]  K.-F. Huang, Technology competencies in competitive environment, Journal of Business Research 64 (2011) 172–179. https://doi.org/10.1016/j.jbusres.2010.02.003

[5]  A. Bauer, K. Kastenhofer, Policy advice in technology assessment: Shifting roles, principles and boundaries, Technological Forecasting and Social Change 139 (2019) 32–41. https://doi.org/10.1016/j.techfore.2018.06.023

[6]  A. Maszke, The analysis of machine operation and equipment loss in ironworks and steelworks, Production Engineering Archives 17 (2017) 45–48. https://doi.org/10.30657/pea.2017.17.10

[7]  K. Mielczarek, M. Krynke, Plastic production machinery – the evaluation of effectiveness, Production Engineering Archives 18 (2018) 42–45. https://doi.org/10.30657/pea.2018.18.07

[8]  G. Aranoff, Competitive manufacturing with fluctuating demand and diverse technology: Mathematical proofs and illuminations on industry output-flexibility, Economic Modelling 28 (2011) 1441–1450. https://doi.org/10.1016/j.econmod.2011.02.016

[9]  S. Takakuwa, I. Veza, Technology Transfer and World Competitiveness, Procedia Engineering 69 (2014) 121–127. https://doi.org/10.1016/j.proeng.2014.02.211

[10] M. Tracey, M.A. Vonderembse, J.-S. Lim, Manufacturing technology and strategy formulation: keys to enhancing competitiveness and improving performance, Journal of Operations Management 17 (1999) 411–428. https://doi.org/10.1016/S0272-6963(98)00045-X

[11] G. Dosi, M. Grazzi, D. Moschella, Technology and costs in international competitiveness: From countries and sectors to firms, Research Policy 44 (2015) 1795–1814. https://doi.org/10.1016/j.respol.2015.05.012

[12] Z. Liao, M.T. Cheung, Do competitive strategies drive R&D? An empirical investigation of Japanese high-technology corporations, J. High Tech. Manag. Res. 13 (2002) 143–156. https://doi.org/10.1016/S1047-8310(02)00052-4

[13] D. Jolly, The issue of weightings in technology portfolio management, Technovation 23 (2003) 383–391. https://doi.org/10.1016/S0166-4972(02)00157-8

[14] J. Frishammar, U. Lichtenthaler, M. Kurkkio, The front end in non-assembled product development: A multiple case study of mineral- and metal firms, Journal of Engineering and Technology Management 29 (2012) 468–488. https://doi.org/10.1016/j.jengtecman.2012.07.001

[15] Depenbrock, T. Balint, J. Sheehy, Leveraging design principles to optimize technology portfolio prioritization, IEEE Aerospace Conference (2015) 1–10. https://doi.org/10.1109/AERO.2015.7119203

[16] P.H. Nguyen, K.-J. Wang, Strategic capacity portfolio planning under demand uncertainty and technological change, Flex Serv Manuf J 31 (2019) 926–944. https://doi.org/10.1007/s10696-019-09335-w

[17] H. Ajjan, R.L. Kumar, C. Subramaniam, Information technology portfolio management implementation: a case study, Journal of Enterprise Info Management 29 (2016) 841–859. https://doi.org/10.1108/JEIM-07-2015-0065

[18] M. Ingaldi, Use of the SWOT ANALYSIS and 3x3 matrix to determine the technological posistion of the chosen metal company, Acta Metall. Slovaca – Conf. 4 (2014) 207–214. https://doi.org/10.12776/amsc.v4i0.248

[19] S. Borkowski, R. Ulewicz, J. Selejdak, M. Konstanciak, D. Klimecka-Tatar, The Use of 3x3 Matrix to Evaluation of Ribbed Wire Manufacturing Technology, METAL 2012: 21st Int. Conf. on Metallurgy and Materials, Ostrava, TANGER, 2012, 1722–1728.

[20] P. Lowe, Management of Technology: Perception and Opportunities, Chapman & Hall, London, 1995.

[21] R. Dwornicka, N. Radek, M. Krawczyk, P. Osocha, J. Pobedza, The laser textured surfaces of the silicon carbide analyzed with the bootstrapped tribology model. METAL 2017 26th Int. Conf. on Metallurgy and Materials, Ostrava, Tanger, 2017, 1252-1257.

[22] N. Radek, M. Scendo, I. Pliszka, O. Paraska, Properties of Electro-Spark Deposited Coatings Modified Via Laser Beam. Powder Metall. Met. Ceram. 57 (2018) 316-324. https://doi.org/10.1007/s11106-018-9984-y

[23] S. Wojciechowski, P. Twardowski, T. Chwalczuk, Surface Roughness Analysis after Machining of Direct Laser Deposited Tungsten Carbide, Journal of Physics Conference Series 483 (2014) art. 012018. https://doi.org/10.1088/1742-6596/483/1/012018

[24] I. Pliszka, N. Radek, A. Gadek-Moszczak, P. Fabian, O. Paraska, Surface improvement by WC-Cu electro-spark coatings with laser modification. Materials Research Proceedings 5 (2018) 237-242. https://doi.org/10.5604/01.3001.0010.5906

[25] L. Wojnar, A. Gadek-Moszczak, J. Pietraszek, On the role of histomorphometric (stereological) microstructure parameters in the prediction of vertebrae compression strength. Image Analysis and Stereology 38 (2019) 63-73. https://doi.org/10.5566/ias.2028

[26] P. Krawiec, A. Marlewski, Spline description of non-typical gears for belt transmissions. J. Theor. Appl. Mech. 49 (2011) 355-367.

[27] G. Filo, E. Lisowski, M. Domagala, J. Fabis-Domagala, H. Momeni, Modelling of pressure pulse generator with the use of a flow control valve and a fuzzy logic controller. AIP Conference Proceedings, vol. 2029, art. 020015-1.

[28] A. Pacana, K. Czerwinska, R. Dwornicka, Analysis of non-compliance for the cast of the industrial robot basis, METAL 2019 28th Int. Conf. on Metallurgy and Materials (2019), Ostrava, Tanger 644-650. https://doi.org/10.37904/metal.2019.869

Terotechnology XI                                                                    Materials Research Forum LLC
Materials Research Proceedings **17** (2020) 72-78                        https://doi.org/10.21741/9781644901038-11

[29] J. Pietraszek, E. Skrzypczak-Pietraszek, The optimization of the technological process with the fuzzy regression. Adv. Mater. Res-Switz. 874 (2014) 151-155. https://doi.org/10.4028/www.scientific.net/AMR.874.151

[30] L. Skrzypczak, E. Skrzypczak-Pietraszek, E. Lamer-Zarawska, B. Hojden, Micropropagation of Oenothera-Biennis L. and an assay of fatty-acids. Acta Soc. Bot. Pol. 63 (1994) 173-177. https://doi.org/10.5586/asbp.1994.023

[31] L.J. Orman, R. Chatys, Heat transfer augmentation possibility for vehicle heat exchangers. 15th Int. Conf. on Transport Means, Kaunas (2011) 9-12.

[32] T. Styrylska, J. Pietraszek, Numerical modeling of non-steady-state temperature-fields with supplementary data. ZAMM 72 (1992) T537-T539.

[33] J. Pietraszek, M. Kolomycki, A. Szczotok, R. Dwornicka, The Fuzzy Approach to Assessment of ANOVA Results. ICCCI 2016 8th Conf. Comp. Coll. Intell. LNAI 9875 (2016) 260-268. https://doi.org/10.1007/978-3-319-45243-2_24

Terotechnology XI
Materials Research Proceedings **17** (2020) 79-85

Materials Research Forum LLC
https://doi.org/10.21741/9781644901038-12

# Selection and Verification of Camouflage Colors

PLEBANKIEWICZ Ireneusz[1]*, MAZURCZUK Robert[1]
and SZCZODROWSKA Bogusława[1]

Military Institute of Engineer Technology, ul. Obornicka 136, 50-961 Wrocław, Poland

plebankiewicz@witi.wroc.pl

**Keywords:** Military, Camouflage, Image Processing

**Abstract** The paper describes a methodology used for analyzing data generated in the natural environment and in real time based on proprietary computer program developed by the authors. Directions for further work geared towards determining the number of colors in camouflage. The ones which would ensure the highest level of camouflaging efficiency for a soldier's uniform and for a given area were indicated. The necessity was demonstrated of conducting research aimed at developing a specific color palette for the purpose of perfecting camouflage patterns dedicated to a specific climatic- and geographical zone in which the object would not be distinguished from the background of the terrain.

## Introduction

The dynamic development of reconnaissance systems operating in various ranges of electromagnetic radiation has still not eliminated the importance of reconnaissance, in particular the $400 \div 700$ nm wavelength range. Optical reconnaissance in the visible range is based on the motion, shape and color detection. Due to the specific structure of the human eye, movement is detected immediately after it occurs, and that is why it is an "unmasking" feature. Many predators utilize this fact, and despite them having advantage of speed during the attack, they purposefully freeze to become imperceptible, waiting for the right moment to attack as it gives them the best chance of success. Later in the article, the authors focus on the selection of camouflage colors as an element reducing the contrast of the object in relation to its background.

The color is mathematically described by three components: L – brightness (luminance); a – component determining the shade from green to magenta; b – component determining the shade from blue to yellow.

The CIELAB color space is described by the following equations:

$$L = 116 \sqrt[3]{\frac{Y}{Y_0}} - 16 \qquad (1)$$

$$a = 500 \left( \sqrt[3]{\frac{X}{X_0}} - \sqrt[3]{\frac{Y}{Y_0}} \right) \qquad (2)$$

$$b = 200 \left( \sqrt[3]{\frac{Y}{Y_0}} - \sqrt[3]{\frac{Z}{Z_0}} \right) \qquad (3)$$

where:

$X_0=94.81$
$Y_0=100.00$

$Z_0 = 107.30$

are the color coordinates of a standard white body.

The method of describing color via Lab coordinates is currently the most popular one and forms the basis of color management systems.

In order to analyze the possibilities of color discrimination, one can use a mathematical relationship describing the difference between the location of points of different color within the CIELAB space:

$$\Delta E = \sqrt{(\Delta L)^2 + (\Delta a)^2 + (\Delta b)^2} \qquad (4)$$

where:

$0 < \Delta E < 1$ – observer does not notice any difference;

$1 < \Delta E < 2$ – only experienced observer notices any difference;

$2 < \Delta E < 3.5$ – also inexperienced observer notices the difference;

$3.5 < \Delta E < 5$ – observer notices a clear color difference;

$5 < \Delta E$ – observer perceives two different colors.

Another important element of great impact on the ability to distinguish an object from its background is the optical contrast between any two colored surfaces. Optical contrast can be determined analytically, based on the determined spectral characteristics.

The $k$ contrast is calculated from the following formula:

$$k = \frac{\beta_A - \beta_B}{\beta_A} \qquad (5)$$

where:

$k$ – contrast;

$\beta_A$ – luminance of the color whose luminance is higher;

$\beta_B$ – luminance of the color whose luminance is lower.

$$\beta_A = \sum_{\lambda=400nm}^{\lambda=700nm} Y(\lambda) R_A(\lambda) \qquad (6)$$

where:

$Y(\lambda)$ – coefficient dependent on the wavelength;

$R_A(\lambda)$ – reemission coefficient for a specific radiation wavelength of sample A.

$$\beta_B = \sum_{\lambda=400nm}^{\lambda=700nm} Y(\lambda) R_B(\lambda) \qquad (7)$$

where:

$Y(\lambda)$ – coefficient dependent on the wavelength;

$R_B(\lambda)$ – reemission coefficient for a specific radiation wavelength of sample B.

Terotechnology XI
Materials Research Proceedings **17** (2020) 79-85

Materials Research Forum LLC
https://doi.org/10.21741/9781644901038-12

**Analysis of the terrain background based on the determination of the number of camouflage colors and their coordinates**

**Software environment and graphic design of the program.** This section describes a software environment and graphic design of the program for the analysis of the background of the terrain in terms of determining the amount of camouflage colors and their coordinates. Visual Studio 2010 C # programming environment with the .NET Framework ver. 4.0 platform was selected to write a program for the background of the terrain, as far as the determination of the amount of colors, together with new tools and libraries, as offered by AForge.Net and Accord.Net in terms of scientific calculations, image processing and robotics are concerned. The selection of AForge.NET and Accord.NET libraries for writing the above program was not accidental because its development (advancement) is directed towards more complex algorithms for video signal processing (both images and sound signals), e.g. creating histograms, color reduction, edge detection, quantizing, calculations of the Fourier transform, etc.

When designing the user interface (Fig. 1), simplicity, minimization of information available in the application window and placement of interactive elements (widgets) are all taken into account. Individual fields and control buttons are clearly defined by descriptions, which are easily recognized by the user.

*Fig. 1. Graphic interface (software window)*

**Background analysis in order to select representative colors.** The application presented above is used as a tool enabling image processing (in the form of a photograph in any format) captured in the field during analysis of the background of the terrain. It is also used to verify a dedicated camouflage pattern. Proper selection of the main colors using the program is conditioned on initial digital photograph development process using e.g. Adobe Photoshop Lightroom software.

Background analysis is based on a non-linear static mapping of the image color matrix in the $x, y$ plane by limiting the set of their values. The input values for a single color pixel $(R, G, B)$ in the photograph (Fig. 2) are associated with the levels of representation which are assigned to a given range (number of colors). This process is called vector quantization implemented according to the median cut algorithm. As a result of the above operation, a reduced image (visible in Fig. 3 and 4 window on the right hand side) is obtained, with the number of given colors and with simultaneous display of those colors in individual windows in the order of percentage.

Terotechnology XI
Materials Research Proceedings 17 (2020) 79-85

Materials Research Forum LLC
https://doi.org/10.21741/9781644901038-12

*Fig. 2. Background image of the area subjected to color analysis*

*Fig. 3 Background of trees subjected to color analysis*

*Fig. 4. Background of grass subjected to color analysis*

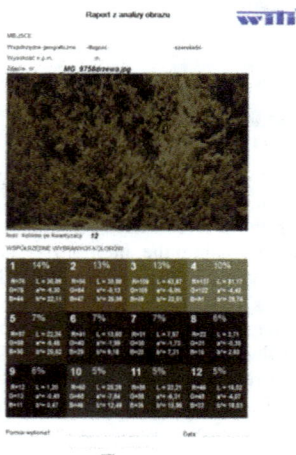

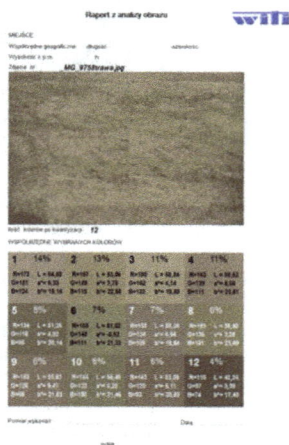

*Fig. 5. Reports generated by the program containing data from color analysis*

Materials Research Forum LLC
https://doi.org/10.21741/9781644901038-12

## Development and verification of camouflage pattern

The stage of developing camouflage for the field uniform is divided into several phases, appearing in succession:

- determining the amount of camouflage colors;
- averaging the color coordinates for different sceneries and seasons;
- determining the percentage of individual colors;
- developing a camouflage pattern.

The exactness of the above- described process is verified as a result of a multi-stage testing of the product in real conditions. Verification during the camouflaging efficiency testing is based on a subjective assessment of the degree of concealment, the object hiding against its background, as determined by the distance between the object tested and the observer. This distance at which the detection occurs (detection in this case means confirmation of the presence of an object of potential military significance in the region under observation) is measured in meters.

The proprietary program described above also allows the user to verify the developed camouflage both in real time – during field tests – and by analyzing the images captured performed during the said tests.

Military uniforms with new camouflage patterns, the so-called spring-summer and autumn-winter underwent verification testing. The current field uniform, as used by the Polish Army, in the *Panthera* camouflage pattern (forest version) was used as a comparative uniform on selected images. All the tested camouflage patterns were developed for use in Central European conditions. Selected colors of these camouflages were chosen on the basis of analysis of forest areas in Poland where they provide the required degree of concealment.

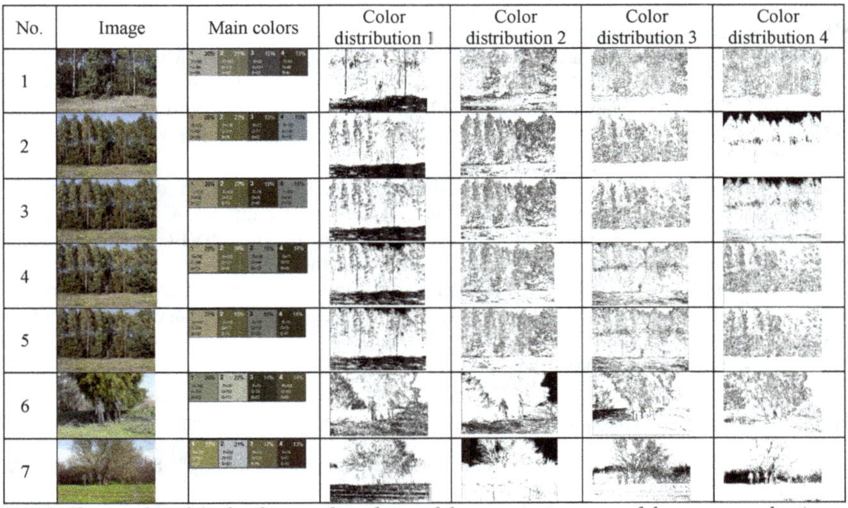

| No. | Image | Main colors | Color distribution 1 | Color distribution 2 | Color distribution 3 | Color distribution 4 |
|-----|-------|-------------|---------------------|---------------------|---------------------|---------------------|
| 1 | | | | | | |
| 2 | | | | | | |
| 3 | | | | | | |
| 4 | | | | | | |
| 5 | | | | | | |
| 6 | | | | | | |
| 7 | | | | | | |

*Fig. 6. The results of the background analysis of the terrain in terms of the correct selection of camouflage colors. Imaging area 1-5 – Poland. Imaging area 6, 7 – Andalusia, Spain.*

Terotechnology XI
Materials Research Proceedings **17** (2020) 79-85

Materials Research Forum LLC
https://doi.org/10.21741/9781644901038-12

The results of the analysis of the images captured during the tests of uniform camouflaging efficiency against the background of representative of the scenery in Poland (items 1 ÷ 5) are presented in Fig. 6. By a selective display of the image after averaging it to a certain number of colors, information on the degree to which the colors and camouflage pattern match the background of the terrain is obtained. A similar series of tests was carried out for another climate zone (Andalusia region in Spain, items 6 and 7). The findings were unambiguous. Camouflage patterns developed for use in the Central European zone exhibited lesser camouflaging efficiency in the Mediterranean climate zone (no matching of the camouflage color to the background of the terrain – the photographs clearly reveal human silhouettes).

**Summary**
The conducted analysis unambiguously confirms the thesis about the necessity of carrying out projects aimed at selecting colors best suited for the uniform(s) used by a soldier during the performance of their official duties in various regions of the world.

Bearing in mind the manner in which the Polish Armed Forces operate during modern conflicts, and considering the possibility of troop deployment, there exists a tangible need to develop a color palette dedicated to be used in the design of camouflage patterns suitable for specific climate zones. Soldiers on peacekeeping missions in other regions of the world should be equipped with uniforms adapted to the place where they are to serve. A dedicated uniform is the fundament of passive protection for a soldier.

On the contemporary battlefield, recognizance systems are increasingly based on infrared detectors. This aspect should be considered when printing the camouflage pattern, dying the uniform fabric and developing dyes used for this particular purpose.

At present, intensive works are underway to develop reconnaissance equipment based on multispectral imaging, which covers the visible-near infrared (NVIR) and short-wave infrared (SWIR) spectral ranges, which is widely used, e.g. in agriculture, to assess the degree of crops vegetation or in forestry, to assess the occurrence of pest infestation. Multispectral imaging allows for the measurement of radiation energy intensity in a given range of coordinates and wavelengths defined as a function of I $(x, y, \lambda)$, where $x$ and $y$ are the coordinates of the imaged surface and $\lambda$ is the wavelength. The results of the analyses are diagrams called re-emission curves.

Military Institute of Engineer Technology devotes a great deal of effort to the analysis of the background of the terrain using advanced image recording techniques, namely multispectral imaging in the range of 380 nm ÷ 2800 nm, including the so-called NVIR and SWIR spectral ranges. The works conducted are aimed at determining the boundaries, within which the re-emission curves of colors reflecting the natural environment in the range of electromagnetic wavelengths not yet defined and not considered when developing camouflage patterns should be contained. These boundaries are vital from the point of view of the features characterizing reconnaissance equipment and their capabilities .

**References**
[1] M. Laprus (ed.), Leksykon wiedzy wojskowej, MON, Warszawa, 1979.

[2] B. Fraser, C. Murphy, F. Bunting, Real World Color Management, 2nd Ed., Peachpit Press, San Francisco, 2004.

[3] Polska Norma Obronna NO-80-A200, Farby specjalne do malowania maskującego. Wymagania i metody badań, 2014.

[3]  M. Pearrow, Web Site Usability Handbook, Charles River Media, Boston, 2006.

[4]  S. Prata, C++ Primer Plus, 5th Ed., Sams, Indianapolis, 2004.

[5]  A. Troelsen, P. Japikse, Pro C# 7 with .NET and .NET Core, Apress, New York, 2017. https://doi.org/10.1007/978-1-4842-3018-3

[6]  AForge.NET, http://www.aforgenet.com/aforge/framework/, (access date 2019.11.08)

[7]  Accord.NET Framework, https://code.google.com/p/accord/, (access date 2019.11.08)

Terotechnology XI
Materials Research Proceedings **17** (2020) 86-93

Materials Research Forum LLC
https://doi.org/10.21741/9781644901038-13

# A Concept of Virtual Reality in Military Camouflage Application

Wojciech PRZYBYŁ*, Ireneusz PLEBANKIEWICZ, Adam JANUSZKO

Military Institute of Engineer Technology, ul. Obornicka 136, 50-961 Wrocław, Poland

przybyl@witi.wroc.pl

**Keywords:** Military, Camouflage, Image Processing

**Abstract:** Camouflage, and in particular one of its forms – military camouflage – is the basic form of protecting friendly forces against the enemy. The article presents the concept of a virtual method, which, based on a photographic simulation for various sceneries and seasons, would allow for testing of camouflaging efficiency in laboratory conditions. The method of image acquisition and attributes which should characterize those images in order for them to be used in photographic simulation was discussed. Furthermore, parameters necessary for a correct visual synchronization of the camouflage patterns applied on objects with the images were presented. Attention was paid to the capabilities and limitations of human vision in the context of military reconnaissance. Methods ensuring colour fidelity during image processing were included, as was selected information on camouflage itself and design of camouflage patterns. The test workstation and the required parameters of the equipment used to conduct camouflaging effectiveness studied in virtual environment were described, and a general algorithm for the virtual method of camouflaging effectiveness assessment, as well as the method of calculating this evaluations were proposed. The method was tested on the example of a Leopard 2A4 main battle tank with a newly-designed camouflage pattern dedicated to autumnal deciduous forest of Central Europe.

## Introduction

The idea of using virtual reality for military purposes, mainly in training, has recently become very popular [1]. The biggest advantage of using the virtual method is lowering the costs, because all the operations will be carried out in place; what will be changing, however, will be the scenery in virtual reality.

One of the factors of key importance for the armies at contemporary battlefield is the capability to conceal their activities from the enemy, i.e. appropriate camouflage [2]. Military camouflage is a kind of safeguard for combat operations, and it consists in hiding forces and material from being identified by opponents or misleading them about the location of its own forces [3]. One of the ways of concealing the forces is to use camouflage, i.e. selecting colours, shapes and sizes of splotches in order make it difficult to distinguish the camouflaged objects from the background of the terrain.

In terms of the camouflage pattern, two main types of patterns can be distinguished – the mimetic pattern and the disruptive (dazzle) pattern (Fig. 1). Mimetic camouflage is a type of camouflage pattern whose main function is to cause the object to blend into its background of the terrain (crypsis). This function results from the need to avoid being detected by the observer who "sweeps" the area using spatial vision. However, the function of the disruptive pattern is to break the shape of a given object, so that it is significantly harder for the observer to recognize the shape and then classify or even identify the object. This effect is achieved by using large splotches of contrasting colours.

*Fig. 1. Plumage pattern – mimetic [⁴] and USS Hobson off Charleston vessel (1942) in dazzle camouflage [⁵]*

The parameters used to determine whether a given camouflage does perform as it should is to assess camouflaging efficiency, which is usually carried out in field observations, during which, a group of observers determines the degree of visibility of objects in the background. This is done for a specified period of time and from exact distances in the environments with defined parameters. The degree of visibility is estimated according to the scale below:

- **No detection** – confirmation of the absence of an object of potential military significance in the area under observation;
- **Detection** – confirmation of the presence of an object of potential military significance in the area under observation;
- **Recognition** – determination that the object detected is a specific type of object, e.g. a human, a wheeled reconnaissance vehicle, camouflage net;
- **Identification** – determination that the object recognized is a specific type of object, e.g. the recognized human is a soldier and the particular tank is Leopard 2A4.

The main "surveillance instrument" analysed in this field is the human eye. Its physical focal length is about 17 mm, but taking into account the fluid filling the eyeball, the commonly accepted value of 22 mm can be assumed. The pupil can change its diameter in the range of approx. $2 \div 8$ mm, which corresponds to brightness equal $f/2.75 \div f/11$.

The angle of view in humans is approx. $150°$ – horizontally and $130°$ – vertically, but sharp vision is about $20°$, and the field of maximum clear vision is $5°$. This stems from the concentration of receptors – suppositories around the macula – which is $200\ 000/m^2$ [6]. Assuming that one suppository is responsible for creating one impulse responsible for creating the impression of vision (this actually happens only for the central part of the retina [7]), where the minimum angle of view $\alpha$ is 10" (arcseconds) [8].

The number of suppositories in one eye is estimated at about 6 million, which makes it possible to distinguish 160 colours and 600 000 shades.

According to Young-Helmholtz's theory, the human eye has three types of suppositories, and each of these types is sensitive to stimuli from a certain range of a visible spectrum of electromagnetic radiation (Fig. 2). These are:

- the long-preferring ones (stimulated mostly by long wavelengths – L – maximum sensitivity of ca. 565 nm);
- the middle-preferring ones (stimulated mostly by medium wavelengths – M – maximum sensitivity of ca. 530 nm);
- the short-preferring ones (stimulated mostly by short wavelengths – S – maximum sensitivity of ca. 420 nm.)

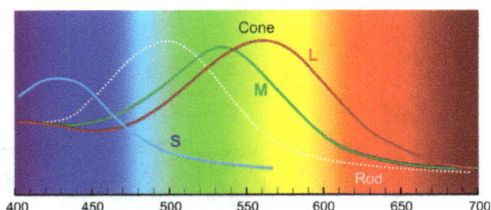

*Fig. 2. Diagram of the wavelengths stimulating individual receptors [9]*

The purpose of the project was to develop a concept for a virtual method of assessing camouflaging efficiency for camouflage patterns, which, taking into account the capabilities of human sense of sight and based on background images for various sceneries and seasons, would allow for testing of this efficiency to be conducted. For this purpose, a proposal has been prepared concerning a laboratory workstation for testing camouflaging efficiency in real conditions which had been imitated. The paper presents examples of photographs of summer scenery with a simulated battle tank contour.

**Methods and materials**

**Images.** The basis of a photographic simulation is a library of images of the natural environment captured in many sceneries and in various seasons. Based on the available image library, sample summer scenery image was selected. The image should be characterised by adequate quality both in terms of the amount of information (image size) and the colour (range and colour accuracy – Fig. 3). The colour and white standard (X-Rite ColorChecker Passport Photo) was used for calibration. Each image contained not only the metadata about the camera with which it was acquired, but also the information on the geographical location, the scenery, the time of year, day and distance from the background.

*Fig. 3. Image capturing (a) using colour standard (b) and white standard (c)*

**Samples.** The test model was a 3-color camouflage pattern. It was generated as a random set of points using Voronoi tessellation for Chebyshev distance and with 33.33% percentage of each colour (Fig. 5a). The contour of the Leopard 2A4 MBT in service with Polish Armed Forces from the 10th Armoured Cavalry Brigade [10] (Fig. 5b) was used as the camouflage carrier.

Before attempting to evaluate camouflaging efficiency, the sample (understood here as camouflage applied onto the vehicle) required visual synchronization with imaging. The scope of synchronization included adjusting the scale, colours, and chiaroscuro using Photoshop (Adobe, San Jose, CA, USA) raster graphics editor and Blender (Blender Foundation, Amsterdam, The Netherlands) 3D creation suite.

The scale was adjusted based on the known equipment dimensions and distances shown in the images (Fig. 4).

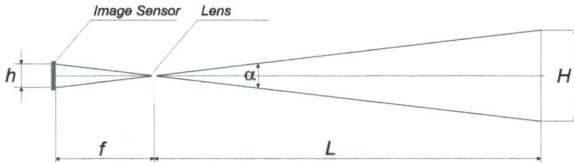

*Fig. 4. Diagram for the proposed captured images. Symbols: H – photographed area, L – distance to the photographed object, h – image sensor size, f – focal length, a – angle of view*

$$H = h \cdot L / f \tag{1}$$

Maintaining uniform colour scheme made the use of one common colour space possible – Adobe RGB and one common colour model – CIE L * a * b *. In addition, all the images were taken with a calibrated camera and had white balance adjusted based on the white standard (Fig. 3). By using a 3D object and illuminating it with virtual light consistent in the direction of the image, the required chiaroscuro was obtained (Fig. 5c and d).

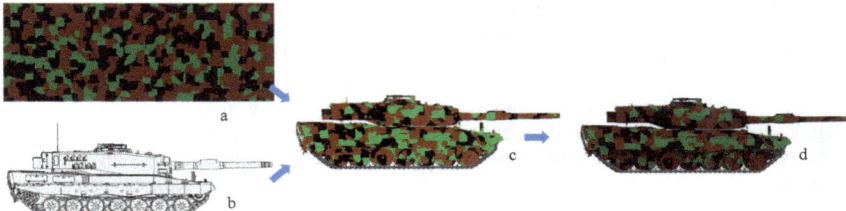

*Fig. 5. Pattern application (a) onto an object (b, c) and generation of chiaroscuro characteristic for the given season and time of day (d)*

**Laboratory workstation.** The proposed laboratory conditions for virtual assessment of the camouflaging efficiency included a projection screen, an observer workstation equipped with a keyboard and a mouse and a supervisor workstation equipped with a control computer (Fig. 6).

*Fig. 6. Drawing of test workstation*

Materials Research Forum LLC
https://doi.org/10.21741/9781644901038-13

Calibration of the presented images (on computers and screens) and the use of the *Color Management System* ensured that the human perception of colour accuracy for images was maintained during the tests (Fig. 7).

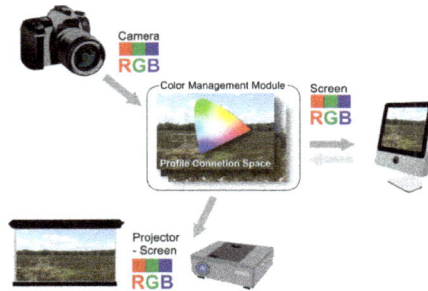

*Fig. 6. Digital Colour Management System laboratory workstation*

**Test results.** The aim of the study was to determine the camouflaging efficiency of a selected object covered with a camouflage pattern in different sceneries. To determine this value, a number of tests should be performed, and the result should be arithmetic **means** from detection, recognition and identification **distances**. In the Polish Defence Standards, the **average** from the recognition distance is usually used.

The first step was to determine the conditions required for a stationary laboratory which would be used to perform tests of camouflaging efficiency using virtual reality. The laboratory design should take into account all the requirements related to the essence of research consisting in observation of a series of images and determination of the degree of visibility of objects against their background in virtual conditions, however, as close to real circumstances as possible.

In order to create the impression of observation of the examined objects in the field as faithfully as possible, it is necessary to specify the distance between the observer's workstation and the screen $L_e$ (2) (Fig. 4). This, in turn, depends on the screen parameters and imaging,

$$L_e = S_e \cdot L/S = H_e \cdot L/H \qquad (2)$$

where:

$H, S$ – height; actual width of the photographed fragment of the background [m],

$H_e, S_e$ – height; width of the image presented on the screen [m],

$L$ – distance of the observer to the object [m],

$L_e$ – distance of the observer from the screen [m].

The rules for presenting images were set out based on the Polish Defence Standard [11], which describes field methods of conducting camouflaging efficiency assessment tests. Therefore it was assumed, in accordance with the said standard, that the order of displayed images should be from the furthest to the nearest distance, and that the time of presenting the image to the observer should be no longer than 2 min. The series of tests is considered to be completed after the imaging has been presented from the shortest distance or after the object has been identified.

Additionally, it was assumed that at least two series of images should be available for a given environment. Furthermore, for one series (including all the simulated distances) the location of objects is not to change, whereas for the second series the object is located in a different place for each distance.

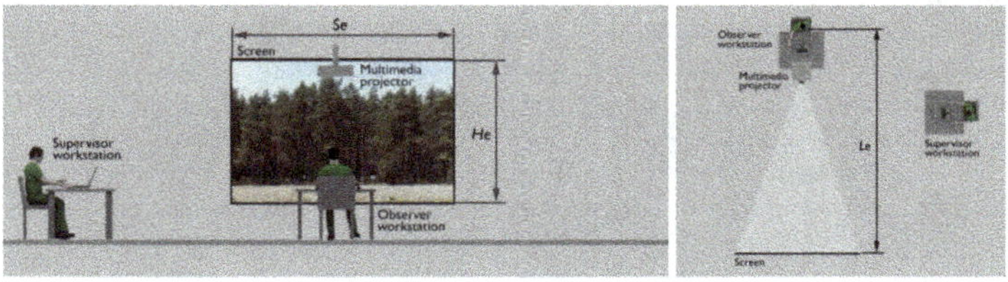

*Fig. 7. Parameters of test workstation*

To support the research process, a computer program was designed which, among others, controls the display of images and records the results (Fig. 9).

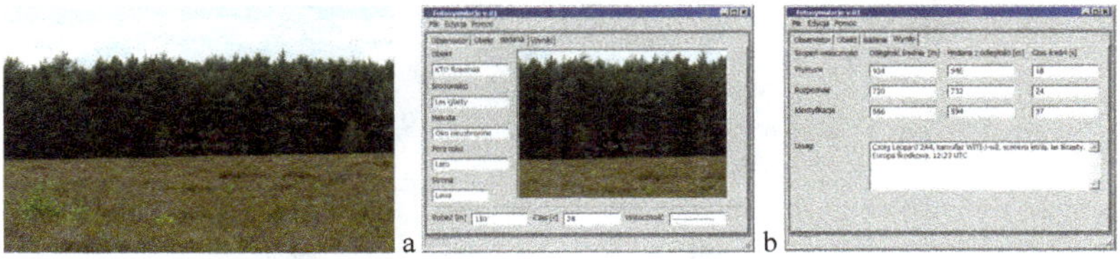

*Fig. 9. View of the observer screen a) and the supervisor b) c) during the test*

The general algorithm of the research methodology using the program is shown in the diagram (Fig. 10).

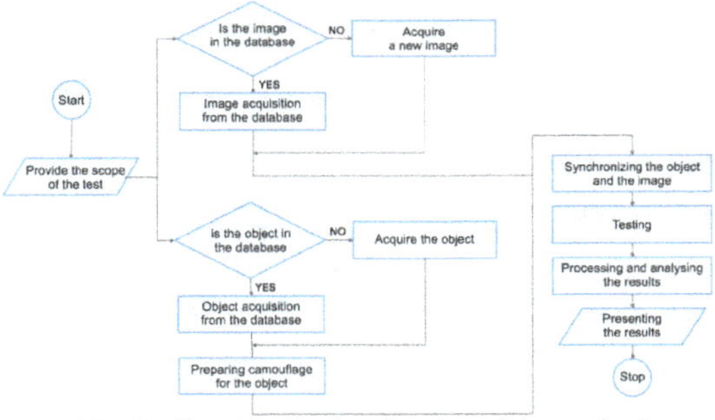

*Fig. 8. Algorithmic visualisation of the method*

Test studies were carried out to verify and validate the concept.

From the image database, an image taken with a Canon EOS 5D camera was downloaded and it was characterised by the following parameters: location – Central Europe; lowlands; latitude = 50.4N, longitude = 21.8E; scenery – deciduous forest; distance to the object observed – 500 m, season – summer; azimuth – 325°, weather condition – no wind, cloud cover – 1/8 octane; resolution 5616 × 3744 px, colour space – Adobe RGB; tonal resolution – 8 bit/channel; white balance – corrected to conform with white standard; lens focal length – 50 mm; matrix size –

Materials Research Forum LLC
https://doi.org/10.21741/9781644901038-13

36×24 mm; exposure – ISO100 sensitivity; aperture 1/9; time 1/400s. For these parameters, the actual image height of H = 0.24·500/0.50 = 240 m was determined.

A camouflage pattern was designed and applied onto the Leopard 2A4 MBT (Fig. 5a-c). The object was visually synchronized with the image. Vehicle dimensions (length, height) were obtained from the tank manufacturer – Krausss-Maffei Wegmann GmbH & Co [12] and were used to adjust the scale of the object and the image. Colour synchronization was ensured via the use of the same colour space – Adobe RGB – and the use of a perceptually homogeneous CIE L*a*b* colour model. Based on the geographic and weather information as well as the imaging date, chiaroscuro for the object was generated, as shown in Fig. 5d.

*Fig. 9. Synchronisation of the sample and image*

The projector was colour-calibrated. Using formula (2), the distance of the observer position from the screen $L_e$ = 1.6 500/240 = 3.3 m was determined as equivalent to the distance to the object (500 m). Due to the perceptive capabilities of the human eye (minimum viewing angle, sharp viewing angle) and the parameters of the image projected by the projector (resolution 1200 × 800 px and dimensions of 3.6 × 2.4 m), the distance $Le'$ = 6.6 m was corrected to avoid perception of individual pixels. A series of scale factors for image $d$ was calculated for the corrected distance $Le'$ and for the corresponding simulated distances to the object ($d$ = ×1 → L = 1000 m, ×1.1 → 900 m, ×1.3 → 800 m, ×1.4 → 700 m, ×1.7 → 600 m, ×2 → 500m).

To properly simulate the real conditions, a group of observers should be furnished with a set of images presenting the equipment to be identified corresponding to different distances from individual distances. Independently, each observer should be presented with an image for max. 2 minutes, starting from the image taken at the greatest distance to the object (1000 m) to the shortest one (or until the object has been identified). Conditions for object visibility, namely *no detection, recognition, identification* will be recorded during tests for each observer. The measurements should include observations for a control sample, i.e. images which do not include the investigated/observed equipment.

*Fig. 10. Examples of images for differing distances of the object to the observer*

easy to segregate, it is possible to cut much thicker machine parts in comparison with laser cutting, the material being cut does not heat up, so there is no contraindication to the treatment of fragile materials with this method.

a)                                                          b)

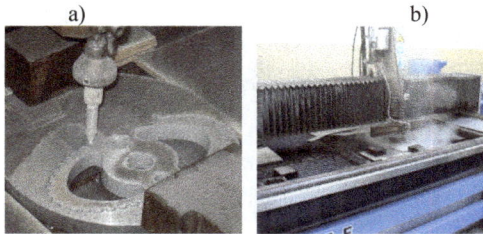

*Fig. 4. Shaping of non-circular cogbelt pulleys: a) cutting a cogbelt pulley with a jet of water with an abradant, b) CNC machine used for machining*

The edges of processed details, unlike laser machining, do not undergo discoloration or thermal hardening and do not undergo structural changes resulting from the thermal effect. The advantages of numerically controlled cutters also result from the possibility of fast and efficient development of the technological process.

The above-described application of methods for shaping non-circular cogbelt pulleys was subjected to verification, one of the main criteria of which was the analysis of the correctness of mapping of geometrical features and surface stereometry. To evaluate the mapping of geometrical features, a Zeiss coordinate measuring machine was used, and the surface quality condition after treatment was checked with a Taylor Hobson profilographometer. Selected cogbelt pulley parameters are given in Table 1. The obtained values meet the requirements for cogbelt pulleys.

*Table 1. Comparison of geometrical features and surface stereometry of non-circular cogbelt pulleys obtained with various shaping methods*

| Geometrical features and surface stereometry | $R_a$ [μm] | Tolerance of outer profile of cogbelt pulleys [mm] | Parallelism tolerance of teeth in relation to cogbelt pulley axis [mm] | Conicity tolerance of cogbelt pulley (for a width of 0.5") [mm] | Pitch errors for cogbelt pulleys (for two adjacent teeth) [mm] | Pitch errors for cogbelt pulleys (for total pitch in an angle range of 90°) [mm] |
|---|---|---|---|---|---|---|
| WEDM | 0.534 | ±0.01 | ±0.02 | ±0.02 | 0.03 | 0.06 |
| Cutting with a water/abradant jet | 2.676 | ±0.05 | ±0.03 | ±0.03 | 0.03 | 0.08 |

**Summary**

The assessment of the surface quality of the cogbelt pulleys indicates that the proposed methods of erosion blasting can be successfully applied. The advantage of the proposed solutions is the relatively low cost of machining and the possibility of making cogbelt pulley shapes of any geometry, the need to design a specialized tool has been eliminated, the program preparation

Terotechnology XI
Materials Research Proceedings **17** (2020) 94-99

Materials Research Forum LLC
https://doi.org/10.21741/9781644901038-14

time is relatively short. Accurate transmission of a cogbelt pulley design to the CAD system is very important in this process. This requires considerable skill and experience of designers.

The obtained results and accumulated experience may also be interesting for other areas of production and processing of machine parts, e.g. production of protective coatings by ESD method [18] and subsequent laser finishing [19, 20], hydraulics of heavy working machines [21], as well as creating realistic simulators of ship engine rooms [22]. The study of the surfaces obtained by erosion blasting may inspire further development of image analysis methods [23].

**References**

[1] M. Kujawski, P.Krawiec, Analysis of Generation Capabilities of Noncircular Cog belt Pulleys on the Example of a Gear with an Elliptical Pitch Line. Journal of Manufacturing Science and Engineering-Transactions of the ASME 133(5) (2011) 051006. https://doi.org/10.1115/1.4004866

[2] P. Krawiec, M. Grzelka, J. Kroczak, G, Domek, A.Kołodziej, A proposal of measurement methodology and assessment of manufacturing methods of nontypical cog belt pulleys, Measurement 132 (2019) 182-190. https://doi.org/10.1016/j.measurement.2018.09.039

[3] P.Krawiec, K.Waluś, Ł.Warguła, J.Adamiec, Wear evaluation of elements of V-belt transmission with the application of optical microscope, MATEC Web of Conf. 157 (2018) art. 01009. https://doi.org/10.1051/matecconf/201815701009

[4] P.Krawiec Analysis of selected dynamic features of a two-wheeled transmission system, Journal of Theoretical and Applied Mechanics 55(2) (2017) 461-467.

[5] K. Gawdzińska, L. Chybowski, W. Przetakiewicz, R. Laskowski, Application of FMEA in the Quality Estimation of Metal Matrix Composite Castings Produced by Squeeze Infiltration. Arch. Metall. Mater. 62 (4) (2017) 2171-2182. https://doi.org/10.1515/amm-2017-0320

[6] Z. Ignaszak, P. Popielarski, J. Hajkowski, E. Codina, Methodology of comparative validation of selected foundry simulation codes, Arch. Foundry Eng. 15(4) (2015) 37-44. https://doi.org/10.1515/afe-2015-0076

[7] Z. Ignaszak, P. Popielarski, J. Hajkowski, Sensitivity of Models Applied in Selected Simulation Systems with Respect to Database Quality for Resolving of Casting Problems, Defect and Diffusion Forum 336 (2013) 135-146. https://doi.org/10.4028/www.scientific.net/DDF.336.135

[8] Ratajczak M., Ptak M., Chybowski L., Gawdzińska K., Będziński R., Material and Structural Modeling Aspects of Brain Tissue Deformation under Dynamic Loads. Materials 12(2) (2019) 271:1-14. https://doi.org/10.3390/ma12020271

[9] K. Gawdzińska, K. Bryll,D. Nagolska, Influence of heat treatment on abrasive wear resistance of silumin matrix composite castings, Arch. Metall. Mater. 61(1) 177-182. https://doi.org/10.1515/amm-2016-0031

[10] Felusiak, T. Chwalczuk, M. Wiciak, Surface Roughness Characterization of Inconel 718 after Laser Assisted Turning, MATEC Web of Conf. 237 (2018) art. 01004. https://doi.org/10.1051/matecconf/201823701004

Terotechnology XI                                                  Materials Research Forum LLC
Materials Research Proceedings **17** (2020) 94-99        https://doi.org/10.21741/9781644901038-14

[11] M. Wiciak, T. Chwalczuk, A. Felusiak, Experimental Investigation and Performance Analysis of Ceramic Inserts in Laser Assisted Turning of Waspaloy, MATEC Web of Conf. 237 (2018) art. 0100. https://doi.org/10.1051/matecconf/201823701003

[12] T. Chwalczuk, D. Przestacki, P. Szablewski, A. Felusiak, Microstructure characterisation of Inconel 718 after laser assisted turning, MATEC Web of Conf. 188 (2018) art. 02004. https://doi.org/10.1051/matecconf/201818802004

[13] S. Wojciechowski, P. Twardowski, T. Chwalczuk, Surface roughness analysis after machining of direct laser deposited tungsten carbide, Journal of Physics: Conference Series 483 (2014) art. 012018. https://doi.org/10.1088/1742-6596/483/1/012018

[14] M. Kawalec, D. Przestacki, K. Bartkowiak, M. Jankowiak, Laser assisted machining of aluminum composite reinforced by SiC particle. (2008) ICALEO 2008 27th Int. Congress on Applications of Lasers and Electro-Optics, 895-900. https://doi.org/10.2351/1.5061278

[15] D. Przestacki, R. Majchrowski, L. Marciniak-Podsadna, Experimental research of surface roughness and surface texture after laser cladding, App. Surf. Sci. 388 (2016) 420-423. https://doi.org/10.1016/j.apsusc.2015.12.093

[16] S. Wojciechowski, D. Przestacki, T. Chwalczuk , The evaluation of surface integrity during machining of inconel 718 with various laser assistance strategies, MATEC Web of Conf. 136 (2017) art. 01006. https://doi.org/10.1051/matecconf/201713601006

[17] M. Kukliński, A. Bartkowska, D. Przestacki, Investigation of laser heat treated Monel 400, MATEC Web of Conf. 219 (2018) art. 02005. https://doi.org/10.1051/matecconf/201821902005

[18] N. Radek, K. Bartkowiak, Laser treatment of electro-spark coatings deposited in the carbon steel substrate with using nanostructured WC-Cu electrodes. Physics Procedia. 39 (2012) 295-301. https://doi.org/10.1016/j.phpro.2012.10.041

[19] R. Dwornicka, N. Radek, M. Krawczyk, P. Csocha, J. Pobedza, The laser textured surfaces of the silicon carbide analyzed with the bootstrapped tribology model. METAL 2017 26th Int. Conf. on Metallurgy and Materials (2017), Ostrava, Tanger 1252-1257.

[20] N. Radek, A. Szczotok, A. Gadek-Moszczak. R. Dwornicka, J. Broncek, J. Pietraszek, The impact of laser processing parameters on the properties of electro-spark deposited coatings. Arch. Metall. Mater. 63 (2018) 809-816.

[21] M. Domagala, H. Momeni, J. Domagala-Fabis, G. Filo, D. Kwiatkowski, Simulation of cavitation erosion in a hydraulic valve. Materials Research Proceedings 5 (2018) 1-6. https://doi.org/10.21741/9781945291814-1

[22] L. Chybowski, K. Gawdzinska, O. Slesicki, K. Patejuk, G. Nowosad, An engine room simulator as an educational tool for marine engineers relating to explosion and fire prevention of marine diesel engines. Scientific Journals of the Maritime University of Szczecin 43 (2015) 15-21.

[23] L. Wojnar, A. Gadek-Moszczak, J. Pietraszek, On the role of histomorphometric (stereological) microstructure parameters in the prediction of vertebrae compression strength. Image Analysis and Stereology 38 (2019) 63-73. https://doi.org/10.5566/ias.2028

Terotechnology XI
Materials Research Proceedings **17** (2020) 100-107

Materials Research Forum LLC
https://doi.org/10.21741/9781644901038-15

# Implementation of Fuzzy Logic in Industrial Databases

KARPISZ Dariusz[1,a*] and KIEŁBUS Anna[1,b]

[1]Cracow University of Technology, Faculty of Mechanical Engineering, Institute of Applied Informatics, al. Jana Pawła II 37, 31-864 Krakow, Poland

[a]dkarpisz@pk.edu.pl, [b]anna.kielbus@pk.edu.pl

**Keywords:** Fuzzy Logic, Industrial Databases, Manufacturing Databases, Production Engineering

**Abstract.** The paper presents selected solutions for the implementation of fuzzy logic in industrial databases. Streaming data processing and classification is one of the most important problems in the Industry 4.0 era. The use of a database engine and appropriate design of the data model for the use of fuzzy logic is a response to expectations of the market. Examples of four types of fuzzy attributes are described. The universal fuzzy data model and its implementation are presented in the article for various internal industry information systems.

**Introduction**

A new approach to building industrial information systems ICT (Information and Communication Technologies) promotes the bottom-up method again instead of the top-down one [1, 2] in building computer applications. It is related to the needs of Industry 4.0 - entry of industrial systems into Big Data problems known in other areas of the market. There are also known problems of processing large amounts of data, where NoSQL class database systems are used as the mid-range layer. Streaming data processing systems may still require a complete or partial transfer of data to relational databases for further analysis. Modern industrial controllers, despite the evolution of the technology, though having the functionality of a computer (Open Controllers) do not have any mechanisms to handle Big Data.

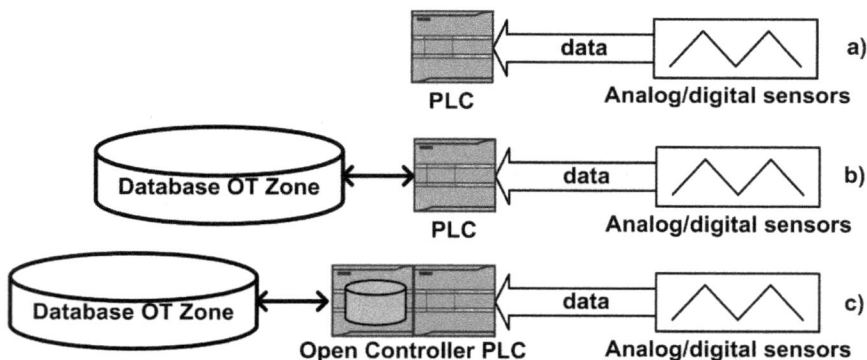

*Fig. 1. Data flow from sensors to databases in OT zone: a) data processing only in PLC, b) using an external industry computer with database, c) using Open Controller.*

Industry 3.0 era, as referred to in Fig. 1a, provided the ability to process selected data on PLC controllers in an OT (Operational Technologies) environment. In this approach, it was possible to respond only to analog or binary values. The use of fuzzy logic was limited only to the use of threshold structures with *if-elseif-else* type of programs without saving data (current data available in one cycle without history or with history limited by a small amount of internal memory). Limitations of data processing in PLCs have forced the use of industrial computers with the ability to store large amounts of data in standard relational databases or NoSQL databases in advanced applications. (Fig. 1b) [3]. With the emergence of Industry 4.0, industrial controllers equipped with a PC class computer with standard operating systems such as MS Windows or Linux as shown in Fig.1c have also been created. The low performance in such a solution is suitable for storing hundreds of thousands of rows of data in classic or NoSQL databases and using more sophisticated mechanisms for data processing and classification. In both cases, according to Fig. 1b and Fig. 1c, it is possible to use fuzzy logic for data classification and processing. It seems that due to the need to install the database server, it may be the best solution to use it as a fuzzy logic processing engine. The database management system (DBMS) uses a large pool of computer resources (CPU and RAM) but does not utilize them. This is the reason for using fuzzy logic in the active database environment in the OT zone.

In many industrial and research areas, there are increasingly imprecise or incomplete requirements, often from end users, that contractors have to deal with when collecting data during research and production. Hence, the presented approach, although strongly formalized, should be interesting for many potential users, including biotechnologists [4, 5], materials [6, 7] and biomaterials [8] engineers, technologists (in heat transfer [9], hydraulics [10] and surface machining [11-13]) or industrial and research data analysts [14-16].

## Methods
**The fuzzy set** has been well described in the literature [17] as a set of pairs according to Eq. 1 as a set A over a universe of discourse X (a finite or infinite interval within which the fuzzy set can take a value). For proper implementation, the $\mu_A(x)$ is named the **membership degree of the element x** to the fuzzy set A, but for universal implementation in technical system it is convenient to assume definition of $\mu_A$ as **member functions** (characteristic functions) without giving an exhaustive list of all the pairs that make the fuzzy set. Also, the **fuzzy domain** as a set of possible values for an attribute was named [3].

$$A = \{\mu_A(x)/x : x \in X, \mu_A(x) \in [0,1] \in \Re\}. \tag{1}$$

However, the problem is a suitable description of the technical object. It is necessary to refer to internal and external standards (legal and industry) and to build the characteristics of the object using a set of coefficients as shown in the simplified Entity Relationship Diagram (ERD) in Fig. 2. The description of a technical object (marked on ERD as *Object of interest*) by linking to the dictionary of usable parameters (marked on ERD as *Factors*) implemented using a connecting entity *Object factors*. The *Factors* typically consist of numerical or boolean coefficients (e.g. TRUE/FALSE values used in electronics as the TTL signal levels). The use of fuzzy factors values has not yet been widely used despite such needs in many modern production management methodologies (e.g. to use in Total Production Management to the assessment of machine condition: good, acceptable, requires immediate intervention).

According to the classification of fuzzy attribute types [16], for the purposes of fuzzy database the following types have been defined:

Terotechnology XI
Materials Research Proceedings **17** (2020) 100-107

Materials Research Forum LLC
https://doi.org/10.21741/9781644901038-15

**Type 1** - Attributes containing precise, non-fuzzy data that allow for the extension of classic databases to use fuzzy queries. In addition to the numerical values, they may also have linguistic labels.
   e.g. measuring instruments - permissible error: *value* (±)2 [μm], *label*: default
   query: *select \* from measuring_instruments where permissible_error < default.*

**Type 2** - Attributes that are an extension of Type 1 allow for storing imprecise data. These factors admit crisp and fuzzy values in the same domain.
   e.g. set *U:(5, 5.8, ~6, 5.5, ~5)*[V] where "~" is refer to "*approximately*".

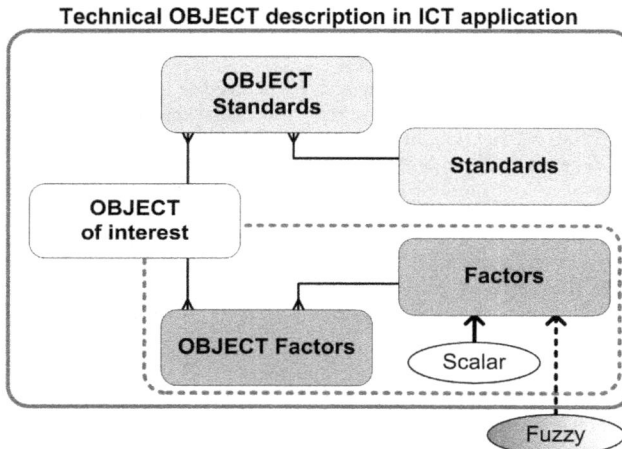

### Technical OBJECT description in ICT application

*Fig. 2. Description of a technical object in standard ICT application without including fuzzy data.*

**Type 3** - Discrete attributes with defined linguistic labels with some degree of similarity. These types of attributes refer to a concave fuzzy set which complies with Eq. 2 with a similarity or proximity relationship. This type allows for examining the similarity of labels. Type 3 also allows for the examination of possibility distributions over the objective attribute domain according to Eq. 3.

$$\forall x,y \in X, \forall \lambda \in [0,1]: \mu_A(\lambda \cdot x + (1 - \lambda) \cdot y) \leq \min(\mu_A(x), \mu_A(y)). \tag{2}$$

$$A = \mu_1/x_1 + \mu_2/x_2 + ... + \mu_n/x_n , \tag{3}$$
   where "+" is the aggregation operator, "/" is the operator of association of both values.

   e.g. Selection of diameter of milling tool from tool storage to cutting a slot with specific dimension (100,100):
   *SlotsMillingTools.Diameter: (mikro2,mikro1,small2,small1,default,large1,large2),*
   *SlotsMillingTools.Dim (100,100).Selection(0.7/small1, 1.0/default).*

**Type 4** - Attributes defined identically to Type 3, but without the need for a similarity relationship between labels or values within the fuzzy domain. This type is suitable for storing non-normalized values, which simplifies this type of attribute compared to Type 3. However, it can be assumed that they are different if we assume a degree of similarity of 0 for each pair of different labels, or attributes are equal if the degree is set to 1.

e.g. *temperature* in case 1 and its impact on tool machining accuracy,

and in case 2, *temperature* and its impact on tool life

(without taking into account the similarity between the two cases).

It is also possible to use fuzzy degrees [18] (associated or nonassociated) instead of (or additional to) fuzzy values of attributes with use either interval values or linguistic labels.

To build the test system, the x86 64-bit server with assigned 8 cores (from AMD Opteron 6344 series physical processor, 2.6 GHz clock) and 80 GB of RAM was used. The database server was located on a virtual machine in the Citrix Server (Xen) environment. As the DBMS the Oracle database 12c was installed and configured with 64 [GB] of System Global Area size. Instance of a fuzzy database was set to use Sort-merge join as primary to optimize all queries joins. The dedicated user schema (*fuzzy_user*) was used to implement the database in DDL SQL language.

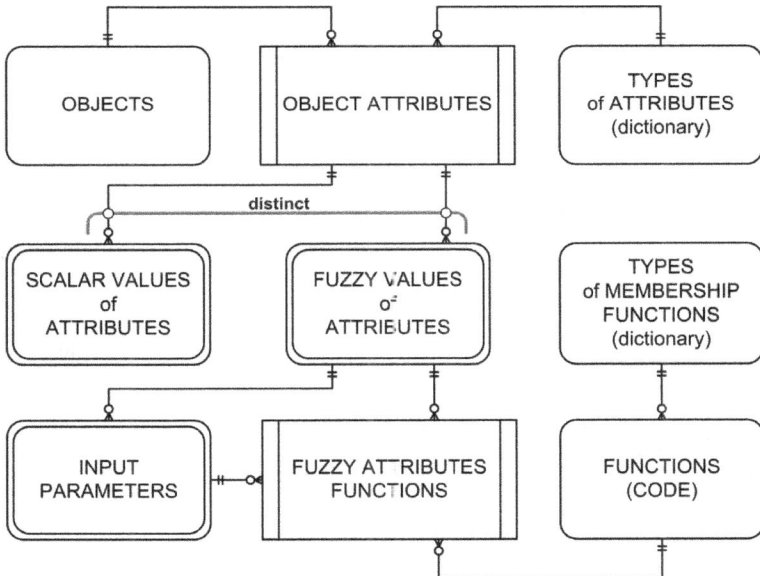

*Fig. 3. A simplified ERD (Martin notation) to describe objects using fuzzy logic.*

**Results**

Based on a study of requirements for a minimum description of technical objects using fuzzy logic, a database design was prepared using the Entity-Relationship Diagram (ERD) as shown in

Fig. 3. This is a simplified diagram, which does not have a complete set of tables and their relationships. In addition, an exclusive condition was introduced between two entities dependent on *OBJECT ATTRIBUTES*, where the attribute can be either scalar (*SCALAR VALUES of ...*) or fuzzy (*FUZZY VALUES of ...*) (for fuzzy type 1-4). In fact, type 1 of attributes can be stored as scalar but cannot be thresholded using linguistic descriptions.

For the proposed project (Fig. 3), an implementation was made in accordance with the database skeleton SK1:

SK1 :=
    { (TA;    { id, name, description }),       /* Types of fuzzy attributes */
     (TF;    { id, name, description }),       /* Types of membership functions */
     (ATFS; {id, label1, label2, degree}),     /* Similarity functions between the labels */
     (OBJ;   { id, name, description }),       /* Object of interest */
     (OBJA; { id, obj_id, name, description }),   /* Attributes of the object*/
     (ATS;  { id, obja_id, value }),       /* Scalar Attributes*/
     (ATF;  { id, obja_id, ta_id, atfs_id, value }), /* Fuzzy Attributes*/
     (ATFF; { id, atf_id, tf_id, input(a,b,c,d,margin){1-4}$^{RegExp}$, function_call})
                                    /* Function for fuzzy attribute */  },

where for each table $T \in$[TA, TF, ATFS, OBJ, OBJA, ATS, ATF, ATFF] each *id* defined as a non-empty subset of the heading of T: ($\forall$t1,t2 $\in$ T: t1↓A = t2↓A => t1=t2),

and condition as above applied to TA(name), TF(name) as unique,
and OBJA(obj_id) $\in$ OBJ, ATS(obja_id) $\in$ OBJA, ATF(obja_id) $\in$ OBJA, ATF(ta_id) $\in$ TA, ATF (atfs_id) $\in$ ATFS, ATFF(atf_id) $\in$ ATF, ATFF(tf_id) $\in$ TF.

For such a fuzzy database schema, it is possible to store any scalar and fuzzy data. Only the dictionary tables (*TA, TF, ATFS*) have predefined tuples with the possibility of extending with new definitions. Defined tuples for *TA* table:

TA := { {(id;1),(name;'Type 1')}, {(id;2),(name;'Type 2')},
        {(id;3),(name;'Type 3')}, {(id;4),(name;'Type 4')} }.

The basic membership functions of triangular, trapezoidal and gaussian sets stored in *TF* dictionary were used for testing. For test cases, the input parameters of implemented programming functions in *ATFF* table were stored. The functions were implemented in the PL/SQL language in Oracle environment.

Example of definition tuples for *ATFS* table
(sets of Labels1 and Labels2={'VeryGood', 'Good', 'Moderate', 'Sufficient', 'Bad', 'VeryBad'} with *degree* step 0,2):

ATFS := { {(id;1),(label1;'VeryGood'),(label2;'VeryGood'),(degree;1.0)}, ..., ...,
(4)
    {(id; 21), (label1;'VeryBad'), (label2;'VeryBad'),(degree;1.0)} }, where *tuples* $\in$ { Label1 x Label2}.

For the exemplary object "*Machine electrical installation*" and the attribute "*Supply voltage*" it is possible to use both specific values (direct readings from the machine measuring system) of **type 1** and fuzzy values of **type 2** ("*about 255*" [V]) and **type 4** (set {"*too low*", "*normal*", "*too high*"} voltage). It should be noted that for such a case it is also possible to define the commonly used "input voltage" attribute and to test the correlation between input and supply voltage. For the attribute defined as type 4, to labels (sets) "*too low*" and "*too high*" it is possible to use the trapezoidal membership function $T(x)$, while for the "*normal*" set it is possible to use the triangular function $A(x)$, where normal is defined as $230 \pm 10\%$ [V], where implementation is:

$\forall x_{input} \in \Re, \forall a,b,c,d \in \Re$:

$T(x,a,b,c,d) = \{$ if $(x \leq a) \Rightarrow y := 0.0$; elsif $(x > a \wedge x \leq b) \Rightarrow y := (x - a)/(b - a)$;
elsif $(x > b \wedge x \leq c) \Rightarrow y := 1.0$; $(x > c \wedge x \leq d) \Rightarrow y := (d - x)/(d - c)$; else $y := 0.0$; return $y; \}$

$A(x,a,b,c) = \{$ if $(x \leq a) \Rightarrow y := 0.0$; elsif $(x > a \wedge x \leq c \wedge x != b) \Rightarrow y := (x - a)/(c - a)$;
elsif $(x > c \wedge x < b \wedge x != b) \Rightarrow y := (b - x)/(b - c)$; elsif $(x = b) \Rightarrow y := 1.0$;
else $y := 0.0$; return $y\}$,

and e.g. "*too low*" set: $\{(a;0),(b;0),(c;207),(d;215)\}$,
"*normal*" set: $\{(a;207),(b;230),(c;253)\}$, /* (d==NULL) */
"*too high*" set: $\{(a;245),(b;253),(c;999),(d;999)\}$.

The mechanisms of active databases [5,6] can, for this defined example, automatically use a trigger of the type "*AFTER INSERT for each row*" for each incoming scalar voltage value (stored in *ATS* table) and perform automatic classification with a fuzzy counterpart defined in *ATF* table. The whole analysis is in this case completely configurable and operated on the database server side. The result of analysis with linguistic label can be sent back to the PLC device and to the operator panel (HMI) and trigger the appropriate alert or alarm.

The example that will be used to describe **Type 3** attributes includes linguistic labels with some degree of similarity. The attribute "*Particulate matter PM10*" will be used to describe the "*Machine exhaust system*" (the object in *OBJ*). For this example, the *ATFS* table (according to Eq. 4 as above) defines a list of similarities between the exhaust gas quality indexes with the PM10 dust for two attributes (in the *OBJA* table) defined for two smokestack or two measuring points:

OBJ := $\{\{(id;1),(name;'Machine exhaust system')\}, \}$
OBJA := $\{ \{(id;1), (obj\_id;1),(name;'PM10\text{-measuring point 1'})\},$
$\{(id; 2), (obj\_id;1),(name;'PM10\text{-measuring point 2'})\}\}$,
ATF := $\{\{(id;1), (obja\_id;1),(ta\_id;3),( value;y)\}, ...,\{(id;n), (obja\_id;1),(ta\_id;3),( value;u)\}$
$\{(id;n+1), (obja\_id;2),(ta\_id;3),( value;v)\}, ...,\{(id;m), (obja\_id;2),(ta\_id;3),($
value;w)$\}\}$,
where $y,u,v,w \in \{$'VeryGood', 'Good', 'Moderate', 'Sufficient', 'Bad', 'VeryBad'$\}$, $n,m \in \mathbb{N}^*$.

In practical application, the defined linguistic labels for *obja_id =1* and *obja_id =2* (as defined above) do not have to correspond to the same scalar values incoming from the measuring systems (e.g. where is setting the measuring points before and after the dedusting cyclone).

## Discussion

The application of fuzzy set theory in classic databases does not have to be complicated. However, there are no standards to describe attributes and analyze their values. The study shows that the high performance database server used to operate with fuzzy sets does not introduce a visible delay in performing queries, despite the needs for joins in queries with additional 2 or 3 database tables. This is a result of the adopted DBMS platform (which is also a standard for many industrial applications such as MES, SCADA, ERP), which operates fast on small data sets of up to 1 million tuples. The proposed data model (ERD in Fig. 3) can be successfully applied directly also on small servers installed in Open Controllers devices as shown in Fig. 1, and the mechanisms of active database systems can classify data in real time using the membership function and similarity matrices data with feedback to the PLC.

Various programming implementations of the proposed data model and functions can be used. What is important for the production implementation of fuzzy database is that variable number of input parameters for membership functions makes it possible to prepare a universal program code using overloaded functions (like in C++ style programming).

Attempts to adapt the proposed data model to issues related to ... [], ... [] and [] have also been undertaken. In all cases it was possible to build structures that fulfill the needs of data analysis requirements in the study area.

## References

[1] D. Karpisz, A. Kielbus, Selected problems of designing modern industrial databases, MATEC Web of Conferences 183 (2018) art. 01017, https://doi.org/10.1051/matecconf/201818301017

[2] D. Karpisz, Design of manufacturing databases, Technical Transactions 113 (10) (2016) 73-77, doi: 10.4467/2353737XCT.16.123.5734

[3] D. Karpisz, A. Kielbus, M. Zembytska, Selected problems of industry databases and information infrastructure security, QPI 1 (1) (2019) 371-377, https://doi.org/10.2478/cqpi-2019-0050

[4] E. Skrzypczak-Pietraszek, J. Pietraszek, Phenolic acids in in vitro cultures of Exacum affine Balf. f. Acta Biol. Crac. Ser. Bot. 51 (2009) 62-62.

[5] E. Skrzypczak-Pietraszek, I. Kwiecien, A. Goldyn, J. Pietraszek, HPLC-DAD analysis of arbutin produced from hydroquinone in a biotransformation process in Origanum majorana L. shoot culture. Phytochemistry Letters 20 (2017) 443-448. https://doi.org/10.1016/j.phytol.2017.01.009

[6] T. Lipinski, The structure and mechanical properties of Al-7%SiMg alloy treated with a homogeneous modifier. Solid State Phenomena 163 (2010) 183-186. https://doi.org/10.4028/www.scientific.net/SSP.163.183

[7] T. Lipinski, Double modification of AlSi9Mg alloy with boron, titanium and strontium. Arch. Metall. Mater. 60 (2015) 2415-2419. https://doi.org/10.1515/amm-2015-0394

[8] D. Klimecka-Tatar, Electrochemical characteristics of titanium for dental implants in case of the electroless surface modification. Arch. Metall. Mater. 61 (2016) 923-26. https://doi.org/10.1515/amm-2016-0156

[9] L. Dabek, A. Kapjor, L.J. Orman, Boiling heat transfer augmentation on surfaces covered with phosphor bronze meshes. MATEC Web of Conf. 168 (2018) art. 07001. https://doi.org/10.1051/matecconf/201816807001

[10] M. Domagala, H. Momeni, J. Domagala-Fabis, G. Filo, M. Krawczyk, J. Rajda, Simulation of particle erosion in a hydraulic valve. Materials Research Proceedings 5 (2018) 17-24.

Terotechnology XI                                                                        Materials Research Forum LLC
Materials Research Proceedings **17** (2020) 100-107                        https://doi.org/10.21741/9781644901038-15

[11] D. Przestacki, M. Kuklinski, A. Bartkowska, Influence of laser heat treatment on microstructure and properties of surface layer of Waspaloy aimed for laser-assisted machining. Int. J. Adv. Manuf. Technol. 93 (2017) 3111-3123. https://doi.org/10.1007/s00170-017-0775-2

[12] S. Wojciechowski, D. Przestacki, T. Chwalczuk, The evaluation of surface integrity during machining of Inconel 718 with various laser assistance strategies. MATEC Web of Conf. 136 (2017) art. 01006. https://doi.org/10.1051/matecconf/201713601006

[13] Radek, N., Kurp, P., Pietraszek, J., Laser forming of steel tubes. Technical Transactions 116 (2019) 223-229. https://doi.org/10.4467/2353737XCT.19.015.10055

[14] J. Pietraszek, A. Gadek-Moszczak, The Smooth Bootstrap Approach to the Distribution of a Shape in the Ferritic Stainless Steel AISI 434L Powders. Solid State Phenomena 197 (2012) 162-167. https://doi.org/10.4028/www.scientific.net/SSP.197.162

[15] J. Pietraszek, A. Gadek-Moszczak, T. Torunski, Modeling of Errors Counting System for PCB Soldered in the Wave Soldering Technology. Advanced Materials Research 874 (2014) 139-143. https://doi.org/10.4028/www.scientific.net/AMR.874.139

[16] A. Pacana, K. Czerwinska, R. Dwornicka, Analysis of non-compliance for the cast of the industrial robot basis, METAL 2019 28th Int. Conf. on Metallurgy and Materials (2019), Ostrava, Tanger 644-650. https://doi.org/10.37904/metal.2019.869

[17] J. Galindo, A. Urrutia, M. Piattini, Representation of Fuzzy Knowledge in Relational Databases, in: Fuzzy databases: modeling, design and implementation, Idea Group Publishing, London, 2005, 145-151. https://doi.org/10.4018/978-1-59140-324-1.ch005

[18] J. Widom, S. Ceri, Active database systems: triggers and rules for advanced database processing, Burlington, Morgan Kaufmann, 1996.

Materials Research Proceedings 17 (2020) 108-113 https://doi.org/10.21741/9781644901038-16

# Simulation and Assessments of Urban Traffic Noise by Statistical Measurands using mPa or dB(A) Units

BĄKOWSKI Andrzej[1, a *] and RADZISZEWSKI Leszek[1,b]

[1]Kielce University of Technology, Aleja Tysiąclecia Państwa Polskiego 7, 25314 Kielce, Poland

[a]abakowski@tu.kielce.pl, [b]lradzisz@tu.kielce.pl

**Keywords:** Urban Noise, Cnossos-EU Method, Heavy Vehicles

**Abstract.** In this paper, some results of heavy vehicle traffic measurements were used to simulate noise measurands by the Cnossos-EU method for this purpose. Heavy vehicle traffic volume and velocity were recorded by a permanent automatic monitoring station. The noise was calculated in octave bands. The results were described using parameters such as the median, average peak noise, average maximum noise, first and third quartiles and relative measures of noise. The values of these parameters were expressed in mPa or dB(A). It was shown that maximum values of the acoustic pressure (mPa) occur for the frequency of $f_0$=500 Hz but of the acoustic pressure level (dB(A)) for $f_0$=1000 Hz. The dispersion of noise and type A uncertainty of the results were evaluated. Depending on the adopted noise unit, different shapes and distribution parameters were obtained.

## Introduction

The ultimate scope of the Directive 2015/996/EC is to enhance the reliability and comparability of noise data in EU [1, 2]. The CNOSSOS-EU method is used for that purpose. Traffic noise and vehicle monitoring systems using permanent monitoring terminals were constructed and installed in Kielce - an example of a medium-size town (a population of approximately 200,000) located in the southern part of central Poland. The measurements results of heavy vehicle traffic flow from two vehicular lanes running towards the town and two lanes running towards Kraków were analysed. Computer simulation of the acoustic pressure in octave bands, in accordance with the CNOSSOS-EU model, were carried out.

## Traffic volume and noise measurements

Traffic noise and volumes analyzed in this study were measured throughout the year by a permanent station located in Krakowska Street in Kielce [3]. The measurements from two vehicular lanes running towards the town and two lanes running towards Kraków were analyzed. The station includes a road radar box, a sound level meter and a weather station. The traffic volume and speed were measured on each lane by a WAVETRONIX digital radar with an operating frequency of 245 MHz. A microphone was positioned at a distance of 4 m from the edge of the lane at a height of 4 m. The measurements were documented in one hour intervals throughout the entire 24 hours of the day (1:00-24:00) throughout the year 2013. The traffic volume and speed data were recorded every 1 minute (buffer) and the averaged results were reported every 1 hour. Counts were used to calculate the traffic flow (understood as the sum of the number of vehicles recorded within a time interval) and average vehicle speed, split into hours. Detailed analyzes were carried out for the day sub-interval (data registered from 6.00 to 18.00) of a 24-hour period because it is the most burdensome time interval of the whole day. The results analyzed contained heavy vehicle traffic flow together with vehicle average speed measurements. In this work, analysis was based on the measurements on all working days of 2013. In this study, calculated noise parameter values are expressed in dB(A) and mPa. Advantages of the noise scale in dB(A) are widely known, but its weakness is non-linearity [4] and difficulties in performing some mathematical functions (operations), e.g. division, multiplication, analysis of variability, analysis of measurement uncertainty [5]. These problems are not there when we express the sound pressure in mPa. However, the scale of noise in mPa also has its weaknesses commonly known in the literature. Interpretative differences in the values of some road traffic noise parameters determined in dB(A) or mPa were pointed out.

## CNOSSOS-EU modeling of traffic noise

In many cities, traffic measurement systems record only traffic volume and speed. To make the full use of data obtained in this way to assess environmental pollution, a noise model is still needed. In the Cnossos-EU model, the sound power level was divided into two parts – propulsion ($L_{WP,i,m}(v_m)$) and rolling ($L_{WR,i,m}(v_m)$) noise [2]. The sound power level emitted by one of the vehicle category $m$ and in octave band number $i$ is the following:

$$L_{W,i,m}(v_m) = 10 \cdot log\left(10^{L_{WR,i,m}(v_m)/10} + 10^{L_{WP,i,m}(v_m)/10}\right) \tag{1}$$

where: $i$ – number of octave bands, from $i$=2 for $f_0$=125 Hz up to $i$=7 for $f_0$=4000 Hz, $m$ – vehicle categories ($m$=1-light motor vehicles, $m$=2-medium heavy vehicles, $m$=3-heavy vehicles, $m$=4-powered two-wheelers), $v_m$ –rolling speed of vehicle category $m$. If a steady traffic flow of vehicles of category $m$ per hour is assumed with an average speed $v_m$ the directional sound power level per 1 m per frequency band $i$ of the source line determined by the vehicle flow is defined by:

$$L_{Weq,i,m} = L_{W,i,m}(v_m) + 10 \cdot log\left(\frac{Q_m}{1000 \cdot v_m}\right) \tag{2}$$

where: $Q_m$ – traffic flow of vehicles of category $m$ per hour with an average speed $v_m$. The acoustic pressure to the second power, measured by microphone, generated by vehicles category $m$ in octave band $i$ we can calculate according to the following formula:

$$p_{i,m}^2 = \sum_{j=1}^{Q_t} p_0^2 \cdot 10^{\left(L_{Weq,i,m} + 10 \cdot log\left(\frac{l_S}{Q_t}\right) - 20\,log(R_j) - 8\right) \cdot 0,1} \tag{3}$$

where: $l_S$ - length of a source line with homogeneous traffic, $Q_t$ – amount of source line segments, $p_0$- reference sound pressure equal to $2 \cdot 10^{-5}$ Pa, $j$ – index of source line segments, $R_j$ – distance of the center of the $j$ source line segments from the measuring microphone. The correction coefficients for deviations from reference conditions were not taken into account in the paper.

The tests [4] for the components contained in the acoustic signals were based on the following percentiles: $C_{10}$, $C_{25}$, $C_{50}$, $C_{75}$, $C_{90}$, and $C_{99}$ defined as the values of noise exceeded by the signal, respectively in 90% (average background noise level), 75%, 50% (median), 25 %, 10% (av. peak level) or 1% (av. maximum noise) of the measurement period. Standard uncertainty of the acoustic pressure, determined in the Type A evaluation, can be calculated from the following relationship:

$$u_A = \sqrt{\frac{1}{n(n-1)} \sum_1^n (p_i - \bar{p}_i)^2} \tag{4}$$

where $n$ is the amount of data.

The authors analyzed acoustic pressure values $p_i$ expressed in units of mPa to be able to easily compare the fixed components (median) and variable components of the acoustic pressure signals. The tests for the variable components contained in the signals were based on the

Terotechnology XI
Materials Research Proceedings **17** (2020) 108-113

Materials Research Forum LLC
https://doi.org/10.21741/9781644901038-16

following measures: coefficient of variation [7], quartile deviation ($Q_{31}$), quartile variation coefficient ($V_q$). The average quartile deviation is the measure of dispersion of the variable:

$$Q_{31} = 0.5 \cdot \left[ C_{75}(p_i) - C_{25}(p_i) \right] \tag{5}$$

By relating it to the median, the positional coefficient of variation is calculated from (6):

$$V_q = \frac{Q_{31}}{Med} \cdot 100\% \tag{6}$$

The positional coefficient of variation and the quartile coefficient of dispersion are positional measures of the data between the first and third quartiles.

It was assumed in this paper that the acoustic source are the following: entry traffic on two lanes that leads from Kraków towards Kielce and exit traffic on two lanes that leads from Kielce towards Kraków. It was assumed that the linear acoustic source is located along the symmetry axis of the respective lanes. Thus, the work analyzed the results of computer simulations of acoustic pressure in the place where the measuring microphone is located, i.e. at a distance of 4 m from lanes and at a height of 4 m for two incoherent acoustic sources using measurements of relevant parameters of road vehicles. The acoustic pressures generated by these sources were also added up, which allowed for the assessment of total noise generated by the examined road section. In [3], the values of the equivalent sound level (for all vehicles category) experimentally measured and calculated according to the Cnossos-EU method, were compared by calculating the RMSE parameter. The calculated value of this parameter is about 1 dB.

**Results**

Shapiro-Wilk and Jarque-Bera tests showed that the acoustic pressure distributions generated by heavy vehicles are not compatible with the normal distribution. Histograms of acoustic pressure distributions for heavy vehicles confirmed deviations from the normal distribution. Examples of histograms in the octave band $f_0 = 500$ Hz are shown in Figure 1. Values of selected data distribution parameters are the following: for figure 3a: skewness is 2.06 and kurtosis is 13.33, for figure 3b: skewness is -0.68 and kurtosis is 10.86. Note the diverse forms of these distributions.

Table 1 compiles the analysis results for acoustic pressure parameters in selected octave bands, on working days for the day sub-interval, generated by heavy vehicles calculated by the Cnossos-EU method. The calculations show that maximum values of median as well as percentiles $C_{10}$ and $C_{90}$ were obtained in an octave band with a central frequency of $f_0 = 500$ Hz. But for the percentile $C_{99}$ the maximum value is in the octave band of $f_0 = 125$ Hz. The minimum values of these parameters were obtained in the octave with a frequency of $f_0 = 4000$ Hz. The values of the parameter $C_{99}$ in relation to the value of $C_{90}$ are as follows: in the octave with a frequency of $f_0 = 125$ Hz they are higher by about 138% and for the frequency of $f_0$=500 Hz, $f_0$=1000 Hz or 2000 Hz by about 80%. Values of positional coefficients of variation are about 8% and uncertainty $u_A$ is less than 0.10 mPa. Statistical analysis of the acoustic pressure values shows that the values of $Q_{31}$ for heavy vehicles fall within the range of 0.4 mPa to 1.6 mPa.

Terotechnology XI                                                    Materials Research Forum LLC
Materials Research Proceedings **17** (2020) 108-113              https://doi.org/10.21741/9781644901038-16

a)                                              b)

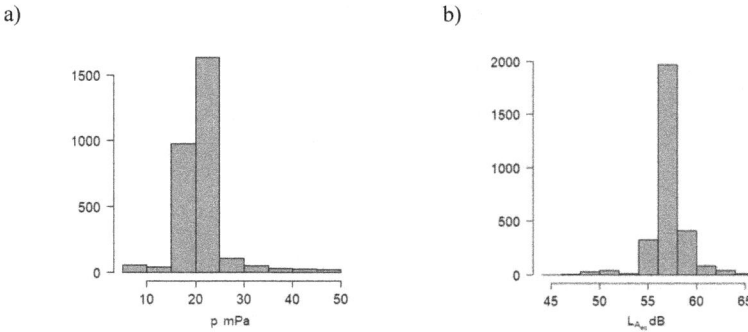

Fig. 1 Histograms of noise distributions in octave band $f_0$=500 Hz for heavy vehicles on working
days of 2013: a) acoustic pressure in mFa, b) acoustic pressure level in dB(A).

Table 1. The values of sound pressure in units mPa calculated by the Cnossos-EU method, for
heavy vehicles on working days for the day sub-interval

| Central frequency band $f_0$ [Hz] | Med. [mPa] | $Q_{31}$ [mPa] | $Vq$ [%] | $u_A$ [mPa] | $C_{10}$ [mPa] | $C_{90}$ [mPa] | $C_{99}$ [mPa] |
|---|---|---|---|---|---|---|---|
| 125 | 16.73 | 1.33 | 7.97 | 0.10 | 14.24 | 20.04 | 47.67 |
| 250 | 18.58 | 1.45 | 7.80 | 0.10 | 15.80 | 22.02 | 46.96 |
| 500 | 20.75 | 1.58 | 7.62 | 0.09 | 17.71 | 24.36 | 43.52 |
| 1000 | 19.04 | 1.44 | 7.54 | 0.08 | 16.23 | 22.16 | 39.09 |
| 2000 | 10.51 | 0.80 | 7.63 | 0.05 | 8.96 | 12.32 | 23.60 |
| 4000 | 5.78 | 0.44 | 7.68 | 0.03 | 4.92 | 6.81 | 13.76 |

Table 2. The values of sound pressure level in units dB(A), calculated by the Cnossos-EU method
for heavy vehicles on working days for the day sub-interval

| Central frequency band $f_0$ [Hz] | Med. [dB(A)] | $Q_{31}$ [dB(A)] | $Vq$ [%] | $u_A$ [dB(A)] | $C_{10}$ [dB(A)] | $C_{90}$ [dB(A)] | $C_{99}$ [dB(A)] |
|---|---|---|---|---|---|---|---|
| 125 | 42.35 | 0.69 | 1.63 | 0.04 | 40.95 | 43.92 | 51.44 |
| 250 | 50.76 | 0.68 | 1.33 | 0.04 | 49.35 | 52.24 | 58.81 |
| 500 | 57.12 | 0.66 | 1.16 | 0.03 | 55.74 | 58.51 | 63.55 |
| 1000 | 59.57 | 0.66 | 1.10 | 0.03 | 58.19 | 60.89 | 65.82 |
| 2000 | 55.61 | 0.66 | 1.19 | 0.04 | 54.23 | 56.99 | 62.64 |
| 4000 | 50.22 | 0.67 | 1.33 | 0.04 | 48.83 | 51.65 | 57.75 |

Table 2 summarizes the results of the data analysis for the values of acoustic pressure level parameters in selected octave bands, on working days for the day sub-interval, generated by heavy vehicles calculated by the Cnossos-EU method. Maximum values of median as well as percentiles $C_{10}$ and $C_{90}$ or $C_{99}$ were obtained in the octave band with a central frequency of $f_0$=1000 Hz. The minimum values of these parameters were obtained in the octave band with a frequency of $f_0$=125 Hz.

The values in octave bands of median and percentile $C_{99}$ of noise for heavy vehicles on working days for the day sub-interval are presented in Figure 2.

a)                                                                  b)

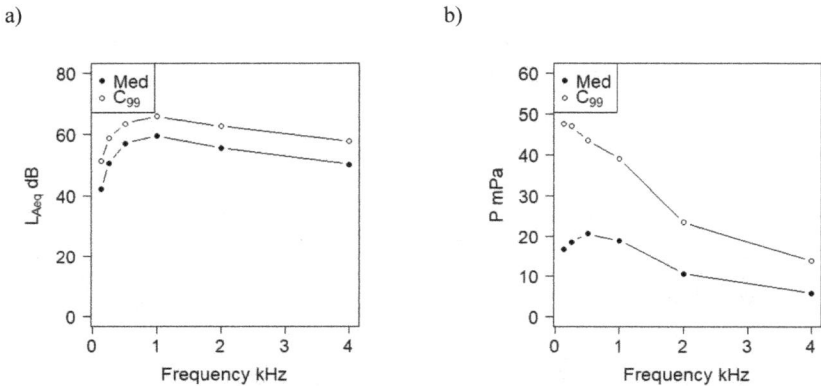

*Fig. 2 The values in octave bands of median and percentile $C_{99}$ of noise calculated by the Cnossos-EU method for heavy vehicles on working days for the day sub-interval: a) acoustic sound pressure level in dB(A) units, b) acoustic sound pressure in mPa units.*

It should be noted that the maximum median value is in the octave band of $f_0$=500 Hz or $f_0$=1000 Hz depending on the adopted scale of units. For the $C_{99}$ percentile expressed in the units of dB(A), the maximum value is for $f_0$=1000 Hz. However, when it is expressed in the units of mPa, its value decreases systematically in each subsequent frequency band. The plots of the $C_{99}$ parameter expressed in dB(A) and mPa units have a slightly different character. In the first case, this waveform has a local extreme for 1 kHz, in the second case there is no such extreme.

**Conclusions**

Statistical tests and histograms confirmed deviations from the normal distribution. The values of selected data distribution parameters depend on the units used. For heavy vehicles, the maximum values in mPa for median, percentiles $C_{10}$ and $C_{90}$ were obtained in the octave band with a center frequency of $f_0$=500 Hz. The minimum values of these parameters were obtained in the octave band with a center frequency $f_0$=4000 Hz. The values of the parameter $C_{99}$ in relation to the value of $C_{90}$ are as follows: in the octave band with a frequency $f_0$=125 Hz they are higher by about 138% and for the frequency of $f_0$=500 Hz or $f_0$=1000 Hz by about 80%. The values of the coefficients of variation $V_q$ are similar in all octave bands but depend on the value of central frequency. For the same parameters but in dB(A) units, the maximum values were obtained in the octave band with a center frequency of $f_0$=1000 Hz and the minimum in the octave band of $f_0$=125

Hz. For the $C_{99}$ percentile expressed in the units of dB(A), the maximum value is for $f_0$=1000 Hz. However, when it is expressed in the units of mPa, its maximum value is in the octave band of $f_0$=125 Hz and decreases systematically in each subsequent frequency band. The research showed that the physical and metrological aspects of noise are more convenient to analyze using mPa but its environmental impact is better described by the dB(A) scale.

**References**

[1]  G. Dutilleuks, B. Soldano, Matching directive 2015/996/EC (Cnossos-EU) and the French emission model for road pavements, Euronoise 2018 Conf. Proc., 2018, 1213-1218.

[2]  S. Kephalopoulos, M. Paviotti, F. Anfosso-Lédée, D. Van Maercke, S. Shilton, N. Jones, Advances in the development of common noise assessment methods in Europe: The CNOSSOS-EU framework for strategic environmental noise mapping, Sci. Total Environ. 482-483 (2014) 400-410. https://doi.org/10.1016/j.scitotenv.2014.02.031

[3]  A. Bąkowski, V. Dekýš, L. Radziszewski, Z. Skrobacki, Validation of Traffic Noise Models, AIP Conf. Proc. 2077 (2019) art. 020005. https://doi.org/10.1063/1.5091866

[4]  W. Batko, B. Stępień, Type A Standard Uncertainty of Long-Term Noise Indicators, Archives of Acoustics 39 (1) (2014) 25–36. https://doi.org/10.2478/aoa-2014-0004

[5]  J. Pietraszek, A. Szczotok, M. Kolomycki, N. Radek, E. Kozien, Non-parametric assessment of the uncertainty in the analysis of the airfoil blade traces, Metal 2017 26th Int. Conf. on Metallurgy and Materials, Ostrava, Tanger, 2017, 1412-1418.

[6]  J. Pietraszek, N. Radek, K. Bartkowiak, Advanced statistical refinement of surface layer's discretization in the case of electro-spark deposited carbide-ceramic coatings modified by a laser beam, Solid State Phenom. 197 (2013) 198-202. https://doi.org/10.4028/www.scientific.net/SSP.197.198

[7]  A. Bąkowski, L. Radziszewski, M. Žmindak, Analysis of the coefficient of variation for injection pressure in a compression ignition engine, Procedia Engineering 177 (2017) 297-302. https://doi.org/10.1016/j.proeng.2017.02.228

Terotechnology XI
Materials Research Proceedings **17** (2020) 114-119

Materials Research Forum LLC
https://doi.org/10.21741/9781644901038-17

# Effect of Gas Laser Beam applied during Machining of Metal Matrix Composites Reinforced by Sic Particle

PRZESTACKI Damian[1,a*], BARTKOWSKA Aneta[2,b], KUKLIŃSKI Mateusz[1,c*], KIERUJ Piotr[1,d*] and SZYMAŃSKI Michał[1,e*]

[1] Faculty of Mechanical Engineering and Management, Institute of Mechanical Technology Poznan, University of Technology, Piotrowo 3, 60-965, Poznań, Poland

[2] Faculty of Mechanical Engineering and Management, Institute of Material Science and Engineering, Poznan University of Technology, Jana Pawła II 24, 60-965, Poznań, Poland

[a] damian.przestacki@put.poznan.pl, [b] aneta.bartkowska@put.poznan.pl, [c] mateusz.kuklinski@doctorate.put.poznan.pl, [d] piotr.kieruj@put.poznan.pl, [e] michal.mari.szymanski@doctorate.put.poznan.pl

**Keywords:** Metal Matrix Composites (MMCs), Laser Assisted Machining, Tool Wear, Turning

**Abstract.** This paper presents a study of Laser Assisted Machining (LAM) when turning the AlSi9Mg alloy reinforced with 20 vol.% particles of SiC. Due to hard ceramic reinforcing, components are difficult to machine using conventional manufacturing processes. The applied LAM process was used to heat the cutting zone. The aluminum matrix becomes softer and easer in plastic deformation, which leads to the reduction of pushing force of the SiC particles on the clearance face of a cutting tool, with is the reason of its wear. This research was carried out for tungsten carbide inserts. The results obtained with the laser assisted machining were compared with results obtained in conventional turning.

## Introduction

The reinforcement of metallic alloys with ceramic particles has generated a new family of material called metal matrix composites (MMCs). Aluminium, titanium and magnesium alloy are used as matrix elements, while silicon carbide (SiC) and alumina ($Al_2O_3$) are popular reinforcements.

The commonly used lightweight materials, e.g. aluminum alloy composites are still of great interest in the field of automotive, aerospace, electronics and medical industries [1-4]. They have outstanding properties like high specific strength, low weight, high modules, low ductility, high wear resistance and high thermal conductivity. As a consequence of the applications of metal matrix composites (MMCs), the machining process has become a very important subject for investigation. The reinforcing materials are characterized by harder and stiffer features than the matrix. This material belongs to the group which is difficult to machine, like tungsten carbide [5], hardened steel [6], Waspaloy [7] or Inconel 718 [8, 9] and therefore machining is much more difficult in comparison with conventional materials like steel [10]. However, due to hard ceramic reinforcing components in MMC, they are difficult to machine using conventional manufacturing processes due to heavy tool wear. As a consequence, hybrids machining processes, e.g. electro discharge machining (EDM), laser [4], numerical method [11-13] and other techniques [14-16], are becoming more popular for the MMC machining. The most LAM hot processes still require further investigations to get all optimum parameters.

The work reported here investigated the effectiveness of laser assisted machining process to machine MMC. A different defocus of laser beam was used to heated the cutting zone. The laser irradiation makes the aluminum matrix softer and easer in plastic deformation, which leads to the reduction of pushing force of the SiC particles on the primary flank face of the cutting tool. This phenomenon is the main reason of increasing tool wear.

The collected results and observations on composite materials and their laser processing may be interesting in the context of similar technological processes, e.g. production of nanocomposites [17] or protective coatings by ESD with subsequent laser machining [18, 19], which is of great importance in the case of hydraulic seals of heavy machines [20] . This work can also inspire the further development of experimental data analysis, such as DOE [21] and image analysis [22].

**Experimental details**
Composite AlSi9Mg (aluminum with 9.2% silicon, 0.6% magnesium, 0.1% iron), reinforced with silicon carbide particles, was the material selected for this study. The SiC particles in the machined composite workpieces were about 20% in volume and about 8-15 μm in diameter. The workpieces had a cylindrical shape of 25 mm length and 38 mm in diameter. The workpieces were painted by absorptive coating each time to increase laser absorption.

Investigation on the wear of SNMG 120408 MS Kennametal tool in the machining of MMC was carried out by turning. The tool was a commercially available physical vapor deposition (PVD) TiCN coated KC5510 tungsten carbide inserts with a fine-grain substrate, with rake angle of 5°, clearance angle of 5°, cutting edge angle of 75°, minor cutting edge angle of 15°, nose radius of 0.8 and cutting edge inclination angle of 6°.

Laser assisted machining was carried out with a 2.6 kW continuous wave (CW) $CO_2$ laser (2.6kW Trumpf, type TLF2600t). After each trial, the machined surface roughness was measured using a Hommel T1000 profilometer with a diamond stylus type instrument set to a 0,8 mm cut-off length and a tracing length of 4.8 mm. Three readings were taken at random points on the machined surface and averaged after each test. Tool wear was measured on the primary flank face by an optical microscope. The measurements of tool wear ($VB_c$) and machined surface roughness ($Ra$) were carried out at regular intervals during a 44 s machining period.

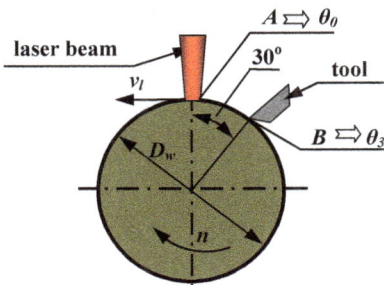

Figure 1. Scheme of the experimental set-up. Designations: A - heating area by laser beam,

B – zone of machining, $\theta_0$, $\theta_3$ - areas of temperature measurement, $D_w$- workpiece diameter.

Figure 1 shows the scheme of a laser assisted machining set-up used during the present study. Surface temperatures were measured with a two RAYTEK pyrometers (model:MA2SC and S5XLT). One of these measured temperatures in A area (Fig.1) and the second in B area (Fig.1) at the same time. Emission was set in the software to the value of ε = 0.3 based upon primarily made calibration tests.

Terotechnology XI                                                                    Materials Research Forum LLC
Materials Research Proceedings **17** (2020) 114-119                    https://doi.org/10.21741/9781644901038-17

The constant parameters were the following: cutting speed ($v_c$ = 107 m/min), depth of cut ($a_p$ = 0.1 mm), feed ($f$ = 0.04 mm/rev), emissivity ($\varepsilon$ = 0.3), cutting time ($t$ = 0.73min) and workpiece diameter ($d_w$ = 38mm). Several others parameters were the following variables: laser power ($P$ = 0.26÷2.34 kW), laser beam diameter ($d_l$ = 2÷4mm).

### Results and discussions

In Figure 2 the effect of laser power on the temperatures in $A$ and $B$ areas are shown. It was noticed that the temperature ($\theta_0$) in $A$ zone (Fig.2a) insignificantly increases with the increasing power of laser beam. The increase of laser power more than three times causes small a increase of temperature to about 200°C. In the case of the temperature ($\theta_3$) in $B$ area (Fig.2b), increasing power of laser by approximately 1kW had distinct effect on increased temperature in this area. On the other hand, increasing power of laser beam above 1 kW practically did not change temperature ($\theta_3$) in $B$ area. It can be brought out of appearance plasma in area heated by laser beam which in this conditions intensively absorbs the energy of laser radiation.

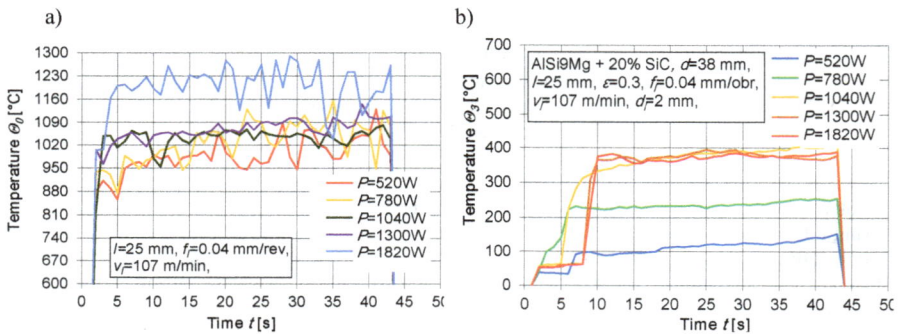

*Figure 2. Comparison of measured temperatures with different power of laser beam, a) measured from $\theta_0$ area, b) measured from $\theta_1$ area*

### Tool wear and surface finished in laser-assisted machining

In comparison with conventional cutting, laser assisted machining increases tool wear but only for 2mm of laser beam diameter, which is depicted in Figure 3. It can be caused by the softening effect of laser heating on the aluminum matrix which reduces the pushing forces of SiC particles on the cutting tool. This effect was observed for severe cutting tool wear in the conventional machining of this composite. Figure 3 shows that 4 mm of laser beam diameter was not sufficient to provide suitable power of laser to soften the matrix and finally wear of wedges had comparable value for conventional and hot assisted machined cutting.

Turning with laser heating improves machined surface roughness in comparison with conventional turning but surface quality decreases with the increase of a laser power beam. In low range value (to 1kW) of laser power, LAM improved surface roughness in both 2mm of laser beam diameter and 4mm of laser beam diameter.

As decrease tool wear (Fig.3) as well increase surface roughness (Fig.5) during higher values of laser power can be explained by the fact that fusion surface causes appearance of plasma which intensively absorbs laser radiation. The result was that the thin layer of composites was heated so deeper layers of workpiece were not soft enough.

Terotechnology XI                                          Materials Research Forum LLC
Materials Research Proceedings **17** (2020) 114-119        https://doi.org/10.21741/9781644901038-17

Abrasion of the deposited workpiece material on both primary and secondary flank faces results in the grooves on the flank face (Fig. 4).

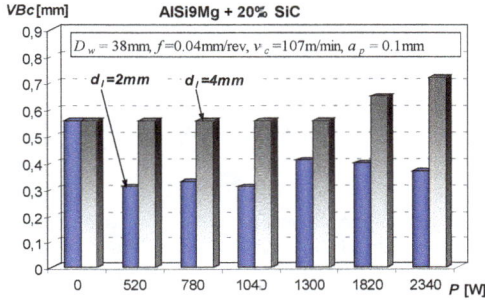

*Figure 3. Tool wear after conventional cutting and laser-assisted hot cutting of AlSi9Mg+20% SiC MMC with diference laser beam diameter*

a)                                    b)

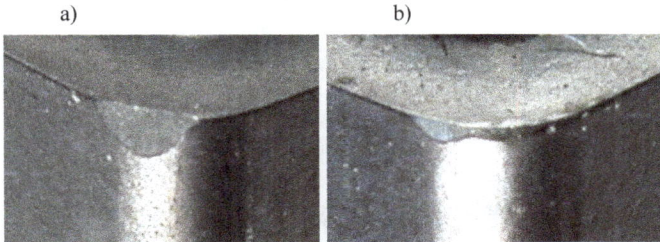

*Figure 4. The images of KC5510 inserts after: a) conventional turning, b) with laser assisted machining. Parameters: $v_c = 107m/min$, $f = 0,04mm/rev$, $a_p = 0,1mm$, $t_s = 0,73min$*

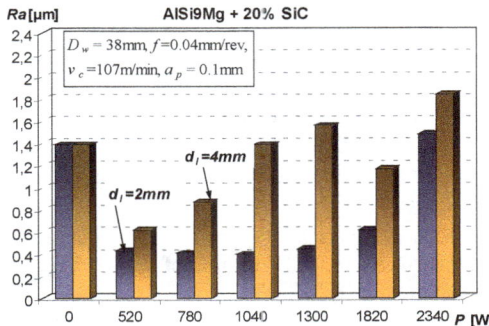

*Figure 5. Comparison of machined surface roughness between conventional and LAM turning*

117

Terotechnology XI
Materials Research Proceedings **17** (2020) 114-119

Materials Research Forum LLC
https://doi.org/10.21741/9781644901038-17

## Summary and conclusions

The results of the experiment show how laser assisted machining improved tool wear resistance and machined surface quality in comparison with conventional cutting. Hence, it is necessary to understand the whole processes during laser assisted machining for this material through further investigation. Conclusions derived from this study are as follows:

1. Laser- assisted machining is a very effective method (in low range value to 1kW) of laser power) in the machining of SiCp/Al composite.
2. LAM of MMC was studied to prove tool life increase by using a $CO_2$ laser to heat spot on a workpiece.
3. In the low value of laser power, the laser-assisted hot machining of SiCp/Al composite improves the machined surface.

## References

[1] Y.Wang, L.J. Yang, N.J. Wang, An investigation of laser-assisted machining of Al2O3 particle reinforced aluminium matrix composite, Journal of Materials Processing Technology 129 (2002) 268-272. https://doi.org/10.1016/S0924-0136(02)00616-7

[2] K. Gawdzińska, L. Chybowski, A. Bejger, S. Krile, Determination of technological parameters of saturated composites based on sic by means of a model liquid, Metalurgija 55 (2016) 659-662.

[3] K. Gawdzińska, L. Chybowski, W. Przetakiewicz, R. Laskowski, Application of FMEA in the Quality Estimation of Metal Matrix Composite Castings Produced by Squeeze Infiltration. Arch. Metall. Mater. 62 (2017) 2171-2182. https://doi.org/10.1515/amm-2017-0320

[4] J. Hajkowski, P. Popielarski, Z. Ignaszak, Cellular Automaton Finite Element Method Applied for Microstructure Prediction of Aluminium Casting Treated by Laser Beam, Arch. Foundry Eng. 19 (2019) 111-118.

[5] S. Wojciechowski, P. Twardowski, T. Chwalczuk, Surface roughness analysis after machining of direct laser deposited tungsten carbide, Journal of Physics: Conference Series 483 (2014) art. 012018. https://doi.org/10.1088/1742-6596/483/1/012018

[6] P. Twardowski, M. Wiciak-Pikuła, Prediction of Tool Wear using Artificial Neural Networks during Turning of Hardened Steel, Materials 12 (2019) art. 3091. https://doi.org/10.3390/ma12193091

[7] M. Wiciak, T. Chwalczuk, A. Felusiak, Experimental Investigation and Performance Analysis of Ceramic Inserts in Laser Assisted Turning of Waspaloy, MATEC Web of Conf. 237 (2018) art. 0100. https://doi.org/10.1051/matecconf/201823701003

[8] A. Felusiak, T. Chwalczuk, M. Wiciak, Surface Roughness Characterization of Inconel 718 after Laser Assisted Turning MATEC Web of Conf. 237 (2018) art. 01004. https://doi.org/10.1051/matecconf/201823701004

[9] T. Chwalczuk, M. Wiciak, A. Felusiak, P. Kieruj, An Investigation of Tool Performance in Interrupted Turning of Inconel 718, MATEC Web of Conf. 237 (2018) art. 02008. https://doi.org/10.1051/matecconf/201823702008

[10] A. Bartkowska, A. Pertek, M. Jankowiak, K. Jóźwiak, Borided layers modyfied by chromium and laser treatment on C45 steel, Arch. Metall. Mater. 57 (2012) 211-214. https://doi.org/10.2478/v10172-012-0012-9

[11] Ratajczak, M. Ptak, L. Chybowski, K. Gawdzińska, R. Będziński, Material and Structural Modeling Aspects of Brain Tissue Deformation under Dynamic Loads. Materials 12 (2019) art. 271. https://doi.org/10.3390/ma12020271

[12] J. Hajkowski, P. Popielarski, R. Sika, Prediction of HPDC Casting Properties Made of Al-Si9Cu3 Alloy, Advances In Manufacturing, Lecture Notes in Mechanical Engineering, Springer 2018, 621-631. https://doi.org/10.1007/978-3-319-68619-6_59

[13] Z. Ignaszak, J.Hajkowski, P. Popielarski, Mechanical properties gradient existing in real castings taken into account during design of cast components, Defect and Diffusion Forum, 334-335 (2013) 314-321. https://doi.org/10.4028/www.scientific.net/DDF.334-335.314

[14] P. Krawiec, M. Grzelka, J. Kroczak, G, Domek, A.Kołodziej, A proposal of measurement methodology and assessment of manufacturing methods of nontypical cog belt pulleys, Measurement 132 (2019) 182-190. https://doi.org/10.1016/j.measurement.2018.09.039

[15] P. Krawiec, A. Marlewski, Profile design of noncircular belt pulleys, Journal of Theoretical and Applied Mechanics 54 (2016) 561-570. https://doi.org/10.15632/jtam-pl.54.2.561

[16] P. Krawiec, K. Waluś, Ł. Warguła, J. Adamiec, Wear evaluation of elements of V-belt transmission with the application of optical microscope, MATEC Web of Conf. 157 (2018) art. 01009. https://doi.org/10.1051/matecconf/201815701009

[17] E. Piesowicz, I. Irska, K. Bryll, K. Gawdzinska, M. Bratychak, Poly(butylene terephthalate/carbon nanotubes nanocomposites part ii. Structure and properties. Polimery 61 (2016) 24-30. https://doi.org/10.14314/polimery.2016.024

[18] R. Dwornicka, N. Radek, M. Krawczyk, P. Osocha, J. Pobedza, The laser textured surfaces of the silicon carbide analyzed with the bootstrapped tribology model. METAL 2017 26th Int. Conf. on Metallurgy and Materials (2017), Ostrava, Tanger 1252-1257.

[19] N. Radek, A. Szczotok, A. Gadek-Moszczak, R. Dwornicka, J. Broncek, J. Pietraszek, The impact of laser processing parameters on the properties of electro-spark deposited coatings. Arch. Metall. Mater. 63 (2018) 809-816.

[20] G. Filo, E. Lisowski, M. Domagala, J. Fabis-Domagala, H. Momeni, Modelling of pressure pulse generator with the use of a flow control valve and a fuzzy logic controller. MSM 2018 14th Int. Conf. Mechatronic Systems and Materials, AIP Conference Proceedings, vol. 2029, art. 020015-1. https://doi.org/10.1007/978-3-642-29347-4_36

[21] J. Pietraszek, Fuzzy Regression Compared to Classical Experimental Design in the Case of Flywheel Assembly. In: Rutkowski L., Korytkowski M., Scherer R., Tadeusiewicz R., Zadeh L.A., Zurada J.M. (eds) Artificial Intelligence and Soft Computing ICAISC 2012. Lecture Notes in Computer Science, vol 7267. Berlin, Heidelberg: Springer, 2012, 310-317. https://doi.org/10.1007/978-3-642-29347-4_36

[22] L. Wojnar, A. Gadek-Moszczak, J. Pietraszek, On the role of histomorphometric (stereological) microstructure parameters in the prediction of vertebrae compression strength. Image Analysis and Stereology 38 (2019) 63-73. https://doi.org/10.5566/ias.2028

Terotechnology XI
Materials Research Proceedings 17 (2020) 120-125

Materials Research Forum LLC
https://doi.org/10.21741/9781644901038-18

# Laser Cutting Methods – Review

KURP Piotr[1,a*]

[1] Kielce University of Technology, Faculty of Mechatronics and Mechanical Engineering Department of Terotechnology and Industrial Laser Systems, Aleja Tysiąclecia Państwa Polskiego 7, PL- 25314 Kielce

[a]pkurp@tu.kielce.pl

**Keywords:** Laser Cutting, Wood Cutting, Steel Cutting, Ceramic Cutting

**Abstract.** Nowadays, sheet and profile cutting services, including cutting in 3D systems, are a significant part of laser technologies used in heavy industry. Laser welding is also becoming more common, mainly in the automotive industry [1]. Due to high power fiber lasers, it is possible to robotize the process. Surface treatment such as hardening, padding, alloying, etc. also has a small market share in laser services [2-6]. However, laser cutting is still the main laser treatment technology applied by heavy and machine industry. In this paper, the author described laser cutting methods and showed examples of various materials laser cutting using the discussed methods.

## Introduction

Laser cutting was the first laser technology applied in industry. The first described use of the laser beam as a cutting tool was in 1967 [7]. In order for the cutting process to take place, the laser beam must be brought to the workpiece surface and sufficiently high power density must be ensured. Depending on the type and thickness of material being cut, this density is about $10^4 \div 10^6$ W/cm$^2$ (max $\sim 10^{11}$). At the same time, a suitable working gas must be applied, whose type depends on the type of material being treated.

By definition, laser cutting is a thermal process that leads to material continuity loss as a result of the laser beam affecting it with the participation of working gas fed under pressure and coaxial with the laser beam. This gas may be inert or reactive. The choice of gas type depends on the cutting method. Its type and pressure significantly affects the quality and speed of a cutting process.

The laser cutting mechanism is complex and depends on physical properties of the material being cut and the parameters of the laser beam, which acts as a linear heat source forming a stable cutting mesh.

The laser cutting process can be briefly described as follows. A focused laser beam falling on material's surface is partly reflected and partly absorbed (the passing wave is negligible). Laser beam photon energy is absorbed by free electrons of the electron cloud surrounding atomic nuclei of the material being cut. Electrons under the influence of absorbed energy go into the state of forced vibrations, which is expressed in the form of thermal energy. When a sufficient amount of laser radiation energy is delivered, the thermal vibrations of the electrons are so intense that there is a decrease in molecular bonds strength leading to the material transition from solid to liquid state. If the energy of laser radiation increases further, then the energy of electron vibrations increases as well. This leads to a significant decrease in molecular bond strength, which results in material transition from liquid to gaseous state [8]. The stream of working gas blows liquid metal and it vapors out of the gap (inert gas) and/or provides additional energy (reactive gas). A diagram of the cutting process is shown in Fig. 1.

Terotechnology XI

Materials Research Forum LLC

Materials Research Proceedings **17** (2020) 120-125

https://doi.org/10.21741/9781644901038-18

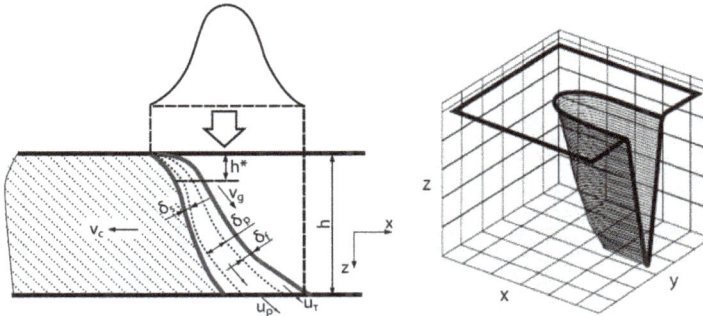

*Fig. 1. Diagram of the laser cutting process and the theoretical shape of the cutting gap [5]: $v_C$ - laser beam speed, $v_g$ - flowing gas speed, $\delta_f$, $u_\tau$ - respectively thickness and velocity of liquid in the boundary layer caused by gas flow, $\delta_p$, $u_p$ - respectively liquid thickness and velocity due to pressure gradient, $\delta_s$ - thickness of boundary layer.*

Laser beam can cut anything. From materials of natural origin (wood, leather, stone, etc.), through metal alloys (ferrous and non-ferrous, pure metals) to all kinds of plastics and laminates. The only limitation is a quality restriction that determines the maximum cutting thickness of a given material. In turn, the type of cut material affects the choice of a laser cutting method.

**Laser cutting methods**
There are five main laser cutting methods separated in the scientific literature [8], [9]. There are methods descriptions below with examples of treated materials.

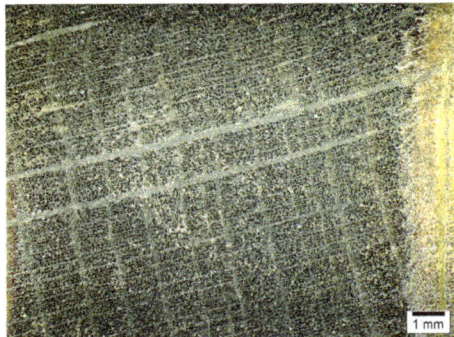

*Fig. 2. View of wood cut edge - 13 mm thick. $CO_2$ laser, working gas: air, working gas pressure: p=2 bar, power: P=1500 W, cutting speed v=2,0 m/min*

*Fig. 3. View of ceramic tile cut edge - 5,0 mm thick. $CO_2$ laser, working gas: air, working gas pressure: p=1,5 bar, power: P=1200 W, cutting speed v=1,0 m/min*

They are as follows:
1. Vaporization cutting - in this method, the material exposed to a focused laser beam is heated to the boiling point, which in consequence leads to its transition from solid to gaseous state. The material evaporates under an inert gas atmosphere. This method is used to cut materials with low thermal conductivity and non-melting materials, such as wood, leather, some plastics and others – Fig.2 and Fig.3.
2. Melt and blow cutting - due to the high power of a laser beam, the material transits from solid to liquid state. Then, due to the high flow pressure of working inert gas, the liquid material is removed from the cutting gap. Due to the fact that this method does not burn the material, the cutting surface is free of oxides. This method is applied for treatment of high-alloy corrosion resistant and stainless steels, aluminum alloys, nickel, titanium, tantalum, and zirconium – Fig.4…Fig.7.

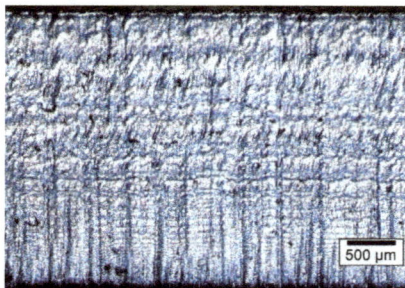

*Fig. 4. View of PA6 aluminum alloy cut edge - 3,0 mm thick. $CO_2$ laser, working gas: $N_2$, working gas pressure: p=14 bar, power: P=5000 W, cutting speed v=2,5 m/min*

3. Reactive cutting - a type of laser cutting in which additional energy is supplied by active gas - oxygen. The material under the action of a focused laser beam is burned out by a stream of oxygen or a mixture of gases containing oxygen in a sufficiently high concentration. The extra

energy from the exothermic reaction allows for cutting materials with less laser radiation power, less working gas volumetric flow and higher process speed. This type of cutting is mainly used for carbon and low alloy steels, some plastics, rubber, quartz, etc. - Fig. 8.

*Fig. 5. View of MO58 brass cut edge - 1,5 mm thick. $CO_2$ laser, working gas: $N_2$, working gas pressure: p=15 bar, power: P=4000 W, cutting speed v=5,3 m/min*

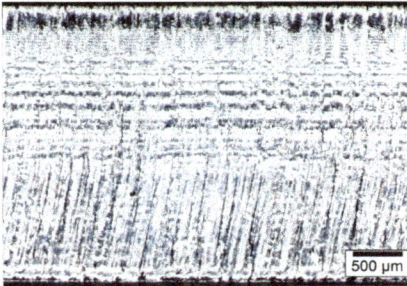

*Fig. 6. View of X5CrNi18-10 stainless steel cut edge - 3,0 mm thick. $CO_2$ laser, working gas: $N_2$, working gas pressure: p=17 bar, power: P=4000 W, cutting speed v=4,0 m/min*

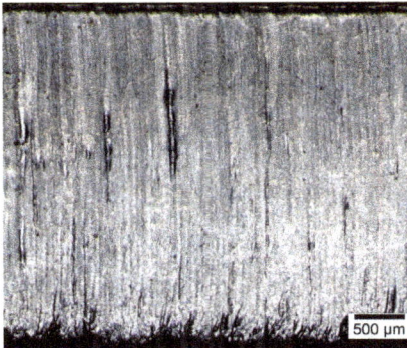

*Fig. 7. View of K303 powder steel cut edge - 3,0 mm thick. $CO_2$ laser, working gas: $N_2$, working gas pressure: p=8 bar, power: P=3500 W, cutting speed v=3,3 m/min*

4. Notching can be considered as a hybrid cutting method. In the first phase, as a result of laser beam action, the material is incised or hollowed out in a series of small holes arranged in a straight line. In the second phase, the material is mechanically broken off. In this case, the

Terotechnology XI                                                          Materials Research Forum LLC
Materials Research Proceedings 17 (2020) 120-125                https://doi.org/10.21741/9781644901038-18

incise or series of holes act as a notch. This type of cutting is mainly used for brittle materials (e.g. ceramics), laminates (e.g. used for the production of printed circuit boards) and plastics.

5. Cold cutting may occur due to the interaction of laser radiation with matter, resulting in chemical degradation of molecular bonds. As a result, the material loses its integrity here. This process takes place without temperature. This type of cutting is used for special plastics and organic materials. Photochemical ablation (also called cold ablation) is another type of cold cutting. It is a transition of material from solid to gaseous state bypassing the liquid phase. In this process, sufficiently high laser radiation energy leads to the disruption of interatomic bonds of the material, and limited thermal effects appear only on boundary zones of the laser beam impact on the material.

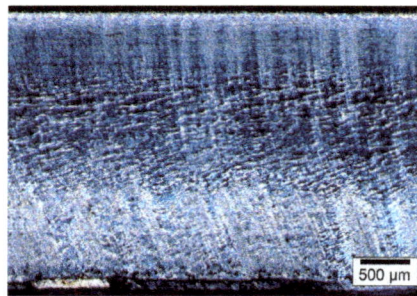

*Fig. 8. View of 50HF spring steel cut edge - 3,0 mm thick. CO$_2$ laser, working gas: O$_2$, working gas pressure: p=1 bar, power: P=1800 W, cutting speed v=3,6 m/min*

Depending on the cutting method, a suitable gas with sufficient pressure is applied. Examples of materials and types of gases used to cutting are shown in Table 1.

*Table 1. Commonly used types of working gases applied for cutting typical construction materials and examples of pressure values.*

| Gas | air (p=0,5÷2 bar) | oxygen (p=0,5÷1 bar) | nitrogen (p=8÷20 bar) | argon (p=8÷20 bar) |
|---|---|---|---|---|
| Material | plastics, wood, composites, ceramics, rubber, quartz, aluminum | carbon and low-alloyed steels, copper alloys | austenitic steels, high-alloy steels, powder steels, aluminum alloys, nickel alloys | titanium alloys, silicon carbide zirconium alloys (zircaloy), tantalum alloys |

**Summary**

The choice of a laser cutting method depends primarily on the type of material, but has an equally significant impact on the edge quality of the material being cut. Therefore, in this case the ratio of cut quality to process costs is the most reasonable criterion. In this case, the expense is working gas.

Terotechnology XI                                                    Materials Research Forum LLC
Materials Research Proceedings 17 (2020) 120-125          https://doi.org/10.21741/9781644901038-18

The cheapest solution is cutting with air shielding. Due to the high nitrogen content, this gas can be used in the melt and blow method. However, due to the oxygen content, the formation of oxides on the cutting edges of metal alloys cannot be avoided. Therefore, air is generally used in the vaporization cutting method as a shielding gas, and the materials are not sensitive to the negative effects of oxygen.

Cutting in pure oxygen (reactive cutting) allows for applying lower laser power and lower working gas expenditure compared to cutting with nitrogen (melt and blow). Oxygen consumption can be up to ten times lower than nitrogen consumption. This affects the treatment costs – the melt and blow method is about twice as expensive. Unfortunately, due to the fact that oxygen is highly reactive, it is not suitable for cutting some materials, e.g. titanium alloys. An oxide film appears on the cutting edges, which is undesirable in some applications. For example, cutting structural carbon steels intended for direct galvanizing or welding should be carried out in an inert atmosphere. Oxides formed on the edges, if not removed, will cause the zinc layer to fall off and may lead to welding incompatibilities.

**Remark**
The figures of cut edges of various materials presented in this paper are the result of research and were made by the author of this publication.

**References**
[1] H. Danielewski, Laser welding of pipe stubs made from super 304 steel. Numerical simulation and weld properties, Technical Transactions 116 (1) (2019) 167–176. https://doi.org/10.4467/2353737XCT.19.011.10051

[2] A. Gądek-Moszczak, N. Radek, S. Wroński, J. Tarasiuk, Application the 3D image analysis techniques for assessment the quality of material surface layer before and after laser treatment, Advanced Materials Research 874 (2014) 133-138. https://doi.org/10.4028/www.scientific.net/AMR.374.133

[3] R. Banak, T. Mościcki, S. Tofil, B. Antoszewski, Laser Welding of a Spark Plug Electrode: Modelling the Problem of Metals with Disparate Melting Points, Lasers in Engineering 38 (2017) 267-281.

[4] N. Radek, J. Pietraszek, B. Antoszewski, The average friction coefficient of laser textured surfaces of silicon carbide identified by RSM methodology, Advanced Materials Research 874 (2014) 29-34. https://doi.org/10.4028/www.scientific.net/AMR.874.29

[5] J. Pietraszek, N. Radek, K. Bartkowiak, Advanced statistical refinement of surface layer's discretization in the case of electro-spark deposited carbide-ceramic coatings modified by a laser beam, Solid State Phenom. 197 (2013) 198-202. https://doi.org/10.4028/www.scientific.net/SSP.197.198

[6] N. Radek, K. Bartkowiak, Laser treatment of electro-spark coatings deposited in the carbon steel substrate with using nanostructured WC-Cu electrodes. Physics Procedia 39 (2012) 295-301. https://doi.org/10.1016/j.phpro.2012.10.041

[7] B. Sullivan, P. Houldcroft, Gas-jet laser cutting. British Welding Journal, Aug. 1967, 443.

[8] A. Klimpel, Podstawy teoretyczne cięcia laserowego metali, Przegląd Spawalnictwa 84 (6) (2012) 2-7. https://doi.org/10.26628/ps.v84i6.289

[9] W. Steen, Laser Material Processing, London, Springer, 2003. https://doi.org/10.1007/978-1-4471-3752

Terotechnology XI
Materials Research Proceedings 17 (2020) 126-131

Materials Research Forum LLC
https://doi.org/10.21741/9781644901038-19

# The Micro Machining of Polypropylene by UV Laser - the Influence of Laser Operating Parameters

TOFIL Szymon

Kielce University of Technology, Faculty of Mechatronics and Mechanical Engineering, Al. 1000-lecia P.P. 7, 25-314 Kielce, Poland

tofil@tu.kielce.pl

**Keywords:** Picoseconds Laser Micro Machining, Polymer Materials, Cold Ablation Phenomenon, Laser Surface Treatment, Polypropylene

**Abstract.** This article presents the research on the impact of operating parameters of a TruMicro 5325c laser device with ultra-short pulses and UV radiation on the polypropylene surface. By changing the frequency of laser pulses, the effect of their impact on the surface of the processed material, which was propylene, was studied. To verify the results, a HIROX KH-8700 confocal digital microscope with software for analyzing the obtained image was used. The efficiency of the performed process was calculated and the optimal working parameters of the device, which do not cause damage to the processed material, were determined.

## Introduction

Constantly evolving industry requires the development of increasingly new construction materials. Even the latest construction materials can be modified in a way that its designers did not anticipate [1-4]. Requirements for the quality of the surface of materials used are one of the most important selection factors in many different technological processes - e.g. in gluing different constructions. These requirements relate, among others, to the condition of a surface - smoothness or surface roughness, including texture. They may also relate to the increased adhesion of a given surface (important in the gluing process) or its resistance to wear. Increasingly, it also has aesthetic aspects. The importance of various surface issues for polymeric materials is very broad. Certainly, the methods of shaping surfaces of various construction materials - including of course polymeric - include laser techniques. The reasons for using these technologies are associated with the provision of specific requirements for the above-mentioned performance properties of the surface. Laser devices are characterized by different properties of the generated laser beam that affects the surface of a workpiece [5-8]. Particular attention should be paid to laser devices giving the possibility of using the phenomenon of cold ablation, which requires high energy and ultra-short time of impact on the material [4].

Laser ablation is a process in which chemical bonds of macromolecules of modified material break under the influence of concentrated laser light. Then, fragments of these macromolecules break away from their surface layer. This applies not only to polymeric materials, but also to metals - including hard-melt alloys (e.g. titanium, iridium, platinum or tantalum). In the case of processing polymer materials, laser ablation is used in the processes of manufacturing micro-modules of micro modules (microlithography), miniaturized machines and their structural elements, contact lenses, as well as for very precise correction of the shape of miniature objects.

Shaping structure and surface properties of materials can be combined with coating the surface of construction materials with polymeric materials and surface treatment. Obtaining the correct properties of processed materials is conditioned by proper surface preparation in order to ensure physical and chemical conditions of adhesion, very often only adhesion between the base material surface and polymer coatings applied to it. In general, the surface treatment of polymeric materials includes cleaning processes that increase adhesion, most often adhesion, and functional modification of the surface. In many cases, these processes combine these functions.

The purpose of laser surface treatment of polymeric materials can improve adhesion, hardness, tribological properties, including scratch resistance, to chemical agents and UV ultraviolet radiation. It should be noted that, above all, surface treatment is aimed at improving the wettability and adhesive properties of many polymeric materials - including polypropylene. Laser techniques allow for pre-treatment on the surface by developing the actual surface of the material, or increasing the free surface energy, so that the difference from the applied surface coverings is not less than 10 mJ / m².

## Material used for research
Polypropylene (PP) is obtained by polymerizing propylene in the presence of metallographic catalysts. This reaction occurs at a temperature of about 100°C in an environment of liquid aliphatic hydrocarbons. Polypropylene is one of the lightest plastics. Its density is about 0.92g/cm³. In its natural form it is a colorless and odorless material and, most importantly, it is non-toxic and harmless to humans.

The physical and chemical properties of polypropylene allow for a wide use of this material in industry and in everyday life. Polypropylene is used, among others, for the production of pharmaceutical packaging, as well as elements of medical apparatus and equipment, e.g. syringes and medicine packaging. Due to high chemical resistance, polypropylene is used to manufacture chemical devices and containers for storing aggressive chemicals. Good mechanical and thermal properties of polypropylene have determined its use with great success as a construction material in the industry for machine components, covers and housings. Good electrical properties in combination with other properties mean that polypropylene is widely used in the electronics industry for the production of various elements of apparatus and equipment for the needs of this industry. The first positive tests of galvanic coating of plastics with metal coatings were carried out on polypropylene products.

## Experimental part
Experimental research was carried out on a test bench existing at the Laser Processing Research Center of the Kielce University of Technology.

A test stand is a laser machine for micro machining with automatic axis. The scheme of the test stand is shown in Fig.1.

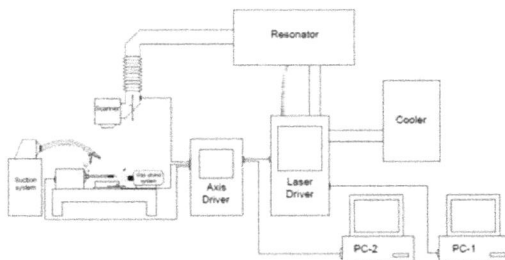

*Fig.1 The scheme of the test stand with TruMicro 5325c laser.*

Terotechnology XI                                        Materials Research Forum LLC
Materials Research Proceedings 17 (2020) 126-131        https://doi.org/10.21741/9781644901038-19

The characteristics of the basic units of the laser machine for micro machining are as follows: Laser TruMicro 5325c - type of laser: pulsed diode impulse laser disk with 3 harmonic generation,- wavelength: 343 nm, - average power: 5 W, - minimum pulse duration: 6.2 ps, - 400 kHz pulse frequency with the possibility of dividing by natural numbers from 1 to 10000, - maximum pulse energy: 12.6 µJ, - mod: $TM_{00}$, - $M^2=1.3$ - maximum fluency: 4.8 J/cm$^2$.

The purpose of the research was to determine how changing the operating parameters of the laser device used affects the quality of the polypropylene surface. In the presented research, only one parameter of the micro machining process changes, which was the pulse frequency. Tests were carried out for the following laser device operating parameters: pulse energy - 12.6 µJ, laser beam scanning speed - 1000 mm/s and pulse frequency from 12.5 kHz to 400 kHz. The results of the work are presented below. A HIROX KH-8700 digital microscope with built-in software for analyzing the obtained image was used to analyze the results of the micro machining.

**Research results and analysis**
The results of the carried out tests are presented below (Table 1). Each test was carried out with an interval of 1 minute in order to obtain repetitive initial conditions of the tested sample - temperature stabilization. In addition, this time was needed to position the sample and change the operating parameters of the laser device. The performed laser micromachining process was repeated five times for each frequency used.

*Table 1. Measurement results of the effects of the laser beam on the material being tested.*

| Pulse repetition [kHz] | Measured values | Results |
|---|---|---|
| 12,5 | Depth [µm] | minimal impact on the material - unmeasurable values |
| | Volume [mm$^3$] | |
| | Machining efficiency [g/s] | |
| 25 | Depth [µm] | minimal impact on the material - unmeasurable values |
| | Volume [mm$^3$] | |
| | Machining efficiency [g/s] | |
| 50 | Depth [µm] | 24.802 |
| | Volume[mm$^3$] | 0.00661656 |
| | Machining efficiency [g/s] | 38.75 |
| 100 | Depth [µm] | 39.140 |
| | Volume [mm$^3$] | 0.0095722 |
| | Machining efficiency [g/s] | 56.06 |
| 200 | Depth [µm] | 75.628 |
| | Volume [mm$^3$] | 0.0296643 |
| | Machining efficiency [g/s] | 173.74 |
| 400 | Depth [µm] | significant material damage |
| | Volume [mm$^3$] | |
| | Machining efficiency [g/s] | |

Terotechnology XI                                                Materials Research Forum LLC
Materials Research Proceedings **17** (2020) 126-131          https://doi.org/10.21741/9781644901038-19

For samples made with 12.5 kHz (Fig.2) and 25 kHz frequencies, the result on the material was insignificant. Therefore, these results were rejected. However, due to the interesting effect of individual interactions at this frequency, these results may constitute separate material for research.

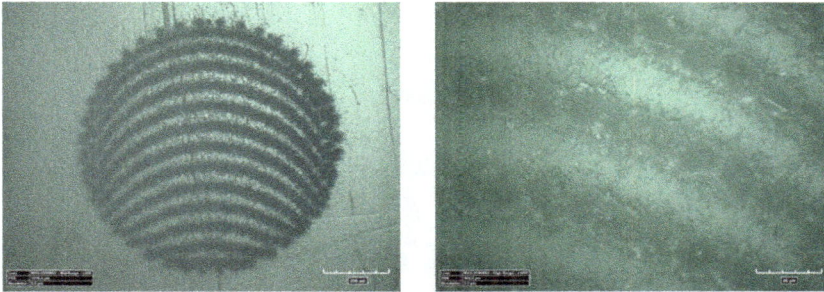

*Fig. 2. View of a single micro texture element made with a pulse repetition rate of 12.5 kHz. Magnification: left – x200 (the marker is 250 μm); right – x1000 (the marker is 50 μm).*

The first measurable effects of the laser radiation allowing for measurements made outside the scale of measurement error could be observed for a pulse repetition rate of 50 kHz. The pulse repetition frequency of 100 and 200 kHz also allowed for the measurement of traces of the laser beam's impact on the tested material. However, for a pulse repetition frequency of 400 kHz, visible material damage was seen - it was melted down, which resulted in the rejection of these results from the result analysis process. Therefore, samples made with 50 kHz, 100 kHz (Fig.2, Fig.4, Fig. 5) sand 200 kHz pulse repetition rates were accepted for testing.

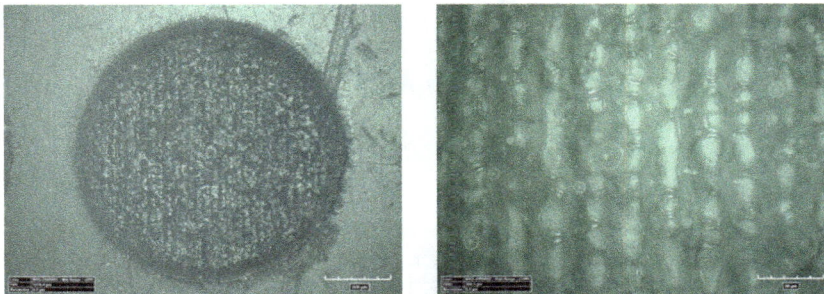

*Fig. 3. View of a single micro texture element made with a pulse repetition rate of 100 kHz. Magnification: left – x200 (the marker is 250 μm); right – x1000 (the marker is 50 μm).*

The diameter of the circle-shaped micro texture was about 1006 μm. The assumed diameter was 1 mm. The small diameter deviation was within the measurement tolerance.

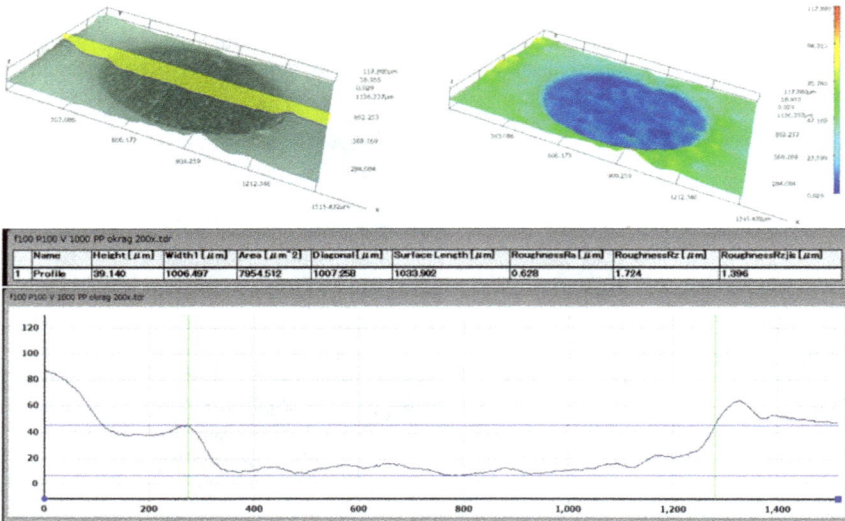

| Name | Height [µm] | Width1 [µm] | Area [µm^2] | Diagonal [µm] | Surface Length [µm] | RoughnessRa [µm] | RoughnessRz [µm] | RoughnessRzjis [µm] |
|------|------|------|------|------|------|------|------|------|
| 1 Profile | 39.140 | 1006.497 | 7954.512 | 1007.258 | 1033.902 | 0.628 | 1.724 | 1.396 |

*Fig. 4. The 3D view of a single element of micro texture (with a pulse repetition frequency of 100 kHz at magnification x200) in a pseudo-color and with a surface profile with a table of values for its width, height, roughness.*

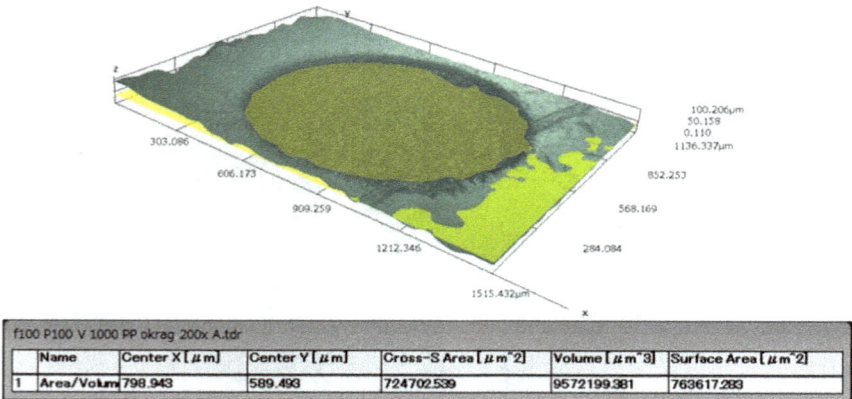

| Name | Center X [µm] | Center Y [µm] | Cross-S Area [µm^2] | Volume [µm^3] | Surface Area [µm^2] |
|------|------|------|------|------|------|
| 1 Area/Volum | 798.943 | 589.493 | 724702.539 | 9572199.381 | 763617.283 |

*Fig. 5. The 3D view of a single element of the micro structure (with a pulse repetition frequency of 100 kHz at magnification x200) and the estimated volume.*

## Conclusions

Tests and observations have shown that a laser emitting UV radiation at 343 nm in picoseconds pulses can be recommended as a tool for the micro machining of elastomeric materials including polypropylene. The micro machining of recesses is efficient and precise and the workpiece does not lose its elasticity. For pulses repetition up to 200 kHz at the site of the treatment and its surroundings, no charring and other signs of overheating were observed. At 200 kHz, a slight remelting of the material could be observed, but not yet significantly damaging the surface. At 12.5 kHz and 25 kHz, there was no significant processing effect but only low traces of individual pulses. At 400 kHz, significant material remelting could be observed. This indicates that the heat supplied to the surface of the material was too high. Dimensions of single micro pockets retained their characteristic dimensions with a +/- 5% deviation along the entire diameter, which demonstrates high repeatability and accuracy of the machining.

Detailed microscopic examinations are required to evaluate structural changes at the work site. The conducted research confirms the possibility of applying a developed technology to shape the surface of elastomeric materials. This opens new prospects for improving the properties of elastomeric materials, for example to glue it with other materials.

**Acknowledgment**
The research was supported by the National Centre for Research and Development (NCBiR), project No. LIDER/30/0170/L-8/16/NCBR/2017.

**References**
[1] N. Stankova et al., Fs- and ns-laser processing of polydimethylsiloxane (PDMS) elastomer: Comparative study, App. Surf. Sci. 336 (2015) 321–328. https://doi.org/10.1016/j.apsusc.2014.12.121

[2] C. Dupas-Bruzek et al. Transformation of medical grade silicone rubber under Nd:YAG and excimer laser irradiation: First step towards a new miniaturized nerve electrode fabrication process, App. Surf. Sci. 255 (2009) 8715–8721. https://doi.org/10.1016/j.apsusc.2009.06.025

[3] H. B. Liu, H. Q. Gong, Templateless prototyping of polydimethylsiloxane microfluidic structures using a pulsed $CO_2$ laser, J. Micromech. Microeng., 19, 3 (2009) 1-8. https://doi.org/10.1088/0960-1317/19/3/037002

[4] B. Antoszewski, Sz. Tofil, M. Scendo, W. Tarelnik, Utilization of the UV laser with picosecond pulses for the formation of surface microstructures on elastomeric plastics, IOP Conf. Series: Materials Science and Engineering 233 (2017) 1-6. https://doi.org/10.1088/1757-899X/233/1/012036

[5] N. Radek, J. Pietraszek, B. Antoszewski, the average friction coefficient of laser textured surfaces of silicon carbide identified by RSM methodology, Advanced Materials Research 874 (2014) 29-34. https://doi.org/10.4028/www.scientific.net/AMR.874.29

[6] N. Radek, K. Bartkowiak, Laser treatment of Cu-Mo electro-spark deposited coatings, Physics Procedia 12 (2011) 499-505. https://doi.org/10.1016/j.phpro.2011.03.061

[7] Gądek-Moszczak A., Radek N., Wroński S., Tarasiuk J., Application the 3D image analysis techniques for assessment the quality of material surface layer before and after laser treatment, Advanced Materials Research 874 (2014) 133-138. https://doi.org/10.4028/www.scientific.net/AMR.874.133

[8] Radek N., Bartkowiak K., Laser treatment of electro-spark coatings deposited in the carbon steel substrate with using nanostructured WC-Cu electrodes, Physics Procedia 39 (2012) 295-301. https://doi.org/10.1016/j.phpro.2012.10.041

Terotechnology XI
Materials Research Proceedings **17** (2020) 132-137

Materials Research Forum LLC
https://doi.org/10.21741/9781644901038-20

# The Influence of Plasma Cutting Parameters on the Geometric Structure of Cut Surfaces

RADEK Norbert[1, a *], PIETRASZEK Jacek [2,b], RADEK Mateusz [3,c]
and PARASKA Olga[4,d]

[1]Kielce University of Technology, Faculty of Mechatronics and Mechanical Engineering,Al. 1000-lecia Państwa Polskiego 7, 25-314 Kielce, Poland

[2]Cracow University of Technology, Institute of Applied Informatics, Faculty of Mechanical Engineering, Al. Jana Pawła II 37, 31-864 Cracow, Poland

[3]The Catholic Public Middle School, Stanisława Kostki 17, 25-341 Kielce, Poland

[4]Khmelnitskiy National University, Department of Chemical Technology, Instytutska 11, 29016, Khmelnickiy, Ukraine

[a] norrad@tu.kielce.pl, [b] jacek.pietraszek@mech.pk.edu.pl, [c] mateusz6676@wp.pl, [d] olgaparaska@gmail.com

**Keywords:** Plasma Cutting, Surface Geometric Structure, Various Materials

**Abstract.** The paper presents the results of microgeometry measurements of cut surfaces. The cutting was carried out using a Hypertherm hand plasma cutter. Samples made of copper, 0H18N9 stainless steel and S355E low-alloy steel were the used tests. The current values at which the smallest roughness of the cut surface was obtained for a given material were determined.

**Introduction**

In modern industries, technologies using a concentrated energy stream are increasingly dominating [1-4]. These technologies include, among others, the plasma cutting method, which, due to high technological capabilities and low operating costs, is one of the most common thermal cutting processes [5, 6].

Plasma arc cutting, commonly known as plasma cutting, is a modification of the GTA plasma welding process. Plasma cutting was introduced to the industry in the 1950s to enable cutting of corrosion resistant steels and independent metals. The plasma cutting process involves melting and removing metal from the cutting gap with a highly concentrated plasma electric arc glowing between the non-fusible electrode and the workpiece. In a plasma electric arc, gas is strongly ionized with a high concentration of kinetic and thermal energy. Then the gas travels from the plasma nozzle, narrowing towards the cutting gap, at a speed close to the speed of sound. The temperature of the plasma stream is in the range of 10,000÷30,000 K and depends on the current intensity, the degree of arc narrowing and the type and composition of plasma forming gas [5].

It is possible to cut all electrically conductive construction materials. Non-metallic materials can only be cut with independent arc plasma torches. Unlike oxygen cutting, plasma cutting allows for cutting materials such as aluminum and its alloys as well as high-alloy steels [7].

The plasma cutting process is used for manual, mechanized and robotic cutting of steel and independent metals at high speeds in all positions. Thanks to the high temperature of the plasma arc, cutting begins immediately, without heating. The disadvantage of the process is very high noise levels, electric shock hazard, strong arc light radiation, large amounts of gases and fumes.

Terotechnology XI
Materials Research Proceedings **17** (2020) 132-137

Materials Research Forum LLC
https://doi.org/10.21741/9781644901038-20

Currently, a dynamic development of hand-held devices and numerical machines for plasma cutting is observed. The main development directions concern improvement of technical parameters of cutting devices, obtaining high quality of cutting surface and increasing cutting accuracy [8, 9]. It may be of particular interest in the preparation of hydraulic accessories requiring high tightness [10-12], difficult-to-cut materials [13-15] and precise processing of smart plates [16, 17]. This method of cutting should significantly reduce the number of failures and non-conformities, which will affect the efficiency of management systems [18, 19]. Thanks to the cutting with minimal heating, damage leading to corrosion is avoided, which may be beneficial in the production of agricultural machinery parts that are exposed to contact with chemically aggressive biological agents [20]. Last but not least, plasma cut edges have a very interesting geometry and microstructure, which should inspire the development of theoretical and practical image analysis methods used in materials science [21, 22].

**Materials and cutting parameters**
50 x 5 x 250 mm samples made of copper, 0H18N9 stainless steel and S355E low-alloy steel were the subject of the study. Five samples of each of the aforementioned material were prepared for testing. The thermal cutting was carried out on a stand with a Hypertherm mechanized plasma cutter, model Powermax 1650. The plasma cutting holder was guided by a welding carriage. The scheme of the plasma cutting process is shown in Fig.1. The cutting parameters were selected experimentally. The following parameter values were adopted:
− cutting speed 0.8 m/min (for 0H18N9 and S355E steel) and 0.5 m/min (for copper),
− cutting current 40÷80 A,
− 60% duty cycle,
− plasma gas - air with a pressure of 0.62 MPa,
− rated air flow of 250 l/min.
The cut samples were subjected to the following tests:
− observation of cut surfaces with a stereoscopic microscope,
− microgeometry measurements.

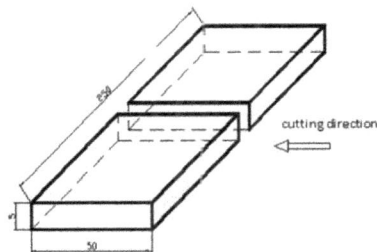

*Fig. 1. The scheme of the plasma cutting process*

**Results and discussion**
Surface geometric structure (SGS) substantially influences many processes that occur in the outer layer. A lot of publications deal with measurement methods and the assessment of surface roughness and waviness [23, 24].

Measurements of surface geometric structure were carried out at the Laboratory of Computer Measurements of Geometric Quantities of the Kielce University of Technology.

Terotechnology XI                                                    Materials Research Forum LLC
Materials Research Proceedings **17** (2020) 132-137              https://doi.org/10.21741/9781644901038-20

Selected results of microgeometry measurements for individual materials are presented in the form of graphs, which are presented in Figures 2÷4. Exemplary protocols for measuring the microgeometry parameters of the tested samples are presented in Figures 5÷7.

Figure 2 shows that the smallest roughness of the cut surface was obtained at a cutting current of I = 70 A and was Ra = 7.76 μm, the largest roughness was obtained at a current I = 40 A and was Ra = 13.7 μm.

In the case of cut samples of OH18N9 stainless steel, it appears that the smallest roughness of the cut surface Ra = 20.2 μm was obtained at a cutting current of I = 60 A (Fig. 3). The largest roughness of the cut surface was created at the current I = 50 A and was Ra = 27.7 μm.

While analyzing Figure 4, it was found that the smallest roughness of the cut surface was obtained at a cutting current of I = 60 A where the roughness was Ra = 2.95 μm, and the largest Ra = 13.1 μm at a cutting current of I = 40 A.

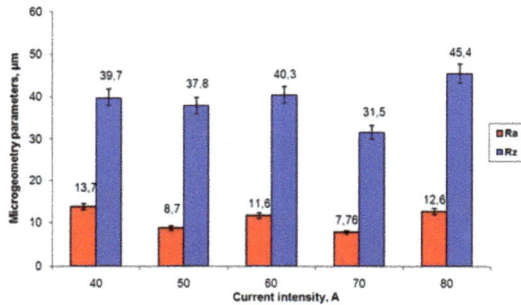

*Fig. 2. Selected microgeometry parameters for copper samples*

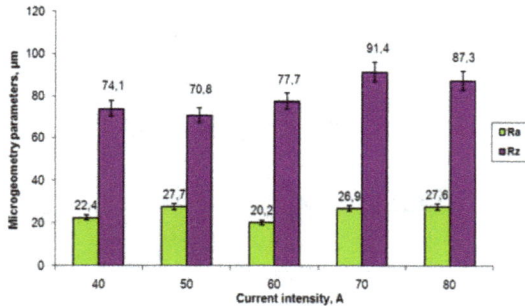

*Fig. 3. Selected microgeometry parameters for 0H18N9 steel samples*

From the graphs presented in Fig.2, Fig.3 and Fig.4, we can read that the smallest values of roughness of the cut surfaces of the tested materials occur at a cutting current of 60 to 70 A.

In the further part of the research, stereoscopic observations of the cut surfaces were made using a set consisting of an OLYMPUS stereoscopic microscope and a digital camera. Examples of photographs of cut surfaces of samples of copper, OH18N9 stainless steel and S355E low-alloy steel are presented in Figures 5a–5c.

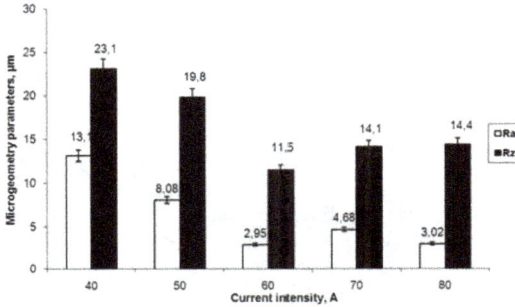

*Fig. 4. Selected microgeometry parameters for S355E steel samples*

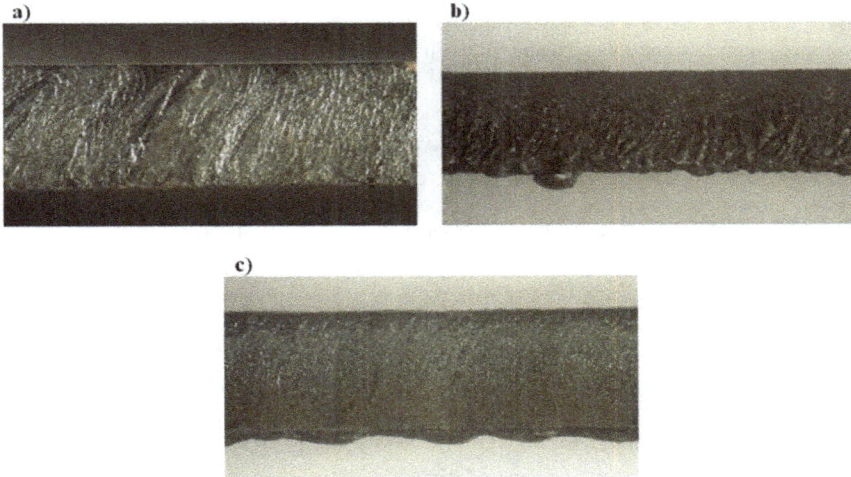

*Fig. 5. Stereoscopic photographs of surfaces cut with plasma (x8 magnification):*
*a) sample from Cu - I = 60 A, b) sample of OH18N9 steel - I = 40 A,*
*c) sample of S355E steel - I = 80 A*

While analyzing stereoscopic photographs of cut surfaces, we can observe that at a cutting current of 80 A, the copper sample is not cut, only melted. The lower edge shows the oxidized form and slag. By reducing the current to 70 A, we get a cutting surface with a lower roughness with a small amount of slag on the bottom edge. The most accurate surface was obtained at a

Terotechnology XI                                                    Materials Research Forum LLC
Materials Research Proceedings 17 (2020) 132-137        https://doi.org/10.21741/9781644901038-20

current of 60 A. A reduction of the cutting current to 50 A caused the cut surfaces to be characterized by considerable unevenness. At 40 A, the material was not cut completely.

When analyzing stereoscopic photographs of cut surfaces of OH18N9 steel samples, we can see that the most favorable surface quality effect occurred when cutting the material at a current of 40 A.

However, observing stereoscopic photographs of the cut surfaces of S355E steel samples, we can conclude that the best cutting results were obtained at a current of 80 A. The most adverse cutting occurred at a current of 50 A, as evidenced by slag on the lower edge of the cut surface.

## Summary

The lowest roughness values of the cut surfaces of samples of copper, OH18N9 steel and S355E steel were obtained at a current of 60÷70 A. While analyzing stereoscopic photographs of the cut surfaces of the samples, we can see traces of cut in the form of oblique grooves and a characteristic slag overhang on the bottom edge of the cut material. When assessing the geometric structure of the cut surfaces, at a later stage of the study, chemical composition and structural changes that occur in the material as a result of the plasma arc should be performed. It also seems advisable to take measurements of the cut surface hardness and the heat affected zone.

## References

[1] N. Radek, K. Bartkowiak, Laser treatment of electro-spark coatings deposited in the carbon steel substrate with using nanostructured WC-Cu electrodes, Physics Procedia 39 (2012) 295-301. https://doi.org/10.1016/j.phpro.2012.10.041

[2] P. Kurp, D. Soboń, The influence of laser padding parameters on the tribological properties of the Al2O3 coatings, METAL 2018 27th Int. Conf. Metallurgy and Materials, Ostrava, Tanger, 1157-1162.

[3] N. Radek, Determining the operational properties of steel beaters after electrospark deposition, Eksploatacja i Niezawodność – Maintenance and Reliability 4 (2009) 10-16.

[4] A. Gądek-Moszczak, N. Radek, S. Wroński, J. Tarasiuk, Application the 3D image analysis techniques for assessment the quality of material surface layer before and after laser treatment, Advanced Materials Research 874 (2014) 133-138. https://doi.org/10.4028/www.scientific.net/AMR.874.133

[5] A. Klimpel, Welding and cutting of metals, Warszawa, WNT, 1999.

[6] J. Czech, J. Dworak, Welding plasma techniques in the domestic industry, Bulletin of the Welding Institute 5 (1995) 54-55.

[7] T. Pfeifer, Plasma cutting of high-alloy steel, Polish Welding Review 4 (2001) 15-19.

[8] T. Pfeifer, Modern systems of automated plasma cutting, Polish Welding Review 7-8 (2000) 22-25.

[9] ZAKMET company homepage – plasma, oxygen, laser CNC cutting. http://www.zakmet.pl (access 23.04.2020)

[10] P. Walczak, A. Sobczyk, Simulation of water hydraulic control system of Francis turbine. Proc. 8th FPNI Ph.D Symposium on Fluid Power, 2014, art. V001T04A001. https://doi.org/10.1115/FPNI2014-7814

[11] M. Domagala, H. Momeni, J. Domagala-Fabis, G. Filo, D. Kwiatkowski, Simulation of cavitation erosion in a hydraulic valve. Materials Research Proceedings 5 (2018) 1-6. https://doi.org/10.21741/9781945291814-1

[12] J. Krawczyk, A. Sobczyk, J. Stryczek, P. Walczak, Tests of new methods of manufacturing elements for water hydraulics. Materials Research Proceedings 5 (2018) 200-205.

[13] M. Hebda, S. Gadek, J. Kazior, Influence of the mechanical alloying process on the sintering behaviour of Astaloy CrM powder mixture with silicon carbide addition. Arch. Metall. Mater. 57 (2012) 733-743. https://doi.org/10.2478/v10172-012-0080-x

[14] A. Szczotok, K. Rodak, Microstructural studies of carbides in MAR-M247 nickel-based superalloy. IOP Conference Series: Materials Science and Engineering 35 (2012) art. 012006. https://doi.org/10.1088/1757-899X/35/1/012006

[15] M. Mazur, K. Mikova, Impact resistance of high strength steels. Materials Today-Proceedings 3 (2016) 1060-1063. https://doi.org/10.1016/j.matpr.2016.03.048

[16] M.S. Kozien, J. Wiciak, Passive structural acoustic control of the smart plate - FEM simulation. Acta Phys. Pol. A 118 (2010) 1186-1188. https://doi.org/10.12693/APhysPolA.118.1186

[17] I. Dominik, J. Kwasniewski, K. Lalik, R. Dwornicka, Preliminary signal filtering in self-excited acoustical system for stress change measurement. CCC 2013 32nd Chinese Control Conf. (2013) 7505-7509.

[18] A. Pacana, M. Pasternak-Malicka, M., Zawaca. A. Radon-Cholewa, Decision support in the production of packaging films by cost-quality analysis. Przem. Chem. 95 (2016) 1042-1044.

[19] A. Pacana, K. Czerwinska, R. Dwornicka, Analysis of non-compliance for the cast of the industrial robot basis, METAL 2019 28th Int. Conf. on Metallurgy and Materials (2019), Ostrava, Tanger 644-650. https://doi.org/10.37904/metal.2019.869

[20] T. Lipinski, D. Karpisz, Corrosion rate of 1.4152 stainless steel in a hot nitrate acid. METAL 2019: 28th Int. Conf. on Metallurgy and Materials, Ostrava, TANGER, 2019, 1086-1091. https://doi.org/10.37904/metal.2019.911

[21] A. Gadek-Moszczak, History of stereology. Image Anal. Stereol. 36 (2017) 151-152. https://doi.org/10.5566/ias.1867

[22] A. Gadek-Moszczak, P. Matusiewicz, Polish stereology – a historical review. Image Anal. Stereol. 36 (2017) 207-221. https://doi.org/10.5566/ias.1808

[23] Radek, N., Kurp, P., Pietraszek, J., Laser forming of steel tubes. Technical Transactions 116 (2019) 223-229. https://doi.org/10.4467/2353737XCT.19.015.10055

[24] S. Adamczak, W. Makieła, Analyzing variations in roundness profile parameters during the wavelet decomposition process using the Matlab environment, Metrology and Measurement Systems 1 (2011) 25-34. https://doi.org/10.2478/v10178-011-0003-6

Terotechnology XI
Materials Research Proceedings **17** (2020) 138-145

Materials Research Forum LLC
https://doi.org/10.21741/9781644901038-21

# Influence of Electrode Material on Surface Roughness during Die-Sinking Electrical Discharge Machining of Inconel 718

CHWALCZUK Tadeusz[1,a], FELUSIAK Agata[1,b*], WICIAK-PIKUŁA Martyna[1,c] and KIERUJ Piotr[1,d]

[1] Poznań University of Technology, Piotrowo 3 St., 60-965 Poznan, Poland

[a] tadeusz.chwalczuk@put.poznan.pl, [b*] agata.z.felusiak@doctorate.put.poznan.pl,
[c] martyna.r.wiciak@doctorate.put.poznan.pl, [d] piotr.kieruj@put.poznan.pl

**Keywords:** Die-Sinking Electric Discharge Machining, EDM, Surface Roughness, Inconel 718, Electrode Materials

**Abstract.** The article presents an analysis of the impact of electrode material and its roughness on the surface roughness of a machined surface. An Inconel 718 element was machined with four different electrodes with constant machining parameters. It was proved that under certain conditions there is a relationship between the roughness of an electrode and a machined surface. The research results show that considering the quality of a machined surface, the copper alloy electrode gives the highest predictability of roughness parameters.

**Introduction**

Nickel-based alloys are difficult to machine with traditional methods such as turning or milling [1]. Difficulty in machining, among others, is connected with the creation of adhesive bonds with the tool material, which results in build-ups, low thermal conductivity compared to carbon steel and tendency to strengthen during machining [2,3]. With traditional machining of such materials, it is difficult to ensure adequate surface roughness [4]. Ensuring adequate roughness and properties of manufactured parts is extremely important from the point of view of performance [5,6].

Due to the difficulties which occur during machining with traction methods, unconventional methods such as laser assisted machining are being increasingly used (LAM) [7]. In [8], the authors used different laser powers to study the effect on stainless steel microstructure. The use of a laser, however, involves the possibility of changing the properties of the surface layer [9]. The use of electrical discharge machining (EDM) is another method. It involves removing material due to controlled electrical discharges which occur between the electrode (tool) and the workpiece. The method is used for the production of machine parts such as noncircular cogbelt pulleys [10]. However, this is not a high-performance manufacturing method [11, 12]. The authors [13] point out that the right choice of an electrode is important in the EDM process. In their research, the machining stainless steel used three electrode materials: aluminum, copper and graphite, at different levels of current. According to their research, the best result (proportion of removed material to electrode wear) was obtained for the current of 6.5 [A] for a graphite electrode. In [14], the authors examined two electrode materials, brass and tungsten carbide, when machining stainless steel and tungsten carbide, with variable process parameters. As a result of the research, they noticed that the brass electrode wears more than the WC electrode, however, it also removes the workpiece faster, due to higher electrical conductivity. In [15], the authors studied electrodes made of aluminum, stainless steel, brass, copper and graphite. Researchers have found that graphite has the best properties as an electrode material. This electrode ensures the fastest material removal and has the lowest wear due to the highest melting point among the mentioned materials. In [16], the authors compared two electrode materials, namely brass and tungsten carbide, when machining aluminum, stainless steel and tungsten carbide. It was noticed that the electrode material has an effect on roughness. With low discharge energy, when machining stainless steel using tungsten carbide, the electrode obtained about 2.5 times lower parameter value $Ra$ (0,23 μm) than with the brass electrode (0,57 μm), while with the increase of the discharge energy, the value of the roughness parameter $Ra$ for the tungsten carbide electrode increased rapidly (1,1 μm) and was almost twice as high as for the brass electrode (0,67 μm). In [17], the authors considered three types of electrodes, such as copper, brass and graphite. They studied the material removal rate (MRR) and surface roughness and found that the material removal rate was higher for the copper electrode than for the brass one. The surface finishing was better for brass than for the copper electrode, and MRR and roughness parameters obtained using a graphite electrode were average. Electrodes and machine parts should be manufactured from a high-quality semi-product, without the separation of alloyed elements and devoid of shrinkage. This is ensured by the use of numerical methods [18].

The obtained results may be interesting for other industries that use EDM or use similar methods, e.g. heavy working machines [19, 20] or applying protective coatings by electro-spark deposition [21, 22]. The effects can also be inspiring for image analysis methods [23] and for health and safety issues [24].

**Experimental details**
The tests were carried out on a sample made of Inconel 718 using an Agie Charmilles Cabinet SP1U spark machining machine. A synthetic hydrocarbon fluid EDM fluid 108 MP-SE was used during machining. The tests were carried out with constant parameters shown in Table 1. Four electrodes made of different materials, the following were used in the experiment: copper impregnated graphite, copper alloy, small grain graphite (grain size <5 μm), large grain graphite (grain size 14 μm). Table 2 presents the percentage of chemical composition of the electrode materials.

*Table 1. Experimental settings.*

| parametr | Peak current [A] | spark gap [mm] | Electrode polarity | cutting depth [mm] |
|---|---|---|---|---|
| value | 10 | 0.2 | + | 0.5 |

*Table 2. Percentage chemical composition of electrode materials*

| Component | Copper impregnated graphite | Copper alloy | Large grain graphite | Small grain graphite |
|---|---|---|---|---|
| Si | 0.08 | 0.62 | 0.10 | 0.14 |
| Cr | 1.22 | 12.99 | 3.01 | 3.38 |
| Fe | 3.12 | 39.44 | 7.08 | 8.58 |
| Ni | 0.33 | 3.66 | 0.99 | 0.97 |
| Cu | 6.26 | 42.62 | 0.00 | 0.00 |
| S | 0.20 | 0.67 | 0.35 | 0.37 |
| C | 88.79 | 0.00 | 88.46 | 86.56 |

Terotechnology XI                                                                Materials Research Forum LLC
Materials Research Proceedings **17** (2020) 138-145                  https://doi.org/10.21741/9781644901038-21

Copper impregnated graphite formation based on technologies of the liquid matrix consists in infiltrating a porous structure of the composite reinforcement phase with a liquid technical alloy [25, 26]. Four operations were performed with each electrode. After each operation, the roughness of treated surface and electrode surface were measured. Roughness parameters were estimated from the area dimensions of 1,5x1,5mm. For the surface topography assessment, stationary profilographometer T8000 was carried out. The following roughness parameters were measured: the texture direction $St$, arithmetic mean height of surface $Sa$, the maximum value peak height on surface $Sp$, the maximum value valley depth $Sv$, the maximum height of surface $Sz$, the root mean squared height $Sq$, the kurtosis of 3D surface texture $Sku$, Skewness $Ssk$.

**Results analysis**
Figure 1 shows the electrodes after four passes. On the basis of the obtained results, the change in electrode roughness and workpiece rosughness over time (Fig.2) and the correlation between electrode roughness and values of workpiece surface roughness parameters were analyzed (Fig. 3).

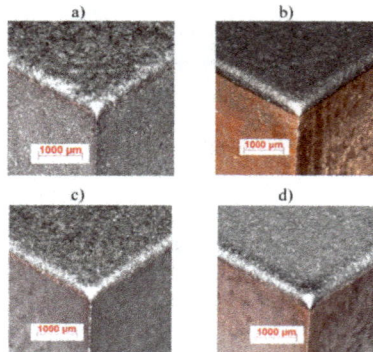

*Fig.1. Electrodes after a completed series of tests: a) large grain graphite, b) copper alloy, c) small grain graphite, d) copper impregnated graphite*

The next picture (Fig.4) shows the topography of the surface worked after the last pass. The graphs in Figure 4 show the relationships between some parameters of the machining surface roughness and the working surface of the electrodes: the texture direction $St$, arithmetic mean height of surface $Sa$, the maximum height of surface $Sz$, the kurtosis of 3D surface texture $Sku$.

Analyzing the graphs of $Sa$, the parameters change (Fig.2a and Fig.2b) over time for the electrodes and workpiece. A satisfactory correlation can be seen only for the copper impregnated graphite electrode. Very high correlation occurs for the surface treated with the Large grain graphite electrode, but the $Sa$ values themselves have a large spread. The opposite situation can be observed for the electrolytic copper electrode. For small grain graphite, no correlation can be found over time.

Analyzing the $Sz/St$ graph (Fig.2c), it can be stated that the surface is equalized for the copper alloy electrode and the tips are worn during machining. In the case of other electrodes, the tendency is reversed as there are more vertices and / or valleys, and the dispersion of results is large, which may indicate a random removal of graphite grains from the electrodes. The analysis of the skewness of the electrode roughness profile (Fig.2d) indicates the increase in the proportion of vertices in the electrodes containing graphite over time, while in the case of copper electrodes they are sheared.

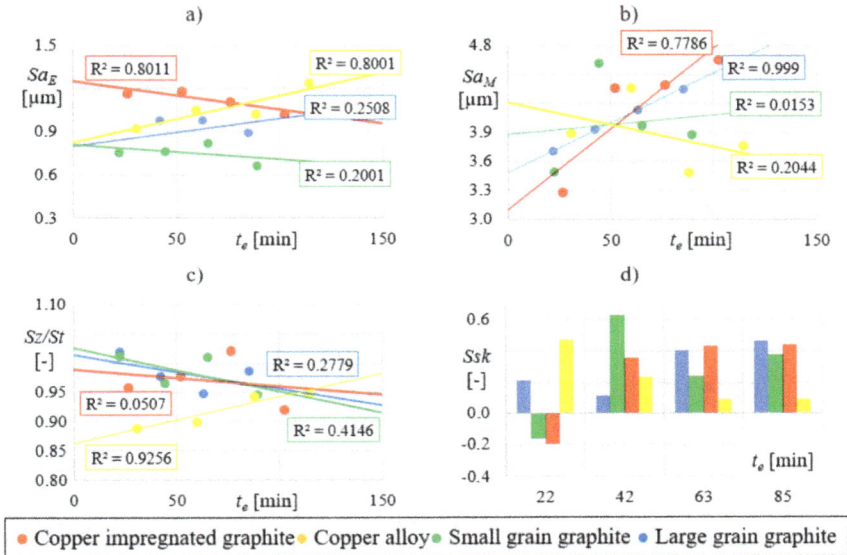

*Fig.2. Course of roughness parameters in time a) Sa of the electrode surface, b) Sa of the treated surface, c) ratio of Sz to St of the electrode surface, d) Ssk of the electrode surface*

It can be seen in the above graphs that the correlation between the roughness of the electrode working surface and the machining surface roughness is ambiguous and does not occur in a satisfactory correlation for some electrode materials.

In the case of graphite electrodes, grain size has a significant impact on this relation. For the large grain graphite, the correlation is very high, usually above 0.9. However, with some parameters for the small grain graphite, there is no relation between the roughness of the treated surface and the electrode. In the case of the *Sa* parameter (Fig. 3d), no relationship can be demonstrated for the copper alloy and the small grain graphite electrodes.

Analyzing these graphs and Table 3, it can be seen that the roughness of electrode surface affects the directivity of the treated surface and kurtosis regardless of the electrode material used. In the case of height parameters, the correlation occurs for the large grain graphite electrode and copper impregnated graphite electrode. For amplitude parameters, the best correlation is for copper and the copper impregnated graphite electrode. It should be noted that the high correlation of valleys on the electrode surface with the peaks on the treated surface occurs (Fig.3b). The inverse relation does not occur (Table 3), which suggests that the peaks on the working surface of the electrode wear during discharge.

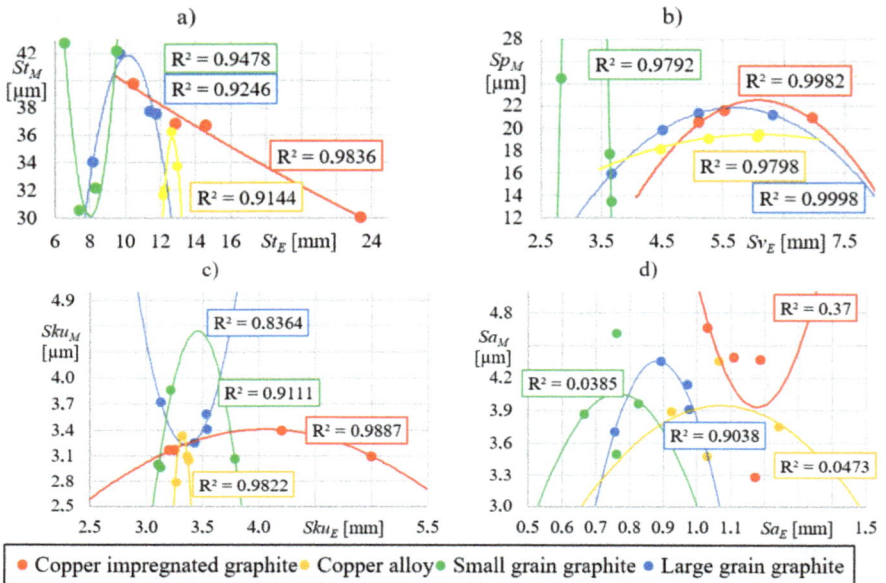

*Fig.3. Correlation between the roughness parameters of machined surface and electrode roughness a) St, b) Sa, c) Sz, d) Sku*

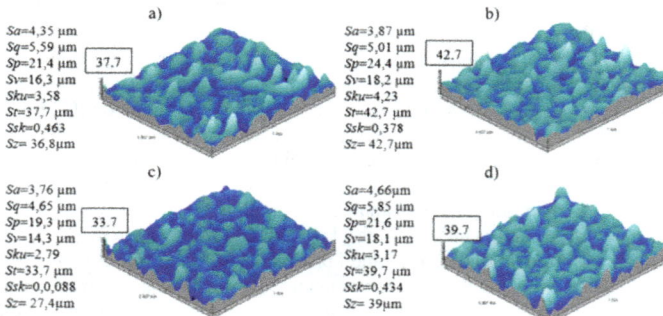

*Fig. 4. Topography of the worked surface for the electrode: a) large grain graphite, b) small grain graphite, c) copper alloy, d) copper impregnated graphite.*

*Table 3.      Polynomial regression equations and correlation coefficient values for individual electrodes*

| parametr | | LGG | SGG | CIG | EA |
|---|---|---|---|---|---|
| $Sq_M$ | eql. | $-34.343\ Sq_E^2 +$ $77.13\ Sq_E - 37.735$ | $-10.09\ Sq_E^2 + 19.192$ $Sq_E - 3.9968$ | $-12.314Sq_E^2+$ $32.326Sq_E-15.385$ | $-3.6128Sq_E^2 +$ $9.3599Sq_E-1.1131$ |
| $Sq_E$ | $R^2$ | 0,84 | 0,011 | 0.98 | 0.07 |
| $Sv_M$ | eql. | $-2.7775\ Sp_E^2+$ $33.208\ Sp_E- 77.579$ | $-5.9657\ Sp_E^2 +$ $57.001\ Sp_E - 111.05$ | $0.1044\ Sp_E^2-$ $1.8771Sp_E+24.631$ | $= 1.7945Sp_E^2-$ $23.108Sp_E +89.22$ |
| $Sp_E$ | $R^2$ | 0.99 | 0.34 | 0.29 | 0.15 |
| $Ssk_M$ | eql. | $-9.8908Ssk_E^2+$ $0.915Ssk_E+ 0.4213$ | $= -30.641Ssk_E^2 +$ $8.566Ssk_E + 0.2357$ | $-15.462Ssk_E^2+$ $2.723Ssk_E+0.8418$ | $= 5.012Ssk_E^2+$ $0.810Ssk_E+0.115$ |
| $Ssk_E$ | $R^2$ | 0.31 | 0,67 | 0.92 | 0.99 |
| $Sz_M$ | eql. | $-1.5604\ Sz_E^2 +$ $30.967\ Sz_E - 116.35$ | $= 0.929\ Sz_E^2 -$ $16.857\ Sz_E + 108.81$ | $-0.0204Sz_E^2-$ $0.167\ Sz_E+ 42.049$ | $-2.2886Sz_E^2+$ $46.606Sz_E- 204.83$ |
| $Sz_E$ | $R^2$ | 0.98 | 0.41 | 0.97 | 0.41 |

As can be seen in Fig.3, only for machining using the copper alloy electrode, roughness parameters *Sa* have a lower value after the last pass than after the first pass. The surface quality after the treatment with small grain graphite electrode is difficult to predict. Due to the high correlation of results, the prediction of the surface roughness treated with the copper impregnated graphite and large grain graphite electrode is the easiest. For some parameters it is also possible to predict the results for the other two electrodes.

**Conclusions**
On the basis of conducted research, the following conclusions were formulated:
- Due to the quality of a machined surface, it is most reasonable to use copper alloy electrodes. The use of this electrode allowed for obtaining the lowest values of *Sa* parameters in the last pass.
- There is a correlation between the roughness of a machined surface and the roughness of an electrode, but it is not a linear relation. Its shape depends on the electrode material. This relation does not exist for all tested electrode materials. This should be noted when choosing the electrode.
- For graphite electrodes, the size of graphite grains is significant. The roughness of a surface machined with the large grain graphite electrode depends on the roughness of the electrode itself. The increase of roughness parameters in subsequent passes is stable. With the small grain graphite electrode, the correlation between the roughness of the electrode and the machined surface is less often obtained, it is also more difficult to predict the roughness in subsequent passes.

**References**

[1] M. Kukliński, A. Bartkowska, D. Przestacki, Investigation of laser heat treated Monel 400, MATEC Web of Conferences 219 (2018) 02005-1-8. https://doi.org/10.1051/matecconf/201821902005

[2] A. Bartkowska, A. Pertek, M. Popławski, D. Przestacki, A. Miklaszewski, Effect of laser modification of B-Ni complex layer on wear resistance and microhardness, Optics and Laser Technology 72 (2015) 116-124. https://doi.org/10.1016/j.optlastec.2015.03.024

[3] P. Kieruj, N. Makuch, M. Kukliński, Characterization of laser-borided Nimonic 80A-alloy, MATEC Web of Conferences 188 (2018) 02003-1-8. https://doi.org/10.1051/matecconf/201818802003

[4] D. Przestacki, R. Majchrowski, L. Marciniak-Podsadna, Experimental research of surface roughness and surface texture after laser cladding, App. Surf. Sci. 388 (2016) 420-423. https://doi.org/10.1016/j.apsusc.2015.12.093

[5] Z. Ignaszak, P. Popielarski, J. Hajkowski, E. Codina, Methodology of comparative validation of selected foundry simulation codes, Arch. Foundry Eng. 15 (2015) 37-44. https://doi.org/10.1515/afe-2015-0076

[6] P. Twardowski, M. Tabaszewski, S. Wojciechowski, Turning process monitoring of internal combustion engine piston's cylindrical surface, MATEC Web of Conferences 112 (2017) 10002-1-6. https://doi.org/10.1051/matecconf/201711210002

[7] M. Kawalec, D. Przestacki, K. Bartkowiak, M. Jankowiak, Laser assisted machining of aluminum composite reinforced by SiC particle, ICALEO Congress Proc. (2008) 895-900. https://doi.org/10.2351/1.5061278

[8] D. Przestacki, A. Bartkowska, M. Kukliński, P. Kieruj, The effects of laser surface modification on microstructure of 1.4550 Stainless steel, MATEC Web of Conferences 237 (2018) 02009-1-5. https://doi.org/10.1051/matecconf/201823702009

[9] J. Hajkowski, P. Popielarski, Z. Ignaszak, Cellular Automaton Finite Element Method Applied for Microstructure Prediction of Aluminum Casting Treated by Laser Beam, Arch. Foundry Eng. 19 (2019) 111-118.

[10] P. Krawiec, M. Grzelka, J. Kroczak, G. Domek, A. Kołodziej, A proposal of measurement methodology and assessment of manufacturing methods of nontypical cog belt pulleys, Measurement 132 (2019) 182-190. https://doi.org/10.1016/j.measurement.2018.09.039

[11] M. Kujawski, P.Krawiec, Analysis of Generation Capabilities of Noncircular Cog belt Pulleys on the Example of a Gear with an Elliptical Pitch Line, Journal of Manufacturing Science and Engineering-Transactions of the ASME 133 (5) (2011) 051006-1-7. https://doi.org/10.1115/1.4004866

[12] P. Krawiec, A. Marlewski, Spline description of not typical gears for belt transmissions, Journal of Theoretical and Applied Mechanics 49 (2) (2011) 355-367.

[13] M.H.F. Al Hazza, A.A. Khan, M.Y. Ali, S.F. Hasim, M.R.C. Daud, A study on capabilities of different electrode materials during electrical discharge machining (EDM), IIUM Engineering Journal 18 (2) (2017) 189-195. https://doi.org/10.31436/iiumej.v18i2.755

[14] G. D'Urso, G. Maccarini, C. Ravasio, Influence of electrode material in micro-EDM drilling of stainless steel and tungsten carbide, International Journal of Advanced Manufacturing Technology 85 (2016) 2013-2025. https://doi.org/10.1007/s00170-015-7010-9

[15] G. Zhu, Q. Zhang, K. Wang, Y. Huang, J. Zhang, Effects of Different Electrode Materials on High-speed Electrical Discharge Machining of W9Mo3Cr4V, Procedia CIRP 68 (2018) 64-69. https://doi.org/10.1016/j.procir.2017.12.023

[16] G. D'Urso, C. Ravasio, Material-Technology Index to evaluate micro-EDM drilling process, Journal of Manufacturing Processes 26 (2017) 13-21. https://doi.org/10.1016/j.jmapro.2017.01.003

[17] S. Choudhary, K. Kant, P. Saini, Analysis of MRR and SR with Different Electrode for SS 316 on Die-Sinking EDM using Taguchi Technique, Global Journal of Researches in Engineering Mechanical and Mechanics Engineering 13 (2013) 14-21.

[18] P.Krawiec, K.Waluś, Ł.Warguła, J.Adamiec, Wear evaluation of elements of V-belt transmission with the application of optical microscope, MATEC Web of Conferences 157 (2018), 01009-1-8. https://doi.org/10.1051/matecconf/201815701009

[19] M. Domagala, H. Momeni, J. Domagala-Fabis, G. Filo, D. Kwiatkowski, Simulation of cavitation erosion in a hydraulic valve. Materials Research Proceedings 5 (2018) 1-6. https://doi.org/10.21741/9781945291814-1

[20] A. Pacana, K. Czerwinska, R. Dwornicka, Analysis of non-compliance for the cast of the industrial robot basis, METAL 2019 28th Int. Conf. on Metallurgy and Materials (2019), Ostrava, Tanger 644-650. https://doi.org/10.37904/metal.2019.869

[21] R. Dwornicka, N. Radek, M. Krawczyk, P. Osocha, J. Pobedza, The laser textured surfaces of the silicon carbide analyzed with the bootstrapped tribology model. METAL 2017 26th Int. Conf. on Metallurgy and Materials (2017), Ostrava, Tanger 1252-1257.

[22] N. Radek, A. Szczotok, A. Gadek-Moszczak, R. Dwornicka, J. Broncek, J. Pietraszek, The impact of laser processing parameters on the properties of electro-spark deposited coatings. Arch. Metall. Mater. 63 (2018) 809-816.

[23] A. Gadek-Moszczak, J. Pietaszek, B. Jasiewicz, S. Sikorska, L. Wojnar, The Bootstrap Approach to the Comparison of Two Methods Applied to the Evaluation of the Growth Index in the Analysis of the Digital X-ray Image of a Bone Regenerate. New Trends in Comp. Collective Intell. 572 (2015) 127-136. https://doi.org/10.1007/978-3-319-10774-5_12

[24] L. Chybowski, K. Gawdzinska, O. Slesicki, K. Patejuk, G. Nowosad, An engine room simulator as an educational tool for marine engineers relating to explosion and fire prevention of marine diesel engines. Scientific Journals of the Maritime University of Szczecin 43 (2015) 15-21.

[25] K. Gawdzińska, L. Chybowski, A. Bejger, S. Krile, Determination of technological parameters of saturated composites based on sic by means of a model liquid, Metalurgija 55(4) (2016) 659-662.

[26] K. Gawdzińska, L. Chybowski, W. Przetakiewicz, R. Laskowski, Application of FMEA in the Quality Estimation of Metal Matrix Composite Castings Produced by Squeeze Infiltration, Arch. Metall. Mater. 62 (4) (2017) 2171-2182. https://doi.org/10.1515/amm-2017-0320

Terotechnology XI
Materials Research Proceedings **17** (2020) 146-151

Materials Research Forum LLC
https://doi.org/10.21741/9781644901038-22

# Impact of Decarburization on the Hardness of the Rails Running Surface

MIKŁASZEWICZ Ireneusz[1,a]

[1] Instytut Kolejnictwa (The Railway Research Institute), Materials & Structure Laboratory, 50, Chlopicki Street, 04-275 Warsaw, Poland

[a] imiklaszewicz@ikolej.pl

Keywords: Rail Defects, Decarburization, Rail Grinding

**Abstract.** Possible causes of rail head defects related to decarburization of the rolling surface are presented. The surface of the rail head was tested directly from the manufacturers and the decarburization of this surface was determined by measuring the hardness and observing the microstructure of the surface layer. The tests were carried out on the basis of requirements included in PN-EN 13674-1 + A1: 2017 [1] and Technical Conditions Id-106: 2010 [2].

**Introduction.**

Rails are one of the most important components of the railway superstructure [4], so the requirements for the quality of the rails and their producers are high. Qualification and acceptance tests of rails must meet the requirements of EN-PN 13674-1 + A1:2017, i.e. have the appropriate chemical composition and the level of gas content, including oxygen of up to 20 ppm and hydrogen of up to 2.5 ppm. Appropriate strength properties and hardness HBW depend on the type of steel, proper level of non-metallic inclusions determined by the K3 index, pearlitic structure of the steel without traces of bainitic-martensitic microstructure, decarburization of the surface of the head up to max. 0.50 mm, proper profile, straightness and dimensions of the rails, and also should not have defects in the rolling surface with a depth of exceeding 0.30 mm, not allowed by the above standard.

The size of decarburization of the rail head on the running surface plays quite a significant role in the quality of rails, which is often overlooked. This phenomenon occurs during the production of rails, mainly at the stage of heating and heating of slabs in heating furnaces at a temperature of about 1150 °C, and during cooling of rails in a cold store. The process causing surface decarburization, i.e. changing the concentration of carbon content in steel, involves gas corrosion at high temperature in an oxidizing atmosphere, causing oxidation of the surface with the formation of a scale surface, simultaneously connected with the diffusion of carbon dissolved in the steel from the surface layers of the material into the environment in the gaseous form. Due to the high carbon content in rail steel, high strength and hardness are obtained. At the same time, a higher carbon content may increase the decarburized steel layer, while the thickness of the decarburized layer depends on many factors, including mainly the atmosphere in the heating furnaces [6, 7].

The initial operation of new rails that have not undergone preventive grinding, i.e. removal of the decarburized layer from the running surface of the rail head is deeply interesting [5]. Rails having decarburization of the surface to the limit permitted by the abovementioned standard and technical conditions are vulnerable to the formation of all types of surface defects as well as defects in the rail head profile during their service life. This is due to the reduced carbon content in the surface layer, resulting in lower hardness and lower surface strength. Thus, the material is softer, prone to wear of the rail running surface, formation of shape defects and, as the track life extends, subjected to strong strengthening of the rail head surface. During this period, the surface layer wears out as a result of wheel-rail interaction, while creating defects in the form of rolling edge cracks occurring mainly in track curves with small radii and straight sections in places with unstable track surface. This kind of defect occurs frequently during operation, to a varying degree connected with decarburization, which can be classified as a fatigue-type defect. Minor cracks in the rail head edges arise in the track curves as a result of wheel flange action, which develop in further operation. These cracks are formed as a result of strong deformation of the softer surface layer of the rail head (Fig.1).

At this stage, there also appears a phenomenon of wave formation and wear of the rail surface, partly linked to decarburization (Fig. 2). This particularly applies to the formation of short waves with low amplitude which probably originate from the uncushioned masses of rolling stock in reaction with the decarburized surface of the rail [8].

Then defects of squat (Fig. 3) and shelling (Fig. 4) types can be distinguished. They are also connected with the surface decarburization, usually arising as a result of the local delamination of the surface layer of the rails and its chipping, as the rails previously underwent so-called spinning or sudden braking of the wheels of the power unit. Also the sticking and flaking of the running surface is related to the micro-slip and friction energy in the wheel-rail interaction and the quality of the rail surface structure.

*Fig. 1. Rail edge defect*

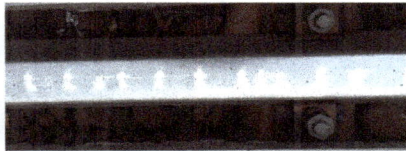

*Fig. 2. Defect in rail waviness*

*Fig. 3. Squat rail defect*

*Fig 4. Shelling rail defect*

**Material for research**

The research material were samples of unused rails from the R260 grade of five major European manufacturers, which were marked with letters from A to E. The chemical composition of the tested rails is presented in Table 1.

Terotechnology XI                                                                          Materials Research Forum LLC
Materials Research Proceedings 17 (2020) 146-151                      https://doi.org/10.21741/9781644901038-22

Samples were cut from the rail sections, whose running surface of the rail head was ground from 0.10 mm to 0.50 mm using 0.10 mm increments. HV5 and HBW hardness measurements were made on samples prepared in this way [3]. The prepared test samples are shown in Fig. 5. Then, after grinding, polishing and etching samples in 4% nital, the decarburized layer of the selected rail manufacturer was shown, indicating the lowest and highest hardness.

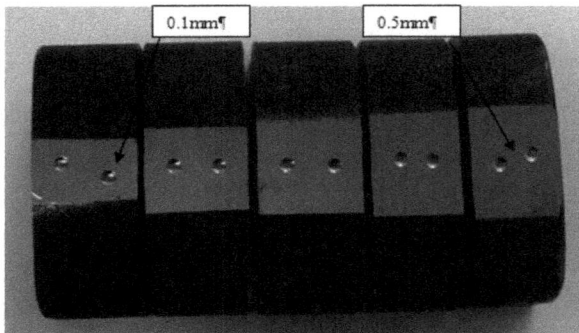

*Fig. 5. C-smelt rails with a ground running surface*

The chemical composition of all tested samples of new R260 rails is within the range of elements provided for in the PN EN 13674-1: 2017 standard. Due to the obligatory vacuum treatment of grades intended for the production of rails, the level of gas content in the tested samples was not determined, assuming that they are in accordance with the abovementioned standard.

The level of carbon content in the R260 grade is highly responsible for strength parameters and the wear rate of rail running surface. The content of carbon is in the range of 0.696% to 0.763%, which indicates the possibility of a difference in hardness of individual rail samples. The contents of the other elements in the melts show slight differences within the limits provided for this grade, which do not have a significant effect on the rail operating parameters.

Figures 6 and 7 present the results of measuring the Vickers HV5 and Brinell HBW hardness depending on the thickness of the ground layer from which surface decarburized was removed. The graph (Fig. 6) shows that the tested rail samples showed hardness in the range of 260 to 310 HBW depending on the thickness of the ground layer. Therefore, the minimum thickness of the layer removed from the rail running surface, guaranteeing the achievement of the required hardness of minimum 260 HBW, should be 0.30 mm.

HV5 measurements, i.e. 260 to 363 HV5 hardness units, show a greater difference in hardness of the tested samples (Fig. 7). Also in this case, the minimum thickness of the ground layer that guarantees the required hardness of the rail running surface should be 0.30 mm.

Figure 8 shows the microstructure of the decarburized surface of the manufacturer's rail sample marked with the letter A. The line indicates the permissible decarburization for the rail of 0.50 mm, and the minimum 0.30 mm securing the required hardness, and Fig. 9 shows the microstructure of the sample decarburized layer 0.30 mm of the rail marked with the letter E. Bainitic-martensitic microstructure was not found in the tested samples.

Terotechnology XI                                                Materials Research Forum LLC
Materials Research Proceedings **17** (2020) 146-151          https://doi.org/10.21741/9781644901038-22

*Table 1. Chemical composition of tested rails in the R260 grade*

| Sample determination | Content of elements in [%] by weight | | | | | | | | | | |
|---|---|---|---|---|---|---|---|---|---|---|---|
| | **C** | **Mn** | **Si** | **P** | **S** | **Cr** | **Ni** | **Cu** | **Al** | **Mo** | **V** |
| **A** | 0.96 | 1.08 | 0.255 | 0.018 | 0.017 | 0.048 | 0.019 | 0.033 | 0.000 | 0.004 | 0.000 |
| **B** | 0.712 | 0.95 | 0.332 | 0.016 | 0.014 | 0.055 | 0.033 | 0.018 | 0.000 | 0.009 | 0.001 |
| **C** | 0.763 | 1.04 | 0.309 | 0.025 | 0.014 | 0.061 | 0.020 | 0.029 | 0.000 | 0.007 | 0.000 |
| **D** | 0.699 | 1.07 | 0.356 | 0.016 | 0.017 | 0.026 | 0.026 | 0.036 | 0.000 | 0.003 | 0.001 |
| **E** | 0.749 | 0.98 | 0.271 | 0.015 | 0.C15 | 0.069 | 0.025 | 0.022 | 0.000 | 0.006 | 0.000 |
| **R260 acc. PN EN 13674-1:2011** | 0.60 – 0.82 | 0.65 – 1.25 | 0.13 – 0.60 | max 0.030 | 0.C08 – 0.C30 | max 0.15 | - | - | max 0.004 | - | max 0.030 |

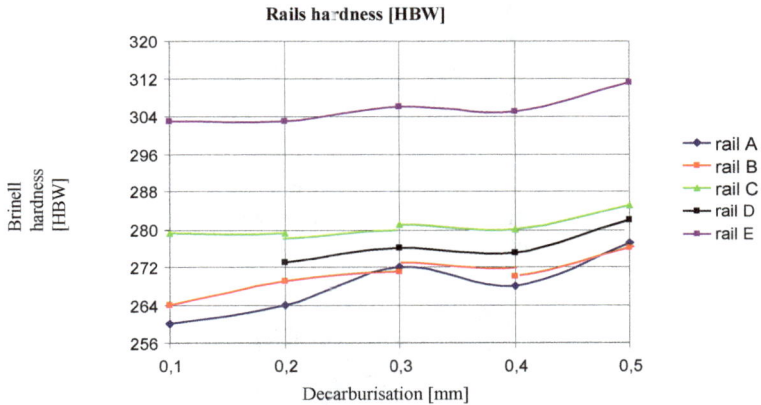

*Fig. 6. HBW hardness distribution of tested rails*

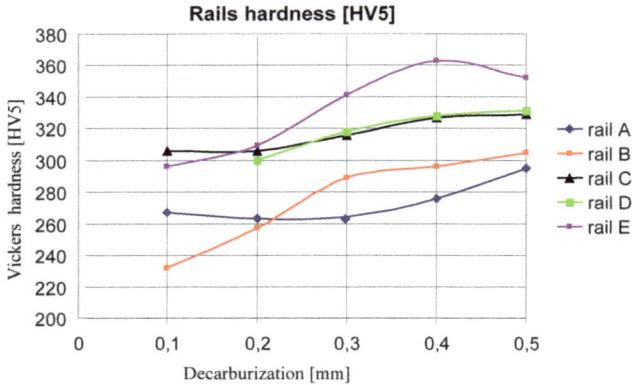

Fig. 7. HV5 hardness distribution of tested rails

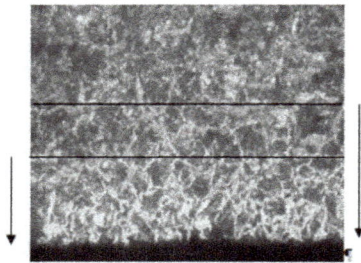

Fig. 8. Sample of rail no. A with 0.50 mm decarburization, area 100x

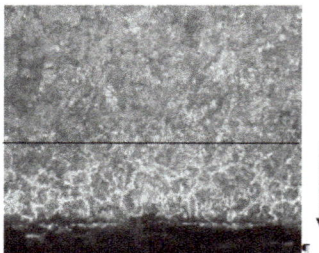

Fig. 9. Sample of rail no. E with 0.30 mm decarburization, area 100x

Materials Research Forum LLC
https://doi.org/10.21741/9781644901038-22

## Summary

The presence of a decarburized layer was confirmed on all the tested rail samples, which is vulnerable to the formation of rail head defects due to its lower hardness. The resulting defects of rails caused by surface crumbling in the form of cracks in the rolling surface and edge of the rails lying at a small depth, short-wavelength waves, a surface white layer appearing on the rails as well as chipping and laminations are the result of microstructure changes occurring on the rail surface.

Currently, the grinding operations of the rail running surface are carried out in running conditions in the tracks using a grinding train. Tests on the volume of the decarburized layer of rails of five European producers confirmed the need of rail grinding. Tests on the hardness and microstructure of new rails showed the thickness of the decarburized rolling surface of the rail head recommended to date for grinding, of minimum 0.30 mm.

However, in the case of rails operated for a longer period, the stage preceding the reprofiling of rails by the grinding method is to check the condition of the track surface and repair the track in order to enable even removal of the defective layer and limit the further development of other rail defects.

## References

[1] PN EN 13674-1 + A1:2017. Railway applications - Track - Rail - Part 1: Vignole railway rails 46 kg/m and above.

[2] WTWiO Id-106:2010. Conditions for making and receiving railway rails - Requirements and tests.

[3] PN-EN ISO 3887:2005. Steel. Determination of decarburization depth.

[4] H. Bałuch, Threats in the railway surface, Railway Research Institute, Warsaw 2017.

[5] H. Bałuch, Advisory system in the assessment of the desirability of rail grinding, Railway Reports, Issue 118, CNTK Warsaw 1995.

[6] G. Parrish, Carburizing: Microstructures and Properties, 1999 ASM International.

[7] G. F. Vander Voort, Understanding and measuring decarburization, Advanced Material and Processes, February 2015.

[8] B Bogdaniuk, A. Massel, Measurements of unevenness of rail rolling surfaces on PKP lines, Railway Reports, Issue 120, CNTK Warsaw 1995.

Terotechnology XI
Materials Research Proceedings **17** (2020) 152-158

Materials Research Forum LLC
https://doi.org/10.21741/9781644901038-23

# Influence of Surface Substrate Texture on the Properties of Al$_2$O$_3$ / IF-WS$_2$ Surface Layers

KORZEKWA Joanna[1, a] * and GĄDEK-MOSZCZAK Aneta[2, b]

[1]University of Silesia, Faculty of Computer and Materials Science, Zytnia str. 10, 41–200 Sosnowiec, Poland, EU

[2]Cracow University of Technology, Faculty of Mechanical Engineering, Al. Jana Pawla II 37, 31-864 Krakow, Poland, EU

[a]joanna.korzekwa@us.edu.pl, [b]aneta.gadek-moszczak@mech.pk.edu.pl

**Keywords:** Surface Layer, Nanomaterials, Nanoparticles, Tribological Properties

**Abstract.** Anodic oxidation of aluminum alloy in a ternary solution of SAS (sulfuric, adipic and oxalic acids) with inorganic fullerene-like tungsten disulfide (IF-WS$_2$) is named in the article Al$_2$O$_3$/IF-WS$_2$. The thickness, geometric structure of the surface (SGP) and the tribological properties such as friction coefficient of Al$_2$O$_3$/IF-WS$_2$ junction with polieteroeteroketon filled with graphite, carbon fiber and PTFE (named PEEK/BG) were investigated. The influence of electrolysis time and temperature on the tribological properties of coatings was studied using $2^k$ factorial design.

## Introduction

Surface texture is generally understood to mean the nature of a surface. In the literature, surface structure tends to be used to refer to surface roughness, characteristics of layer and waviness [1]. It is common knowledge that surface topography is one of the most important factors that control friction and transfer layer formation during sliding. For this reason, a great research effort has been directed towards understanding the effect of surface texture on properties of different kind of materials. Due to their properties, aluminum and its alloys have become one of the most commonly used materials in different kinds of industry. Aluminum alloy has many uses for front and functional surfaces, which come into contact with other materials and weather conditions, therefore processes for its surface treatment have become a critical issue in research works. Kadleckova and co-workers [2] reported that with the multistep etching process, the aluminum-alloy substrate can be effectively modified and textured to the same morphology, regardless of the initial surface roughness. In [3] the authors investigated the effects of surface roughness on the tribological properties of a textured surface. Their investigations show that in contrast with dimple textures, surface roughness is a texture at the micro-level, which will essentially influence the load-bearing capacity of lubricant film. In recent years there has been a growing interest in roughness measurement problems which are important for tribological tests [4]. The hard anodizing process of aluminum alloy is a commonly discussed type of surface technology [5-7]. Thin aluminum-oxide films have a great importance for applications in kinematic friction nodes in dry lubrication (i.e. pneumatic cylinders).

Statistical experimental design methods have been shown to be an efficient technique to different coatings property description [8, 9]. One of these methods is a two-level factorial design which involves simultaneous adjustment of experimental factors at high and low levels. It is well known that a full factorial design may also be called a fully crossed design. Such an experiment allows for analyzing the effect of each factor on the response variable, as well as the effects of interactions between factors on the response variable. The most widely-used type of correlation coefficient is Pearson $r$, also called a linear or product-moment correlation and it is allow for the measurement of the degree of relationship between linearly related variables.

In this paper, the influence of electrolysis time and temperature on the tribological properties of coatings was studied using a $2^k$ factorial design. The thickness, geometric structure of the surface (SGP) and the tribological properties such as friction coefficient of $Al_2O_3/IF\text{-}WS_2$ junction with polieteroeteroketon filled with graphite, carbon fiber and PTFE (named PEEK/BG) were investigated.

The rationale for this study is the importance attached to the durability of coatings, especially those subjected to tribological loads. It is of interest to many industrial sectors, in particular power hydraulics used in heavy working machines [10-12], including industrial robots [13], as well as heat energy transport [14, 15] and parts working in chemically aggressive biotechnological environments [16, 17]. The methodology used should also be interesting for related surface improvement methods, e.g. electro-spark deposition [18, 19] and laser machining [20]. It can also significantly affect the methods of image analysis [21] and the decision inference schemes [22, 23].

**Experimental part**
**Methodological bases.** Macroscopic images of the samples were made with an Olympus BX60M optical microscope with a Motic camera. The thickness of the layers was measured with a Dualscope MP40 by Fischer, using the eddy current method. 10 measurements were performed along the length of the sample and then the average value was calculated. The structural geometry parameters (SGP) measurements of oxide layers were made by the Taylor Hobson Talysurf 3D pin profilometer with the accuracy of 2%. The results of the parameters were developed by means of the TalyMap Universal 3D software. The stereometric analysis was performed on an area of 2 mm x 2 mm. A 2k factorial design with one repetition was applied for analyzing the influence of the surface substrate texture and temperature on thickness, friction coefficient, wear intensity of PEEK/BG plastic and roughness of $Al_2O_3/IF\text{-}WS_2$ coatings. The statistical analysis of the result was performed using the STATISTICA 12.0 software. Tribological measurements were performed on a T17 tester (made by ITE BIP Radom), a pin-on-plate in a reciprocating motion, at room temperature, at the humidity of 30±5%, using 0.5 MPa pressure at an average sliding speed of 0.2 m/s in dry friction conditions. The tribological test was conducted for the sliding distance of 15 km. The commercial PEEK/BG plastic pin with a diameter of 9 mm was used as a counter-body. The friction coefficient was measured when a steady state was reached in the friction test. The wear quantity of the PEEK/BG plastic was studied by means of a WA32 analytical balance with the accuracy of 0.1 mg, after each friction process.

**Sample preparation.** EN-AW-5251 aluminum alloy was the starting material for the process. The samples were etched sequentially with 5% KOH solution for 45 minutes, and 10% $HNO_3$ solution for 10 minutes, at room temperature. After each step of etching, a sample was placed in distilled water to remove residual acid. The electro-oxidation of the aluminum alloy was carried out in a SAS ternary solution (18% sulfuric (33 ml/l), adipic (67 g/l) and oxalic acids (30 g/l)) with admixture of 15 g of the commercially available $IF\text{-}WS_2$ nanoparticles (NanoMaterials Ltd) per 1 liter of electrolyte. The hard anodizing process was performed at 3 $A/dm^2$ current density. In order to ensure homogeneity of the suspension and to prevent the settling of $IF\text{-}WS_2$ nanopowder, mechanical stirring was applied during the electrolysis process. The details concerning the serial number of samples according to $2^k$ factorial design are presented in Table 1. It was decided that the independent variables for this investigation were two levels of temperature of electrolysis and surface of substrate and dependent variables were the thickness, friction coefficient, Ra surface roughness parameter and wear intensity of PEEK/BG. For the purpose of this article, the type of surface substrate was called "horizontal" for samples whose surface addicted by the production of aluminum alloy was parallel to the direction of motion in the tribological test and called "vertical" for perpendicular direction of motion.

**Results and discussion**

The results of tests can be seen in Table 1. These tests showed that the oxide thickness takes values from 22.50 μm for sample 2 to 27.45 μm for sample 4 (Table 1).

*Table 1. Factor settings and results of test for $2^k$ factorial design*

| Factor/ sample name | Independent variables | | Dependent variables | | | |
|---|---|---|---|---|---|---|
| | Temperature (°C) | Surface of substrate | Oxide thickness (μm) | *Ra* before tribological test (μm) | Friction coefficient $\mu$ | Wear intensity of PEEK/BG |
| 1 | 30 | horizontal | 26.62±0.35 | 0.33±0.05 | 0.129±0.001 | 0.124±0.008 |
| 2 | 25 | horizontal | 22.50±0.44 | 0.75±0.13 | 0.271±0.002 | 0.279±0.014 |
| 3 | 30 | vertical | 24.61±0.43 | 0.19±0.06 | 0.126±0.002 | 0.064±0.009 |
| 4 | 25 | vertical | 27.45±0.33 | 0.31±0.07 | 0.273±0.003 | 0.504±0.008 |

The scatter chart of oxide thickness and *Ra* roughness parameter over type of structure of substrate and temperature of anodizing process are depicted in Figure 1a and b respectively. In order to measure the statistical relationship between two continuous variables, the Pearson's correlation coefficient *r* was used (Eq.1). Pearson's coefficient gives information about the magnitude of the association, or correlation, as well as the direction of the relationship.

$$r = \frac{n \sum xy - \sum x \sum y}{\sqrt{[n \sum x^2 - (\sum x)^2][n \sum y^2 - (\sum y)^2]}} \qquad \text{(Eq. 1)}$$

where: $n$ – number of variable, $x$, $y$- variables.

The correlation coefficient represents the strength of an association and is graded from zero to 1.00. It has no units, but may be positive or negative. The Table 2 provides a rule of thumb scale for evaluating the correlation coefficient. In Table 3, the values of correlation coefficient between the analyzed variables are shown.

Our experiment demonstrated little if any correlation of oxide thickness with the temperature of anodizing process and a low negative correlation with the type of structure surface. Using both temperatures, the oxide with the thickness of above 20 μm could be obtained, which is a satisfactory value for a tribological test. As shown in the results (Table 3) of *r* coefficient for *Ra* roughness parameters before tribological test, one could observe a moderate correlation with temperature and the type of structure substrate. The sample obtained in the lower temperature and with the horizontal structure of substrate exhibited the highest value of *Ra* roughness parameters, while the lowest one value of *Ra* was noticed for the higher temperature and vertical structure of substrate.

The further analysis of tribological properties showed that the higher values of friction coefficient $\mu$ between $Al_2O_3$/IF-WS$_2$ surface layer and PEEK/BG pin were achieved by samples 2 and 4. The graphs which show dependence of friction coefficient $\mu$ versus sliding distance is shown in Figure 2a, the scatter chart of friction coefficient in Figure 2b and wear intensity of

PEEK/BG plastic in Figure 2c. Pearson's correlation coefficient (Table 3) between friction coefficient $\mu$ and temperature equals $r = -0.99$, which means very a high negative correlation. A significant correlation was revealed also for friction coefficient and wear intensity and it equalled $r=0.88$. The most surprising correlation was $r = -0.003$, which means that there is any correlation between friction coefficient $\mu$ and the type of structure of a substrate. There was no correlation between friction coefficient $\mu$ and thickness of oxide, either. The friction coefficient correlates with moderate strength with $Ra$ roughness parameter, therefore it can be assumed that there is certain dependence between those variables.

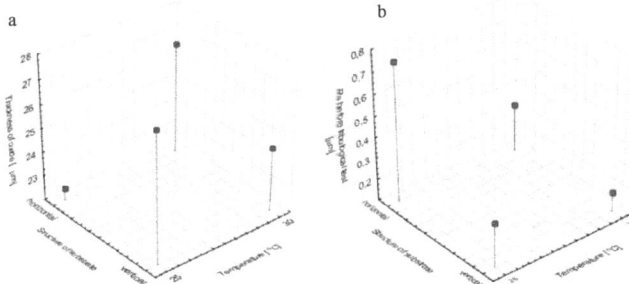

*Figure 1. The scatter chart of (a) oxide thickness and (b) Ra parameter before the tribological test over the structure of substrate and temperature of anodizing process.*

*Table 2. A rule of thumb scale for evaluating the correlation coefficient*

| Strength of Correlation | |
|---|---|
| **Size of r** | **Interpretation** |
| 0.90 to 1.00 (-0.90 to -1.00) | Very high positive (negative) correlation |
| 0.70 to 0.89 (-0.70 to -0.89) | High positive (negative) correlation |
| 0.50 to 0.69 (-0.50 to -0.69) | Moderate positive (negative) correlation |
| 0.30 to 0.49 (-0.30 to -0.49) | Low positive (negative) correlation |
| 0.00 to 0.29 (0.00 to - 0.29) | Little if any positive (negative) correlation |

*Table 3 The values of correlation coefficient between the analyzed variables.*

| Variables / Variables | Strength of Pearson's correlation coefficient r | | | |
|---|---|---|---|---|
| | Friction coefficient | Wear intensity of PEEK/BG | Thickness of oxide | Ra parameter |
| Thickness of oxide | -0.15 | 0.33 | - | - |
| Temperature | -0.99 | -0.87 | 0.17 | -0.64 |
| Structure of substrate | -0.003 | -0.24 | -0.38 | 0.68 |
| Ra parameter | 0.64 | 0.25 | - | - |
| Wear intensity of PEEK/BG | 0.88 | - | - | - |

Pearson's correlation coefficient (Table 3) between the wear intensity of PEEK/BG plastic and temperature was $r = -0.87$, which means a high negative correlation. Such variables as thickness of oxide, structure of substrate and Ra roughness parameter of oxide layers show low or no correlation.

Table 4 shows the pictures of surface coatings and isometric projection of surface before and after tribological tests. The picture of surface coating before and after tribological test depicted the type of a surface substrate which was called in this article "horizontal" for samples whose surface addicted by the production of aluminum alloy was parallel to the direction of motion in tribological tests and called "vertical" for perpendicular direction of motion. The isometric projections clearly show how the roughness of surface is reduced after tribological tests in comparison with the value before tribological test.

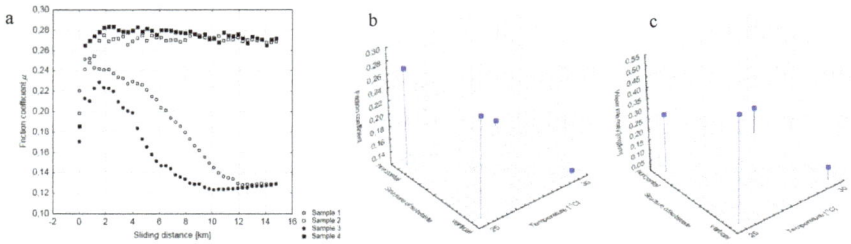

**Figure 2.** *The diagram of friction coefficient versus sliding distance (a), the scatter chart of friction coefficient μ (b) and wear intensity of PEEK/BG plastic (c) over temperature of anodizing process and structure of substrate.*

*Table. 4. Pictures of surface coatings and isometric projection of surface before and after tribological tests.*

## Conclussion

This paper gives an account of the 2k factorial design which was used to determine the influence of electrolysis temperature and the structure of a substrate on the tribological properties of $Al_2O_3$/IF-$WS_2$ coatings. Taken together, these studies indicate that the structure of an aluminium substrate has no influence on the friction coefficient from the point of view of the orientation of samples during a tribological test. The connection of design of experiment and tribological experiment emphasizes the validity of obtained results. The lower values of friction coefficient $\mu$ and wear intensity of PEEK/BG plastic during abrasive wear were obtained for samples obtained in the higher temperature of $30^\circ C$. The analysis of Pearson's coefficient confirms that the friction coefficient $\mu$ and wear intensity of PEEK/BG plastic depend on the temperature of electrolyte during electrolysis and surface roughness parameter $Ra$.

## References

[1] E.P. Degarmo, J.T. Black, R.A. Kohser, (2003), Materials and Processes in Manufacturing, Hoboken, Wiley, 2003.

[2] M. Kadlecková, A. Minarík, P. Smolka, A. Mrácek, E. Wrzecionko, L. Novák, L. Musilová, R. Gajdošík, Preparation of Textured Surfaces on Aluminum-Alloy Substrates, Materials 12 (2019) art. 109. https://doi.org/10.3390/ma12010109

[3] Yuankai Zhou, Hua Zhu, Wenqian Zhang, Xue Zuo, Yan Li and Jianhua Yang, Influence of surface roughness on the friction property of textured surface, Advances in Mechanical Engineering 1–9. https://doi.org/10.1177/1687814014568500

[4] V. Rodriguez, J. Sukumaran, M. Ando, P. De Baets, Roughness measurement problems in tribological testing, Sustainable Construction & Design 2 (2011) 115-121.

[5] P. Kwolek, Hard anodic coatings on aluminum alloys, Advances in Manufacturing Science and Technology 41 (2017) 35-46.

[6] N. Tsyntsaru, B. Kavas, J. Sort, M. Urgen, J.-P. Celis, Mechanical and frictional behaviour of nano-porous anodized aluminium, Materials Chemistry and Physics 148 (2014) 887-895. https://doi.org/10.1016/j.matchemphys.2014.08.066

[7] M. Bara, W. Skoneczny, S. Kaptacz, Tribological properties of ceramic-carbon surface layers obtained in electrolytes with a different graphite content. Maintenance and Reliability 40 (2008) 66-70.

[8] H. Ruiz-Luna, D. Lozano-Mandujano, J.M. Alvarado-Orozco, A. Valarezo, C.A. Poblano-Salas, L.G. Trapaga-Martinez, F.J. Espinoza-Beltran, J. Munoz-Saldana, Effect of HVOF Processing Parameters on the Properties of NiCoCrAlY Coatings by Design of Experiments, Journal of Thermal Spray Technology 23 (2014) 950-961. https://doi.org/10.1007/s11666-014-0121-2

[9] M. J. Anderson, P. J. Whitcomb, Design of Experiments for Coatings, Minneapolis, Stat-Ease, 2006. https://doi.org/10.1201/9781420044089.ch15

[10] P. Walczak, A. Sobczyk, Simulation of water hydraulic control system of Francis turbine. Proc. 8th FPNI Ph.D Symposium on Fluid Power, 2014, art. V001T04A001. https://doi.org/10.1115/FPNI2014-7814

[11] M. Domagala, H. Momeni, J. Domagala-Fabis, G. Filo, M. Krawczyk, J. Rajda, Simulation of particle erosion in a hydraulic valve. Materials Research Proceedings 5 (2018) 17-24. https://doi.org/10.21741/9781945291814-4

[12] J. Krawczyk, A. Sobczyk, J. Stryczek, P. Walczak, Tests of new methods of manufacturing elements for water hydraulics. Materials Research Proceedings 5 (2018) 200-205.

[13] A. Pacana, K. Czerwinska, R. Dwornicka, Analysis of non-compliance for the cast of the industrial robot basis, METAL 2019 28th Int. Conf. on Metallurgy and Materials (2019), Ostrava, Tanger 644-650. https://doi.org/10.37904/metal.2019.869

[14] Z. Ignaszak, P. Popielarski, T. Strek, Estimation of coupled thermo-physical and thermo-mechanical properties of porous thermolabile ceramic material using Hot Distortion Plus® test. Defect and Diffusion Forum 312-315 (2011) 764-769. https://doi.org/10.4028/www.scientific.net/DDF.312-315.764

[15] L. Dabek, A. Kapjor, L.J. Orman, Boiling heat transfer augmentation on surfaces covered with phosphor bronze meshes. MATEC Web of Conf. 168 (2018) art. 07001. https://doi.org/10.1051/matecconf/201816807001

[16] E. Skrzypczak-Pietraszek, J. Pietraszek, Phenolic acids in in vitro cultures of Exacum affine Balf. f. Acta Biol. Crac. Ser. Bot. 51 (2009) 62-62.

[17] E. Skrzypczak-Pietraszek, I. Kwiecien, A. Goldyn, J. Pietraszek, HPLC-DAD analysis of arbutin produced from hydroquinone in a biotransformation process in Origanum majorana L. shoot culture. Phytochemistry Letters 20 (2017) 443-448. https://doi.org/10.1016/j.phytol.2017.01.009

[18] S. Wojciechowski, P. Twardowski, T. Chwalczuk, Surface Roughness Analysis after Machining of Direct Laser Deposited Tungsten Carbide, Met & Props 2013, 14th Int. Conf. on Metrology and Properties of Eng. Surf., Journal of Physics Conference Series 483 (2014) art. 012018. https://doi.org/10.1088/1742-6596/483/1/012018

[19] R. Dwornicka, N. Radek, M. Krawczyk, P. Osocha, J. Pobedza, The laser textured surfaces of the silicon carbide analyzed with the bootstrapped tribology model. METAL 2017 26th Int. Conf. on Metallurgy and Materials (2017), Ostrava, Tanger 1252-1257.

[20] Radek, N., Kurp, P., Pietraszek, J., Laser forming of steel tubes. Technical Transactions 116 (2019) 223-229. https://doi.org/10.4467/2353737XCT.19.015.10055

[21] A. Szczotok, D. Karpisz, Application of two non-commercial programmes to image processing and extraction of selected features occurring in material microstructure. METAL 2019: 28th Int. Conf. on Metallurgy and Materials, Ostrava, TANGER, 1721-1725. https://doi.org/10.37904/metal.2019.971

[22] J. Pietraszek, Response surface methodology at irregular grids based on Voronoi scheme with neural network approximator. In: Rutkowski L., Kacprzyk J. (eds) Neural Networks and Soft Computing. Advances in Soft Computing, vol 19. Physica, Heidelberg: 2003, 250-255. https://doi.org/10.1007/978-3-7908-1902-1_35

[23] A. Pacana, M. Pasternak-Malicka, M., Zawada. A. Radon-Cholewa, Decision support in the production of packaging films by cost-quality analysis. Przem. Chem. 95 (2016) 1042-1044.

Terotechnology XI
Materials Research Proceedings **17** (2020) 159-164

Materials Research Forum LLC
https://doi.org/10.21741/9781644901038-24

# The Impact of Residual Stresses on Bogie Frame Strength

KULKA Andrzej[1a]* and BIŃKOWSKI Robert[1b]

[1] Instytut Kolejnictwa (The Railway Research Institute), Materials & Structure Laboratory, 50, Chlopicki Street, 04-275 Warsaw, Poland

*[a]akulka@ikolej.pl    [b]rbinkowski@ikolej.pl

**Keywords:** Rolling Stock, Rail Vehicle Bogie Frame, Residual Stresses

**Abstract** The article describes a method for determining the residual stresses occurring in the structure of a railway vehicle bogie frame. The performed tests involved the application of the destructive strain gauge method. The impact of residual stresses on bogie frame strength has also been determined.

**Introduction**

Residual stresses appear during most technological processes, and their value and distribution down through the material have a significant impact on fatigue strength and wear, etc. Residual stresses were determined for a railway vehicle bogie frame that had been earlier subject to standard static load and fatigue tests, in accordance with [1] and [2]. After removing the frame from the fatigue test stand, it was subject to a cutting process. Stresses were determined for 14 strain gauges stuck to the smallest possible elements of the frame to make it possible to assume that the residual stresses occurring within the frame were completely released before the frame was cut.

Since the tests were confidential, we are not able to include the outline of the arrangement of the strain gauges or any drawings or pictures of the frame.

*Table 1. Strain gauge location on the frame*

| Strain gauge no. | Strain gauge location on the frame |
|---|---|
| T1, T17 | Bottom flange of the longitudinal member, ¼ of the distance from the end point |
| T2, T18 | Bottom flange of the longitudinal member, ¼ of the distance from the end point |
| T3, T11, T19, T27 | Top flange of the end of the I-beam included in the headstock near the joint |
| T4 | Near the joint connecting the cross beam with the longitudinal member - top flange |
| T5 | Near the joint connecting the cross beam with the longitudinal member - top flange |
| T6 | Near the joint connecting the cross beam with the longitudinal member - bottom flange |
| T7 | Near the joint connecting the cross beam with the longitudinal member - bottom flange |
| T8, T24 | Cross beam near the bump stop support |

Terotechnology XI
Materials Research Proceedings **17** (2020) 159-164

Materials Research Forum LLC
https://doi.org/10.21741/9781644901038-24

Attention: tension gauges marked by the same color in the table are symmetrical relative to the longitudinal or the transverse axis of the frame.

## Materials and methods
Description of the procedure for determining residual stresses
**Test method.** An original test program involving the use of the destructive strain gauge method was applied. The program was developed based on the experience gained during tests of the frames of railway vehicle bogies and when determining the stresses occurring in their respective structures. During the frame cutting, the stresses were recorded using a strain gauge system designed for fatigue testing. Fourteen TF-5/120 strain gauges manufactured by TENMEX were used.

## Description of the procedure followed during frame cutting.
Phase 1 – Cutting the headstocks.
Phase 2 – Cutting the headstocks off and cutting the frame into two large parts.
Phase 3 – Cutting a part of the frame using a panel saw. The process resulted in frame-shaped elements, with a maximum of four tension gauges affixed to each of them.
Phase 4 – Cutting the tension gauges from the frames (Fig. 1). The result was a set of small elements, each with an affixed tension gauge.

The tension gauges were reset before each cut. Stresses were recorded throughout the entire duration of cutting process and for some time after it was finished, until the temperature of the frame and the ambient temperature became stable (i.e. until the strain / stress values registered at the bridge stabilized). If it appeared necessary to disconnect the wires between the tension gauges and the bridge, the reading was recorded and the tension gauges were reset again before the next cutting phase. The procedure was repeated for each following cutting operation, regardless of the tools used. The residual stresses for a given tension gauge were calculated as the sum of the stresses following each cutting operation with a reverse symbol. The values of these residual stresses are given in Table 2.

Fig. 1. Phase 4 – the last strain gauge cutting operation. Cutting the strain gauge out of the frame.

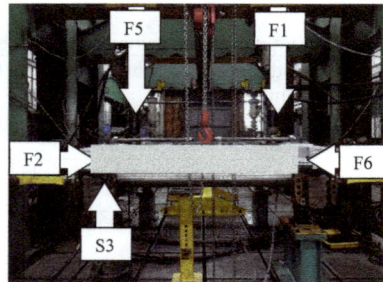

Fig. 2. The frame at the test stand where static load testing was performed

The following tools were used in the course of the above-mentioned operations: an angle grinder, an acetylene torch, a Marvel Armstrong – Blum Mfg.Co Series 8 Mark II panel saw. The occurring strain was recorded using a strain gauge module: NI PXIe-4330 + 4 × 8CH BRIDGE

Terotechnology XI                                                  Materials Research Forum LLC
Materials Research Proceedings **17** (2020) 159-164          https://doi.org/10.21741/9781644901038-24

INPUT NI TB-4330 – National Instruments [9] and the following strain gauge bridges: UPM-100 Hotinger Baldwin  Messtechnik Darmstadt and STRAIN INDICATOR P-3500, Vishay – Measuremets Group – Instruments Division – Raleigh, North Carolina, USA.

Before the frame was cut, nine static load tests – labeled A01 to A09 – were performed. A simplified diagram of the loads and the values of the applied forces are given in Fig. 2 and Table 2. Every load had a corresponding file for which the name included the name of a given load. The files also included the strain gauge numbers and the recorded stress values assigned to the strain gauges. The files were processed following the standard procedure applied in frame static load testing. Each measurement point (strain gauge) had a $\sigma_{max}$ , a $\sigma_{min}$, the average stress value $\sigma_m = (\sigma_{max} + \sigma_{min})/2$, and the UF value determined based on recommendations [5]. A Goodman diagram was developed and safety factors were determined according to [3]. Next, the values of the residual stresses from Table 3 were added to the stresses resulting from frame loading. The operation was performed for each of the nine loads. This produced new files with stress values, labeled B01 to B09. The values were processed in a similar manner in an analogous manner to the previous ones. The data obtained as a result was used as reference material, on the basis of which the impact of residual stresses on the frame strength was determined.

*Table 2. Forces affecting the frame – values like at the $3^{rd}$ stage of fatigue loading.*

| Force labeling according to Fig. 2 | | load 1, A01 | load 2, A02 | load 3, A03 | load 4, A04 | load 5, A05 | load 6, A06 | load 7, A07 | load 8, A08 | load 9, A09 |
|---|---|---|---|---|---|---|---|---|---|---|
| F1 | kN | -123.4 | -123.4 | -154.2 | -154.2 | -92.8 | -92.8 | -61.8 | -61.8 | -100.0 |
| F2 | kN | +103.4 | +103.4 | 0 | 0 | 0 | 0 | 0 | 0 | 0 |
| S3 | mm | -3.2 | +3.2 | -3.2 | +3.2 | -3.2 | +3.2 | -3.2 | +3.2 | 0 |
| F5 | kN | -154.2 | -154.2 | -123.4 | -123.4 | -61.8 | -61.8 | -92.8 | -92.8 | -100.0 |
| F6 | kN | 0 | 0 | +103.4 | +103.4 | 0 | 0 | 0 | 0 | 0 |

Vectors on Fig. 2 indicate the directions of the applied forces and the actuator labels as used in Table 2. The sense of the vector represents negative force (compression) in Table 3.

**Results**

*Table 3. Residual stresses for individual strain gauges*

| Testing gauge no. | | T1 | T17 | T2 | T18 | T3 | T11 | T19 |
|---|---|---|---|---|---|---|---|---|
| Residual stresses within the frame | MPa | +16.8 | -46.3 | -40.1 | -75.5 | -61.1 | +68.8 | -4.1 |
| Testing gauge no. | | T27 | T4 | T5 | T6 | T7 | T8 | T24 |
| Residual stresses within the frame | MPa | +76.2 | +120.8 | -11.1 | -50.7 | -1.6 | +122.3 | +87.5 |

The performed tests showed that the highest and lowest residual stresses were registered by the following strain gauges: T8 (+122.3 MPa), T4 (+120.8 MPa), T24 (+87.5 MPa), T17 (-46.3 MPa), T3 (-61.1 MPa) and T18 (-75.5 MPa). Out of the fourteen examined points, six had tensile stresses (+) determined, and eight had compressive stresses (-) determined. In the

Terotechnology XI
Materials Research Proceedings **17** (2020) 159-164

Materials Research Forum LLC
https://doi.org/10.21741/9781644901038-24

case of two compressive stresses, their values are: - 4.1 MPa for T19 and -1.6 MPa for T7, which means they are below the measurement error threshold. It can be roughly assumed that, in the set of the tested tension gauges, half of the measurement points were subject to positive stresses, and the other half to negative stresses.

It is important to bear in mind that the values of the forces occurring during static loading (A01 ÷ A09) corresponded to the 3$^{rd}$ stage of fatigue loading (at Stage 3, the force amplitudes are 40% larger than at Stage 1) – the principle that UF needs to be below 1 cannot therefore be applied since it pertains to forces corresponding to the 1$^{st}$ stage of fatigue loading. Likewise, the safety factor determined based on the Goodman method cannot be greater than 1. It should be mentioned that the residual stresses occurring within the frame were determined after fatigue testing of the frame. Three stages of fatigue tests in accordance with UIC-615-4 were completed, yielding a total of 10 million loading cycles. This could have an impact on the values of residual stresses (there might have been a relaxation of residual stresses).

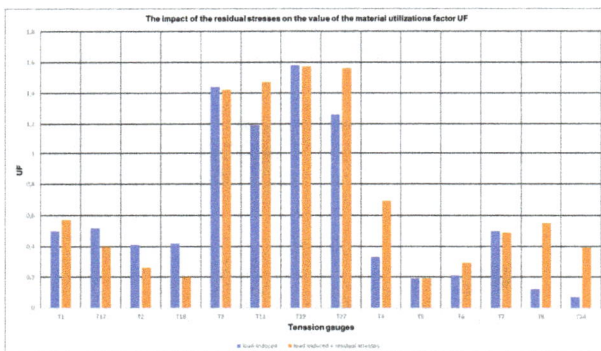

*Fig. 3. Illustration of the change in the UF factor before and after adding residual stresses.*

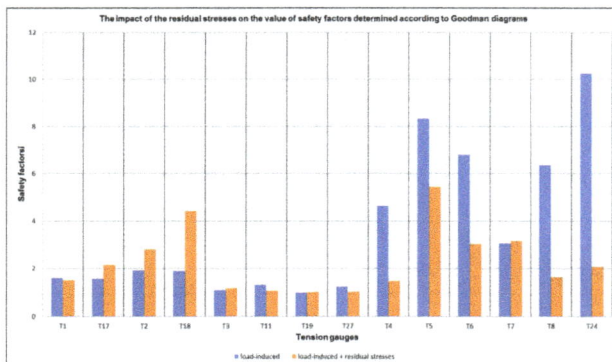

*Fig. 4. Illustration of the change in safety factors determined on the basis of a Goodman diagram.*

Terotechnology XI                                                            Materials Research Forum LLC
Materials Research Proceedings **17** (2020) 159-164                      https://doi.org/10.21741/9781644901038-24

## Analysis

1. Residual stresses were determined for 14 strain gauges fixed to a bogie frame in order to make it possible to determine the stresses generated by static and fatigue loads. Strain gauges are usually fixed to the frame in a way that their directions correspond to the directions of principal stresses generated by loads. It is unlikely for these directions to correspond to principal directions resulting from residual stresses. That is why, in order to make such tests more reliable, it would be reasonable to affix strain gauge rosettes to the frame.

2. It was noticed that the residual stresses recorded for symmetrical strain gauges had different values. This is probably due to the manufacturing technology of the frame (different welding order, different process and different force of restraint during welding).

3. When analyzing Fig. 3. "The impact of residual stresses on the value of the material utilization factor UF", the following was found: an increase in the UF value in 7 out of 14 cases (deteriorated strength), one case of the UF remaining unchanged, and a decrease in the UF value in 6 cases (improved strength).

4. Based on Fig. 4. "The impact of the residual stresses occurring within the frame on the value of safety factors determined according to a Goodman diagram", the following observations could be made: an increase in the safety factor value in 6 cases (improved strength) and a decrease in the safety factor value in 8 cases (deteriorated strength).

5. It was found that T19 was the most strained point out of all fourteen points with strain gauges fixed to them. It was proven in both of the analyses performed according to DVS 1612 and ERRI B12/RP 17, 8th Edition, Test programme for wagons, bogies, bogie frames, UTRECHT, April 1997.

6. At some points, the residual stresses changed the safety factor values to a significant extent – see Fig. 4. But most of those points were not significantly strained and, after the residual stresses were added, they were still on the safe value side of the figure. In the case of the most strained point (tension gauge T19), the safety factor value even grew – from 0.99 to 1.01.

7. Similar observations can be made when we analyze Fig. 3. The material utilization factor value UF for T19 improved from 1.58 to 1.57.

8. In two cases (strain gauges T11 and T27), the strength deteriorated significantly after residual stresses were added. This can pose a threat to the safety of the frame – see Fig. 3 and Fig. 4. This is proven especially by analysis carried out in line with DVS. After 10 million fatigue cycles performed in accordance with UIC-615-4, no cracks were found at those points, though.

## Conclusions

The conducted tests have shown that the values of residual stresses occurring in bogie frames are in the same range as load-induced stresses. Therefore, they have a great impact on the strength of the structure they affect. So far, the tests and calculations for frames have been performed mainly in terms of loads, often disregarding the significance of residual stresses. When analyzing the findings of frame static tests, the impact of residual stresses should be taken into consideration. However, this is not really practicable at present. This is why fatigue tests play such a great part in evaluating frame strength. Strain gauges are fixed to frames to make it possible to control the stresses during static and fatigue loading tests. This is why, out of over thirty frames tested over the past 20 years at the Railway Institute, only one frame has had a crack on it in the location of a tension gauge. Other frames would crack where there were no tension gauges, and where the stresses and their load amplitudes were often minor, which is proven by the earlier MES calculations. The cracks could have been caused by residual stresses combined with load stresses and the impact of the notch effect. The case was similar with the frame tested as described herein. The tests in question were performed when performing the obligatory functional on-site tests of bogie frames, as required under the regulations in force. For this reason, the occurring stresses were determined only at 14 points, i.e. where tension gauges were fixed. This is why the probability of locating a point where significant residual stresses would combine with significant load stresses was very small.

**References**

[1]    PN-EN 13749:2011 Kolejnictwo – zestawy kołowe i wózki – Metody określania wymagań konstrukcyjnych ram wózków; Railway applications – Wheelsets and bogies – Method of specifying the structural requirements of bogie frames.

[2]    UIC CODE 615-4 Motive power units – Bogies and running gear – bogie frame structure strength tests, p. 6. 2nd edition, February 2003.

[3]    ERRI B12/RP 17, 8th Edition, Test programme for wagons, bogies, bogie frames, UTRECHT, April 1997.

[4]    ERRI B12/RP 12, 2nd Edition, Tests to demonstrate the strength of railway vehicles – Regulations for proof tests and maximum permissible stresses, UTRECHT, June 2001.

[5]    DVS1612:2009-08 Technische Regel. Gestaltung und Dauerfestigkeitsbewertung von Schweißverbindungen an Stählen im Schienenfahrzeugbau.

[6]    D. Senczyk, S. Moryksiewicz, D. Senczyk, Naprężenia własne – pojęcia i klasyfikacja. Poznań University of Technology, H. Cegielski Poznań S.A.

[7]    I. Mikłaszewicz, R. Bińkowski, Przypadki pęknięcia szyn w świetle obliczeń MES. Problemy Kolejnictwa 172 (2016) 35-45. https://doi.org/10.36137/1776p

[8]    R. Bińkowski, Influence of the local plasticise zone in the elastic deformation environment on the fatigue strength of the construction elements. Material Research Proceedings 5 (2018) 123-127.

[9]    National Instruments - strain gauge module purchased under the project: "Construction of a testing rig for bogie frames using multifunctional multi-signal control and measurement system, implemented by the proprietary method" (task No 000155).

Terotechnology XI
Materials Research Proceedings **17** (2020) 165-170

Materials Research Forum LLC
https://doi.org/10.21741/9781644901038-25

# Strength Testing of a Composite Mounting Frame for a Multi-Sensor Detection System

KRYSIAK Piotr[1,a] *, SZCZEPANIAK Marcin[1,b], WOJCIESZYŃSKA Patrycja[1,c] and JASIŃSKI Wiesław[1,d]

[1]Military Institute of Engineer Technology, Obornicka 136, 50-961 Wrocław, Poland

[a]krysiak@witi.wroc.pl*, [b]szczepaniak@witi.wroc.pl, [c]wojcieszynska@witi.wroc.pl, djasinski@witi.wroc.pl

**Keywords:** Strength Testing, Strain Gauge Measurements, Glass Fibre-Reinforced Polymer Composites

**Abstract.** The work concerns the design and implementation of a support frame for a multi-sensor detection system and conducting strength tests on it. The system consists of GPR, a metal detector and a non-linear junction detector, so it is vital that the support structure be made of a dielectric material. A commercial frame made of glass fiber reinforced polyester (GFRP) profiles was used to construct the frame. After the system was completed, strength tests were carried out under field conditions for the most adverse load conditions.

## Introduction

At present, fibre-reinforced composites are gaining more and more popularity among constructors, mainly due to the advantages of these particular materials. What gives them this high position in the ranking of construction materials is mainly their high strength and very low weight when compared to other construction materials, including steel. The absence of electrical conduction is also an important parameter determining the use of composites in constructions. This is especially important if, for example, induction detectors are attached to the designed structure [1-3].

## Frame design

The frame design was made on the basis of input data on geometrical parameters and mass of individual system components. The design also takes into account the load and operating conditions of the system in operational conditions, including field conditions and travel off-road. The design of the structure was made using the currently available tools for graphical modelling (Autodesk Inventor) and for strength calculations (Ansys LS-Dyna).

At the beginning, a geometric model of the device was made, taking into account the possibility of attaching it to the base vehicle. Afterwards, strength calculations were conducted within the static range for the position when the frame works without the support of its wheels. In addition, dynamic simulation was performed for the full range of the frame positions and up to 20km/h of operating speed. Based on strength analyses and the available composite profiles, rectangular profiles (stringers and crossbars) made of polyester-glass composites with the dimensions of 100x100x5 mm were used for the designed structure. The properties of the composite used are given in Table 1. Fig. 1 presents the designed frame with elements of the detection system.

Terotechnology XI                                                        Materials Research Forum LLC
Materials Research Proceedings **17** (2020) 165-170        https://doi.org/10.21741/9781644901038-25

*Table 1.*     *Parameters regarding physical- and mechanical properties of composite profiles*
               *[5]. Matrix: isophthalic polyester resin; glass content: 50%-60%.*

| Parameter | Standard | Lengthwise MPa | Crosswise MPa | MPa | [--] |
|---|---|---|---|---|---|
| **Flexural Strength** | EN ISO 14125 | 250 | 30-80 | | |
| **Tensile strength** | EN ISO 527-4 | 250 | 30-80 | | |
| **Compressive strength** | EN ISO 14126 | 240 | 30-80 | | |
| **ShearStrength** | EN ISO 14130 | | | 25 | |
| **E-modulus** | EN 13706 | 25 000 | 9 000 | | |
| **Compressive modulus** | EN ISO 14126 | 10 000 | 4 000 | | |
| **Shear modulus** | EN ISO 14130 | | | 3 000 | |
| **Poisson ratio lengthwise/crosswise** | EN ISO 527-4 | | | | 0.23 |
| **Poisson ratio lengthwise/crosswise** | EN ISO 527-4 | | | | 0.09 |
| **IZOD impact strength [kJm$^2$]** | ASTM D-256 | | | | 300 |
| **Density [kg/dm$^3$]** | ISO 1183 | | | | 1.9 |
| **Barcol hardness** | EN 59 | | | | >30 |

| Application limits | Short-term behaviour | | Long-term behaviour | |
|---|---|---|---|---|
| | Lengthwise [MPa] | Crosswise [MPa] | Lengthwise [MPa] | Crosswise [MPa] |
| **Bending strength** | 135 | 25 | 70 | 20 |
| **Tensile strength** | 135 | 20 | 70 | 15 |
| **Compression strength** | 135 | 25 | 70 | 20 |
| **Shear strength** | 17 | 17 | 8 | 8 |

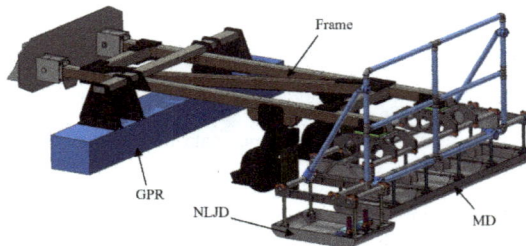

*Fig. 1. Frame model with elements of the detection system*

**Stress tests of a multi-sensor detection system supporting frame under field conditions**
In the next stage, having completed the system, strength tests were carried out using strain
gauges (base: 3 mm, resistance: 120 Ω). The strain gauges were glued on the outer surface of the
beams, in accordance with the diagram shown in Fig. 2 [4].

The tests were conducted using a complete set of sensors being mounted onto the frame.
These sensors included the following: ground-penetrating vehicle, metal detector and non-linear
junction detector.

Six (6) tests in total were performed in order to test the system's operation under varying
operating conditions.

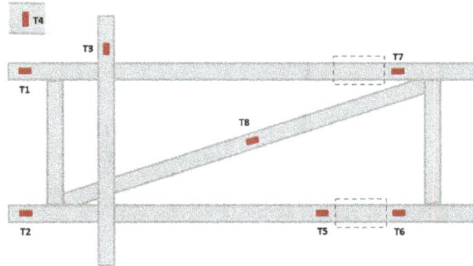

*Fig. 2. Stress pattern as registered by sensors 1÷ 8 for Test No. 1÷6*

Fig. 3. presents a frame with strain gauges and wiring prepared for the tests, as well as the system during testing. The test results are presented in the charts in Figs. 4-9.

*Fig. 3. Loaded frame with strain gauges (on the left) and the frame during tests (on the right)*

During test No. 1, the road was traversed in operational (working) position (the system is wheel-supported). The system was moving forward, down a gentle hill at a speed of 2km/h. During the travel, first a 15 cm-deep and 50 cm-wide hole was encountered, followed by a 15 cm-high and 100 cm-long bump (front right wheel from the UGV perspective).

As part of test No.2, the system was being turned back in operational position (the system was wheel-supported).

Test No. 3 involved the road being traversed in operational (working) position (the system is wheel-supported). The system was moving forward, up a gentle hill at a speed of 2km/h. During the travel, first a 15 cm-deep and 50 cm-wide hole was encountered, followed by a 15 cm-high and 100 cm-long bump (front right wheel from the UGV perspective).

Test No. 4 consisted in the system being turned back with the frame being lifted to a height of ca. 0.5 m above the ground.

Test No. 5 comprised static lifting and of the frame to a height of ca. 0.5 m above the ground followed by lowering of the frame.

As part of test No. 6, static lifting and of the frame to a height of ca. 0.5 m above the ground was followed by impact of a side force equal ca. 300N at the end of the frame on the left, and the identical impact on the right side, followed by lowering of the frame.

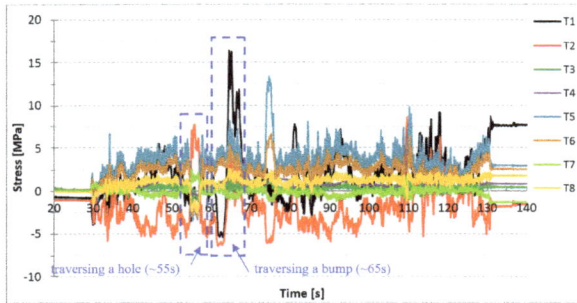

*Fig. 4. Stress pattern as registered by sensors 1÷ 8 for Test No. 1*

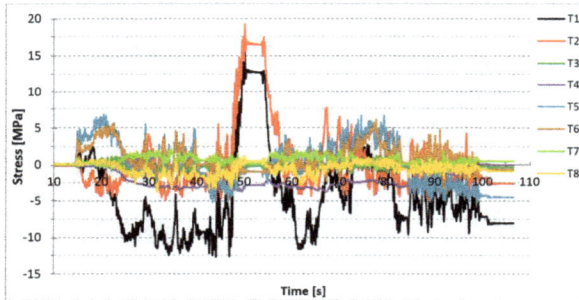

*Fig. 5. Stress pattern as registered by sensors 1÷8 for Test No. 2*

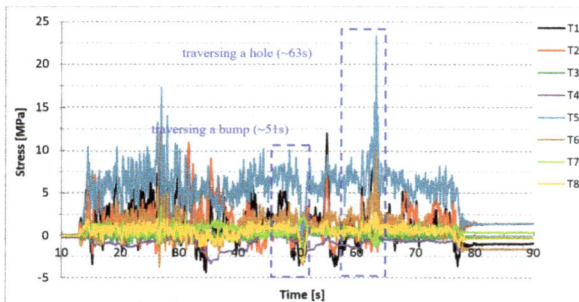

*Fig. 6. Stress pattern as registered by sensors 1÷8 for Test No 3.*

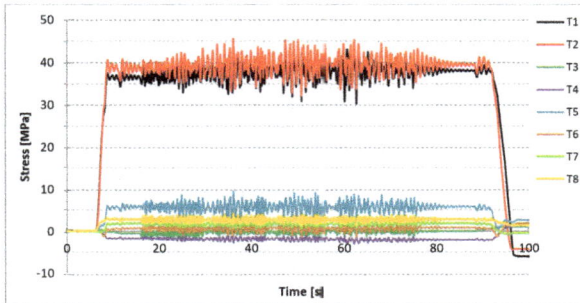

*Fig. 7. Stress pattern as registered by sensors 1÷8 for Test No. 4*

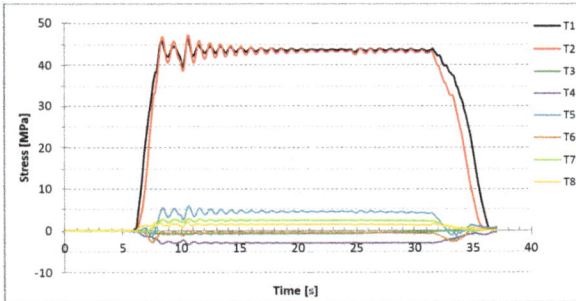

*Fig. 8. Stress pattern as registered by sensors 1÷8 for Test No. 5*

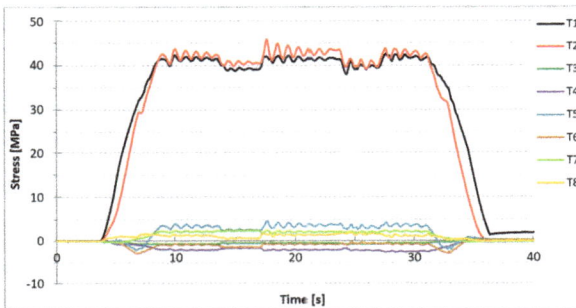

*Fig. 9. Stress pattern as registered by sensors 1÷8 for Test No. 6*

**Summary**

Based on the conducted analyses, the following have been found:

1) The highest normal stress within the main beams occurred during the process of lifting the entire assembly up. In addition, an increase in stress within the beam near the wheel was observed when the system was passing through a hole in the ground, while the passage over a bump did not reveal increased stress.

2) The greatest stress was noted for the support beams near the connection to the attachment area. The maximal values noted equaled 46.5 MPa. What results from the information contained in Table 1 is that the bending stress equals 70 MPa for a prolonged use, which means that the constructed frame operates safely with the coefficient of 1.5. For the short-term stress, the safety coefficient equals 2.9.

3) When the fully-loaded frame operated under field conditions (traversing uneven ground, including, holes and bumps) maximum stress equaled ca. 23 MPa.

4) When the fully-loaded frame operated under field conditions (lifting and lowering of the frame, travel in operating positions, turning), the system was tested multiple times during an 80-hour period. No mechanical damage to the structural elements was detected.

**References**

[1] P. Krysiak, A. Czulak, R. Rybczyński, W. Hufenbach, Dobór elementu układu wieloczłonowego z materiału kompozytowego, in: Polimery i Kompozyty Konstrukcyjne, G. Wróbla (ed.). Cieszyn, Logos Press, 2011, 234-241.

[2] P. Krysiak, A. Błachut, P. Gąsior, J. Kaleta, Influence of fiber type and the layer thickness on the stress distribution in composite pipe, Interdisciplinary Journal of Engineering Sciences 2 (1) (2014) 17-20.

[3] P. Krysiak, J. Kaleta, P. Gąsior, A. Błachut, R. Rybczyński, Identification of strains in a multilayer composite pipe, Journal of Science of the Military Academy of Land Forces 49 (4) (2017) 186. https://doi.org/10.5604/01.3001.0010.7233

[4] Z. Orłoś, Doświadczalna analiza odkształceń i naprężeń, Warszawa, PWN, 1977.

[5] Information contained on http://www.fibrolux.com

Terotechnology XI
Materials Research Proceedings **17** (2020) 171-176

Materials Research Forum LLC
https://doi.org/10.21741/9781644901038-26

# Microstructure and Tribological Properties of DLC Coatings

RADEK Norbert[1, a *], PIETRASZEK Jacek[2,b] , SZCZOTOK Agnieszka[3,c] ,
FABIAN Peter[4,d] and KALINOWSKI Artur[5,e]

[1]Kielce University of Technology, Faculty of Mechatronics and Mechanical Engineering, Al. 1000-lecia P.P. 7, 25-314 Kielce, Poland

[2]Cracow University of Technology, Institute of Applied Informatics, Faculty of Mechanical Engineering, Al. Jana Pawła II 37, 31-864 Cracow, Poland

[3]Silesian University of Technology, Institute of Materials Science, str. Krasinskiego 8, 40-019 Katowice, Poland

[4]University of Zilina, Univerzitna 1, 01026 Zilina, Slovakia

[5]Kielce University of Technology, Faculty of Mechatronics and Mechanical Engineering, Al. 1000-lecia P.P. 7, 25-314 Kielce, Poland

[a]norrad@tu.kielce.pl, [b]pmpietra@gmail.com, [c]agnieszka.szczotok@polsl.pl, [d]fabianp@fstroj.uniza.sk, [e]kalinowski9@outlook.com

**Keywords:** Physical Vapor Deposition, Diamond-Like Carbon, Coatings, Properties

**Abstract.** This paper presents the results of diamond-like carbon coatings deposited using the physical vapor deposition PVD process on 4H13 stainless steel samples. The properties were assessed by analyzing the coating microstructure, nanohardness, roughness and tribological tests. The results obtained during the tests which were carried out showed that the application of diamond coatings considerably improves tribological properties. In addition, coatings, which are desirable in sliding friction pairs, are free of pores and microcracks.

**Introduction**
Wear and abrasion are serious problems in all branches of industry. Although modern technologies permit considerable improvement of properties of the outer layer, they need to be modified or new solutions have to be looked for [1, 2].

Functional properties of many elements of machine parts depend not only on the possibility of transferring mechanical loads through the entire cross-section of the material, but mainly on the structure and properties of surface layers. The increasing demands placed on construction materials have led to constant attempts to produce new types or modifications of wear protection, including changing the chemical composition of a coating or the technology of its production [3-5]. To properly protect surfaces and reduce wear processes, top layers and coatings must perform the following functions [6]:

- reduce consumption,
- prevent and reduce direct contact of metal components,
- facilitate tangential movement by reducing friction forces,
- cause the distribution of normal forces over the largest possible nominal contact area,
- suppress vibrations and oscillatory movements.

Applying protective coatings to machine parts is economically justifiable if the wear is local or if the coating material is expected to display properties different from those of the substrate. Most surface layers are technological surface layers (TSLs) - they are produced before objects are used. Functional surface layers (FSLs), on the other hand, are applied during maintenance.

In recent years there has been great progress in the development of research and application topics related to carbon materials. It includes obtaining diamond-like coatings, DLC (Diamond Like Carbon) and applying them by PVD and CVD methods [7-9]. Carbon creates the most chemical compounds among all elements and has several allotropic varieties.

Due to their special properties, diamond-like carbon coatings are used, among others in sheet metal pressing, steel forming, aluminum foundries, in the production and regeneration of tools and devices, in the production of machine parts and components for the automotive industry. They are also used in the rubber and food industry.

The work discusses the properties of diamond-like carbon coatings deposited using the PVD process. The properties were established based on the results of a microstructure analysis, nanohardness and roughness tests and tribological studies.

There are many alternative technologies for producing coatings and improvements of material properties in relation to PVD technology [10]. The analysis of properties of DLC coating systems requires many methods [11, 12].

Due to its unique and characteristic properties, DLC coatings can be used wherever hardness, wear resistance, slipperiness are required, so primarily hydraulic parts [13], including those heavily mechanically [14-16] and thermally [17-19], as well as exposed to cavitation erosion [20]. The introduction of DLC coatings favorably changes the operational features of machine parts, and in particular reduces the number of failures and non-conformities, which has a great impact on the planning and optimization of production in many industries, such as power plants [21], production of industrial and consumer films [22] and parts of industrial robots [23]. At the same time, these atypical coatings influence the development of research methods in both experimental [24] and data analysis [25].

**Experimental**

Materials with anti-wear functions intended for work in friction nodes covered with diamond-like carbon (DLC) coatings type a-C:H with W and Cr interlayer obtained in physical vapor deposition processes, PVD, were selected for research. The choice of DLC coatings was dictated by their excellent properties and the possibilities of a very wide application in various industries. Amorphous hydrogenated carbon (a-C:H) film, also known as diamond-like carbon coating, is characterized by excellent mechanical properties.

The substrate material was used of 4H13 stainless steel. The elemental composition of the steel used was as follows (wt.%): C: 0.36-0.45, Mn: 0.50-0.80, Cr: 12.0-14.0, Si: 0.60-0.80, Mo: 0.5-0.7, V: 0.2-0.3, Ni: 0.1-0.60, P: max 0.04, S: max 0.03, and the rest is iron.

The processes of applying thin PVD anti-wear coatings take place at elevated temperatures, which causes tempering of the surface layers and reduction of hardness. Individual coatings were obtained in the following processes and temperatures:

- a-C:H in the physical vapor deposition process PVD by ion spray at a temperature < 300° C,
- substrate material temperature of 350° C.

**Results and discussion**

A microstructure analysis was conducted for DLC coating using a *JEOL* JSM-*7100F* scanning electron microscope with field emission. Figure 1 shows the microstructure of DLC coatings. The layer thickness is approximately 1.1 μm. In the photograph, the boundary line between the coating and the substrate is clear. The coating is free of pores and microcracks.

The element maps of an amorphous hydrogenated carbon film on the examined stainless steel surface are shown in Fig. 2. It turned out that the element maps of Fe, Cr, W and C are clearly

Terotechnology XI                                                  Materials Research Forum LLC
Materials Research Proceedings **17** (2020) 171-176              https://doi.org/10.21741/9781644901038-26

visible on diamond-like carbon films. Furthermore, all the element distributions are heterogeneous and some of the grains are oxygen deficient. Moreover, the DLC film covers a significant portion of the specimen surface.

Production of coatings with the required micromechanical properties is a major research challenge. In the production process, it is necessary to specify the controlled parameters and properties that we expect from the resulting coating.

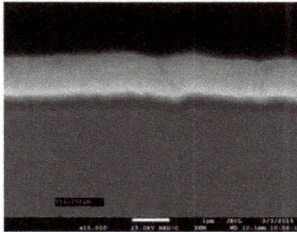

Fig. 1. Microstructure in the DLC coating

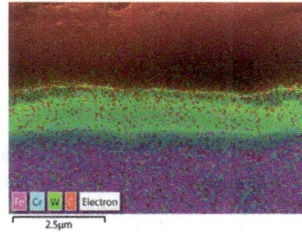

Fig. 2. Maps of elements in DLC coating

The multitude of controlled parameters and their values mean that a large number of combinations results in high costs. In our case, nanohardness and elastic modulus (Young's modulus) of DLC coatings were tested. The hardness and elastic modulus were investigated by nanoitender technique. This Measurement technology was possible due to the development of instruments that continuously measure force and displacement. The tests were carried out with the following parameters: linear load, max. load of 3.2 mN, load and unload speed of 40 mN/min as well as a break time between successive cycles of load and unload of 3 s. On the basis of 10 measurements, the values of average nanohardness and elasticity modulus were determined and placed in Table 1. This table contains the average values of nanohardness and elastic modulus together with the standard error.

Table 1. Value of nonohardness and modulus of elasticity with errors

| Material | Nanohardness [GPa] | Elastic modulus [GPa] |
|---|---|---|
| DLC coating | $7.55 \pm 0.10$ | $92.87 \pm 2.05$ |
| 4H13 steel | $5.60 \pm 0.40$ | $44.00 \pm 4.30$ |

While analyzing Table 1, it can be concluded that the DLC coating nanohardness was approx. 26% higher compared to the 4H13 stainless steel nanohardness. A similar analogy can be observed by analyzing the values of Young's modules for DLC coating and 4H13 stainless steel.

Investigations into dry friction resistances were performed using the T-01M pin-on-disk type tribotester. In Fig. 3, the T-01M principle of operation is shown. The specimens were rings of 4H13 stainless steel, onto which DLC coatings were PVD method deposited. The counter specimen was a ball, $\phi6.3$ mm in diameter, made of 100Cr6 steel. Tribological tests were conducted using the following parameters:
– linear speed, $v = 0.8$ m/s,
– test duration, $t = 3600$ s,
– range of load changes, $Q = 4.9$; 9.8; 14.7 N.

Terotechnology XI
Materials Research Proceedings **17** (2020) 171-176

Materials Research Forum LLC
https://doi.org/10.21741/9781644901038-26

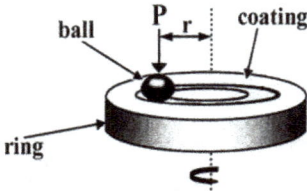

Fig. 3. Operation of the pin on disc type tester

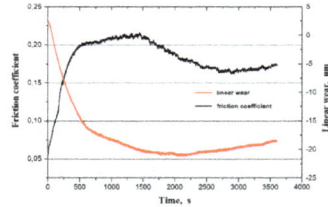

Fig. 4. Friction coefficient and linear wear as
a function of time

An exemplary graph (Fig. 4) shows profiles of friction coefficient and linear wear as a function of time for the load of 4.9 N. The graph presented in Figure 4 refers to the tests on DLC coating. In dry friction, in the examined coating, the technological surface layer (TSL) was transformed into a functional surface layer (FSL). The effect was produced mainly due to sliding stresses and speed and the action of the atmosphere of the environment close to the tested surface. The stabilization of the state of anti-wear surface layer was observed (AWSL). In the profile (Fig. 4) that refers to the DLC coating, it can be seen that the stabilization of the friction coefficient takes place after approx. 2700 seconds, the stabilization value ranges between 0.16-0.18. The course of linear wear is exponential.

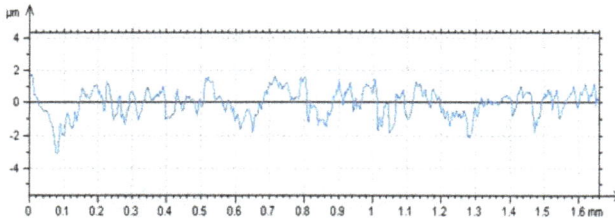

Fig. 5. View of roughness profile of DLC coating

Table 2. Results of roughness profile of DLC coating

| Parameters of roughness | DLC coating | Parameters of roughness | DLC coating |
|---|---|---|---|
| Rp [μm] | 1.81 | Ra [μm] | 0.66 |
| Rv [μm] | 2.33 | Rq [μm] | 0.81 |
| Rz [μm] | 4.14 | Rsk | -0.39 |
| Rc [μm] | 1.91 | Rku | 2.74 |
| Rt [μm] | 4.86 | – | – |

The roughness of the DLC coatings was measured at the Laboratory for Measurement of Geometric Quantities of the Kielce University of Technology using a TALYSURF CCI equipment. The roughness was measured in two directions perpendicular to each other. Then, the average value Ra = 0.60÷0.66 μm was calculated. The steel specimens (4H13) without coatings had a roughness from 0.41 to 0.43 μm. Fig.5 presents an example two-dimensional surface microgeometry measurement of the DLC coating. Table 2 presents the most important average roughness parameters of the tested coating system. Low value of roughness parameters of DLC coating also has influence on mechanical properties.

**Summary**
DLC coatings are characterized by good mechanical properties. They have a homogeneous structure and are free of defects. The layer thickness was approximately 1.1 μm. The average value of the friction coefficient (at the moment of stabilization) obtained during the tribological tests for a DLC coating were between 0.16-0.18. DLC coatings are characterized by low roughness and high nanohardness. Further research will be targeted at the determination of corrosion and erosion resistance.

**References**

[1] N. Radek, A. Sladek, J. Broncek, I. Bilska, A. Szczotok, Electrospark alloying of carbon steel with WC-Co-Al$_2$O$_3$: deposition technique and coating properties, Advanced Materials Research 874 (2014) 101-106. https://doi.org/10.4028/www.scientific.net/AMR.874.101

[2] N. Radek, K. Bartkowiak, Laser treatment of electro-spark coatings deposited in the carbon steel substrate with using nanostructured WC-Cu electrodes, Physics Procedia 39 (2012) 295-301. https://doi.org/10.1016/j.phpro.2012.10.041

[3] R. Ulewicz, Hardening of steel X155CrVMo12-1 surface layer, Journal of the Balkan Tribological Association 21 (2015) 166-172.

[4] A. Dudek, A. Wronska, L. Adamczyk, Surface remelting of 316 L+434 L sintered steel: microstructure and corrosion resistance, Journal of Solid State Electrochemistry 18 (2014) 2973-2981. https://doi.org/10.1007/s10008-014-2483-2

[5] D. Klimecka-Tatar, G. Pawlowska, R. Orlicki, G.E. Zaikov, Corrosion characteristics in alkaline, and ringer solution of Fe68-xCoxZr10Mo5W2B15 metalic glasses, Journal of the Balkan Tribological Association 20 (2014) 124-130.

[6] T. Burakowski, W. Wierzchoń, Surface engineering of metals - principle, equipment, technology, CRC Press, Boca Raton – London – New York – Washington D. C., 1999. https://doi.org/10.1201/9781420049923

[7] M. Madej, The effect of TiN and CrN interlayers on the tribological behavior of DLC coatings, Wear 317 (2014) 179-187. https://doi.org/10.1016/j.wear.2014.05.008

[8] M. Abdollah, Y. Yamaguchi, T. Akao, N. Inayoshi, N. Miyamoto, T. Tokoroyama, N. Umehara, Deformation-wear transition map of DLC coating under cyclic impact loading, Wear 274-275 (2012) 435-441. https://doi.org/10.1016/j.wear.2011.11.007

[9] Y. Liu, E.I. Meletis, Tribological behavior of DLC Coatings with functionally gradient interfaces, Surface and Coatings Technology 153 (2002) 178-183. https://doi.org/10.1016/S0257-8972(01)01688-7

[10] P. Kurp, D. Soboń, The influence of laser padding parameters on the tribological properties of the Al$_2$O$_3$ coatings, METAL 2018 27$^{th}$ Int. Conf. Metallurgy and Materials, Ostrava, Tanger, 1157-1162.

Terotechnology XI                                                                      Materials Research Forum LLC
Materials Research Proceedings **17** (2020) 171-176                    https://doi.org/10.21741/9781644901038-26

[11] L. Radziszewski, M. Kekez,  Application of a genetic-fuzzy system to diesel engine pressure modeling,  Int. J. Adv. Manuf. Tech. 46 (2010) 1-9. https://doi.org/10.1007/s00170-009-2080-1

[12] M. Kekez, L. Radziszewski, A. Sapietova,  Fuel type recognition by classifiers developed with computational intelligence methods using combustion pressure data and the crankshaft angle at which heat release reaches its maximum, Procedia Engineering 136 (2016) 353-358. https://doi.org/10.1016/j.proeng.2016.01.222

[13] P. Walczak, A. Sobczyk, Simulation of water hydraulic control system of Francis turbine. Proc. 8th FPNI Ph.D Symposium on Fluid Power, 2014, art. V001T04A001. https://doi.org/10.1115/FPNI2014-7814

[14] M.S. Kozien, J. Wiciak, Passive structural acoustic control of the smart plate - FEM simulation. Acta Phys. Pol. A 118 (2010) 1186-1188. https://doi.org/10.12693/APhysPolA.118.1186

[15] T. Lipinski, Double modification of AlSi9Mg alloy with boron, titanium and strontium. Arch. Metall. Mater. 60 (2015) 2415-2419. https://doi.org/10.1515/amm-2015-0394

[16] J. Krawczyk, A. Sobczyk, J. Stryczek, P. Walczak, Tests of new methods of manufacturing elements for water hydraulics. Materials Research Proceedings 5 (2018) 200-205.

[17] L.J. Orman, R. Chatys, Heat transfer augmentation possibility for vehicle heat exchangers. 15th Int. Conf. on Transport Means, Kaunas (2011) 9-12.

[18] L. J. Orman, Boiling heat transfer on meshed surfaces of different aperture. AIP Conf. Proc. 1608 (2014) 169-172. https://doi.org/10.1063/1.4892728

[19] L.J. Orman, Boiling heat transfer on single phosphor bronze and copper mesh microstructures. EPJ Web of Conf. 67 (2014) art. 02087. https://doi.org/10.1051/epjconf/20146702087

[20] M. Domagala, H. Momeni, J. Domagala-Fabis, G. Filo, D. Kwiatkowski, Simulation of cavitation erosion in a hydraulic valve. Materials Research Proceedings 5 (2018) 1-6. https://doi.org/10.21741/9781945291814-1

[21] R. Dwornicka, The impact of the power plant unit start-up scheme on the pollution load. Adv. Mat. Res.-Switz. 874 (2014) 63-69. https://doi.org/10.4028/www.scientific.net/AMR.874.63

[22] A. Pacana, L. Bednarova, I. Liberko, A. Wozny, Effect of selected production factors of the stretch film on its extensibility. Przem. Chem. 93 (2014) 1139-1140.

[23] A. Pacana, K. Czerwinska, R. Dwornicka, Analysis of non-compliance for the cast of the industrial robot basis, METAL 2019 28th Int. Conf. on Metallurgy and Materials (2019), Ostrava, Tanger 644-650. https://doi.org/10.37904/metal.2019.869

[24] J. Korzekwa, W. Skoneczny, G. Dercz, M. Bara, Wear mechanism of Al2O3/WS2 with PEEK/BG plastic. J. Tribol.-Trans. ASME 136 (2014) art. 011601. https://doi.org/10.1115/1.4024938

[25] A. Gadek-Moszczak, P. Matusiewicz, Polish stereology - a historical review. Image Anal. Stereol. 36 (2017) 207-221. https://doi.org/10.5566/ias.1808

Terotechnology XI
Materials Research Proceedings **17** (2020) 177-184

Materials Research Forum LLC
https://doi.org/10.21741/9781644901038-27

# Performance Analysis of Diamond Coated End Mill during Machining of Metal Matrix Composite

WICIAK-PIKUŁA Martyna[1, a *], FELUSIAK Agata[1,b] and KIERUJ Piotr[1,c]

[1]Poznan University of Technology, Pl. M. Sklodowskiej-Curie 5, 60-965 Poznan, POLAND

[a]martyna.r.wiciak@doctorate.put.poznan.pl, [b]agata.z.felusiak@doctorate.put.poznan.pl, [c]piotr.kieruj@put.poznan.pl

**Keywords:** Metal Matrix Composites, Performance Analysis, Machinability

**Abstract.** In this paper, a performance analysis of a diamond coated end mill during the milling of a hard-to-cut Duralcan™ metal matrix composite was presented. The conducted tests involved the measurements of cutting force components during milling with one variable parameter, evaluation of the flank wear $VB_B$ and corner wear $VB_C$. Cutting speed $v_c$ in range 300, 500 and 900 m/min was the changeable parameter. Finally, the analysis of cutting forces in time domain, as well as the correlation of the obtained measures with the tool wear values were conducted.

**Introduction**
One of the most popular Metal Matrix Composites (MMCs) is Duralcan™, which is a material based on aluminum cast alloys reinforced with ceramic particles SiC that is suited for high pressure die castings [1]. Its strength and abrasive wear resistance is increased due to reinforcement. In view of high mechanical properties, MMCs are difficult-to-cut materials [2-4]. Machining of MMCs is difficult because of hard ceramic particles content, which simultaneously contributes to the rapid tool wear like in other hard-to-cut materials [5-7]. Nowadays, there are a lot of ideas to improve machinability of difficult-to-cut materials, which improve surface roughness after machining and reduce tool wear. One of such methods is laser assisted machining (LAM), which improves the machining efficiency of hard-to-cut materials [8-10]. One of the example is paper [11], where the laser modification steel specimens with the boronickelized layer (B-Ni complex layer) contribute to the reduction of the microhardness of material and wear resistance. These parameters are reduced in the heating zone by means of a laser beam, which in the case of laser-assisted machining of hard materials improves machining efficiency and tool life. The other method of improving the efficiency of machining of hard-to-cut materials is various kind of prediction models of tool wear. During the monitoring of machining process, the information about tool wear is useful for the tool condition prediction process [12]. There are a lot of methods of edge wear indicator prediction, where the following are the most interesting: genetic algorithm, artificial neural networks or fuzzy logic. The most important stage is the validation of the created model and assessment of its effectiveness in predicting tool wear. Appropriate tool condition assessment will allow for its catastrophic failure detection before cutting edge chipping or surface deterioration occurrences [13, 14].

One of the problems of hard-to-cut materials machining is to simultaneously obtain a satisfactory surface with reduced tool wear and carry out the process in the shortest possible time. Materials such as Duralcan, Inconel or Waspaloy are increasingly used in the automotive or the aviation industry, so it is important to obtain the required accuracy of components. The following are examples of elements that often work at elevated temperatures or difficult conditions: brake rotors, clutch plates, brackets, belt pulley and many others [15-17].

There are still problems with the mechanical processing of these materials. The rationale for this article is the increasing use of materials such as MMC in industry.

The results of this study may be interesting for related technological processes, such as the production of protective coatings [18], including the ESD method [19], later processed with a laser [20], which is of interest to the hydraulics of heavy-duty machines with high operating precision [21]. Further work will include an in-depth analysis of the factors influencing the process [22, 23].

**Materials and method**

The objective of the research involved the performance analysis of metal matrix composite milling, based on the measurement of tool wear in different cutting conditions. The workpiece was a type of MMC with the trade name of Duralcan™. This material exhibits high yield strength, ultimate strength and elastic modulus due to the reinforcement of aluminum matrix with approx. 10% silicon carbide (SiC) particles. This reinforcement allowed for the improvement of mechanical properties and improved abrasion resistance. Table 1 depicts the chemical composition of the Duralcan™ matrix alloy.

*Table 1 The chemical composition of Duralcan™*

| Duralcan™ F3S.10S | Si | Fe | Cu | Mg | Ti | Al |
|---|---|---|---|---|---|---|
| content (%) | 8.50…9.50 | 0.20 max | 0.20 max | 0.45…0.65 | 0.20 max | rest |

Three-edged diamond coated end mills ⌀ 10 mm were selected for machining the composite material. The diamond coating retains its properties up to the maximum of 600°C. The base tool material is a fine-grained carbide grade with a cobalt content of 8%, which significantly improves durability and abrasion resistance.

The cutting tests were conducted on a DECKEL-MAHO DMC 70V machining center, with cutting conditions presented in Table 2. The cutting speed $v_c$ was a variable parameter in the tests. Three repetitions were carried out for each cutting speed.

*Table 2. Cutting parameters*

| Cutting variant | 1 | 2 | 3 |
|---|---|---|---|
| Cutting speed $v_c$ (m/min) | 300 | 500 | 900 |
| Feed per tooth $f_z$ (mm/tooth) | 0.035 | 0.035 | 0.035 |
| Cutting path $L$ (mm) | 122 | 122 | 122 |
| Cutting depth $a_p$ (mm) | 8 | 8 | 8 |
| Cutting width $a_e$ (mm) | 0.2 | 0.2 | 0.2 |

After each milling pass, the tool flank wear $VB_B$ and corner wear $VB_C$ were measured with the use of a microscope. The tool wear was measured for three cutting edges of each tool. The value of tool wear was averaged. The critical tool wear criterion for the milling tool was equaled to 0,3 mm. Additionally, measurements of cutting force components were carried out with the application of piezoelectric force sensor clamped to the machine's worktable. The measurements were conducted in the following directions: X – feed direction $F_f$, Y – feed normal direction $F_{fN}$ and Z – thrust direction $F_z$. Figure 1 presents a simplified diagram of the measurement set-up.

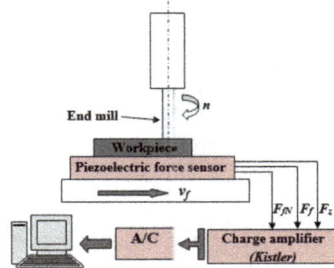

*Figure 1 The scheme of experimental apparatus*

**Analysis of tool wear**

The research results show the measured tool wear $VB_C$ and $VB_B$ generated during the end milling of MMC. Fig.2 depicts the relationship between the flank wear $VB_C$ and the cutting time $t_s$ for all repetitions. To determine the relation, the exponential function $VB_C = a \cdot e^{b \cdot t_s}$ was selected.

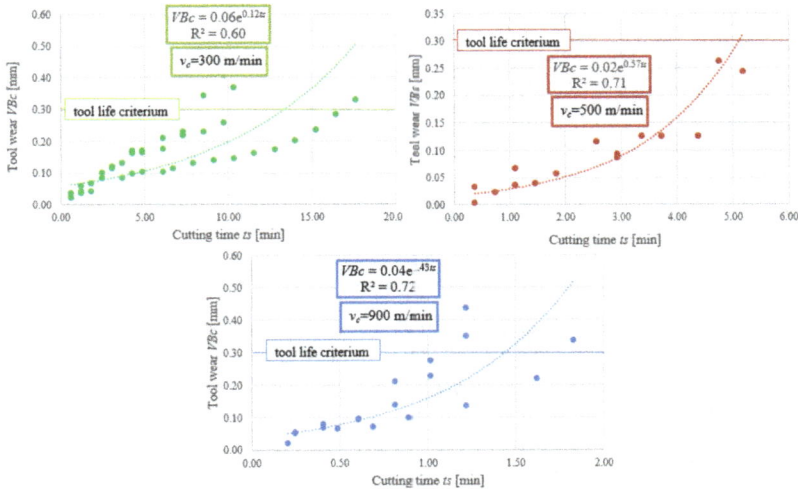

*Figure 2. Tool wear $VB_C$ in function of time for:*
*a) $v_c$ = 300 m/min, b) $v_c$ = 560 m/min, c) $v_c$ = 900 m/min*

On the basis of the selected equations, the average time needed for the excessing the critical tool wear value was calculated and shown in Table 3.

Materials Research Forum LLC
https://doi.org/10.21741/9781644901038-27

Fig.3 presents the tool flank wear $VB_B$ values in function of time $t_s$ for all repetitions. The exponential function $VB_B = a \cdot e^{b \cdot ts}$ was determined similar to the corner wear analysis. The tool life $T$ based on selected equations was calculated and depicted in Table 4. The tool life $T$ based on selected measurement of $VB_B$ was calculated and depicted in Table 4.

*Table 3 Value of tool life T based on measurement of $VB_c$*

| Cutting speed $v_c$ [m/min] | Equation | Tool life $T$ [min] |
|---|---|---|
| 300 | $VB_C = 0.06 \cdot e^{0.12ts}$ | 13.36 |
| 500 | $VB_C = 0.02 \cdot e^{0.57ts}$ | 5.11 |
| 900 | $VB_C = 0.04 \cdot e^{1.43ts}$ | 1.44 |

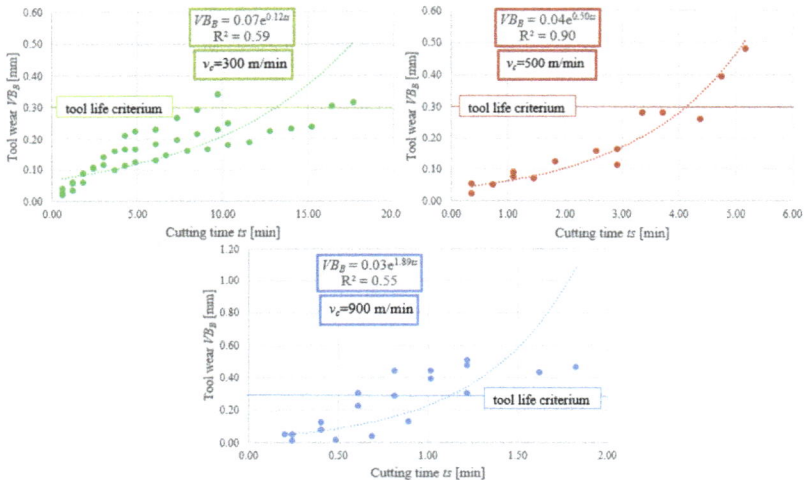

*Figure 3. Tool flank wear $VB_B$ in function of time for:*
*a) $v_c = 300$ m/min, b) $v_c = 500$ m/min and c) $v_c = 900$ m/min*

*Table 4. Value of tool life T based on measurement of $VB_B$*

| Cutting speed $v_c$ [m/min] | Equation | Tool life $T$ [min] |
|---|---|---|
| 300 | $VB_B = 0.07 \cdot e^{0.12ts}$ | 13.12 |
| 500 | $VB_B = 0.04 \cdot e^{0.50ts}$ | 4.41 |
| 900 | $VB_B = 0.03 \cdot e^{1.89ts}$ | 1.15 |

Terotechnology XI
Materials Research Proceedings **17** (2020) 177-184

Materials Research Forum LLC
https://doi.org/10.21741/9781644901038-27

In the next step, the material removal rate (MMR) $Q$ [cm³/min] was calculated to compare these values with tool life $T$ for each cutting tests. The MMR was calculated on the basis of the following equation (Eq.1):

$$Q = \frac{a_p \cdot a_e \cdot v_f}{1000} \ [cm^3/min] \quad (1)$$

where $a_p$ – cutting depth [mm], $a_e$ – cutting width [mm], $v_f$ – fee rate [mm/min].
The summary of tool life and efficiency of metal matrix composite milling is shown in Table 5.

Table 5. Value of tool life T compared with efficiency Q

| Cutting speed $v_c$ [m/min] | Material removal rate $Q$ [cm³/min] | Tool life $T$ [min] |
|---|---|---|
| 300 | 1.61 | **13.12** |
| 500 | 2.68 | 4.41 |
| 900 | **4.82** | 1.15 |

**Analysis of cutting force in time domain**
After the tests, the each milling pass was correlated with measured cutting force components in the three directions.

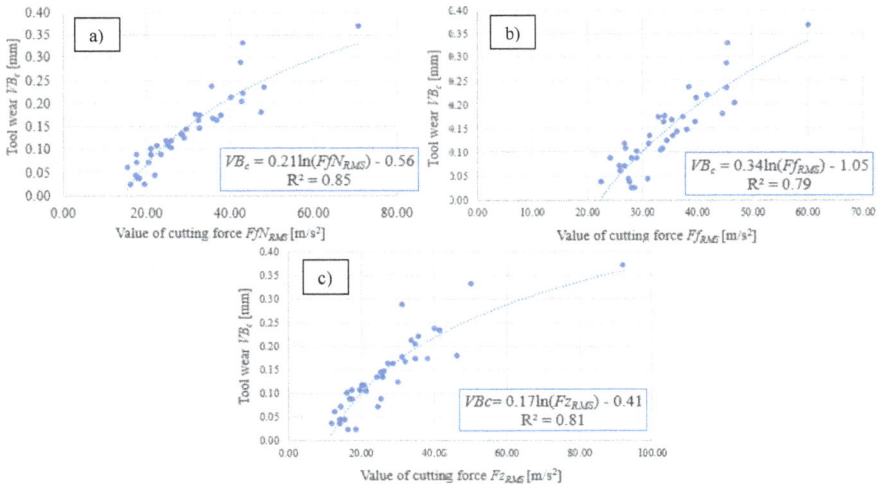

Fig. 4, Relation between VBc and cutting force components: a) Ff, b) Ff, c) Fz

The names of the selected statistical measures are as follows:
- $Ff_{RMS}$ – root mean square value of cutting force measured in the feed direction (along the X axis),
- $FfN_{RMS}$ - root mean square value of cutting force measured in the feed normal direction (along the Y axis),

Terotechnology XI                                                      Materials Research Forum LLC
Materials Research Proceedings **17** (2020) 177-184        https://doi.org/10.21741/9781644901038-27

- *$Fz_{RMS}$* - root mean square and peak value of accelerations of vibrations measured in the thrust direction (along the Z axis),

Figure 4 shows the chart containing the logarithmic equation between the tool wear and cutting forces: $VB_c = A \cdot \ln(x) + b$ for the tests with $v_c$=300 m/min. The conformity of the experimental results with the logarithmic function can be described by the $R^2$ coefficient.

This function reflects the results obtained in the best way, and the coefficient $R^2 = 0.85$ for root mean square value of cutting force measured in the feed normal direction, which indicates high adjustment to the selected function.

## Summary

On the basis of the conducted research, the following conclusions were formulated:

- The analysis of corner tool wear $VB_C$ and flank wear $VB_B$ showed that the milling of Duralcan™ metal matrix composite is difficult due to hard silicon carbide particles in aluminum. The longest tool life $T$ was noted for the smaller cutting speed 300 m/min (approx. 13 min), but the worst results were received for 900 m/min (about 10 times lower than with 300 m/min). The highest parameter of cutting speed did not bring good results.
- The best efficiency of metal matrix composites was achieved for 900 m/min (approx. 5 cm³/min), but compared with tool life there were no satisfying effects. For cutting speed 500 m/min tool life was 4 times bigger than for 900 m/min, but had only two times lower removal rate $Q$ than the highest parameter.
- The analysis of cutting force in time domain allowed for selecting the measure which indicated the highest matching with tool wear during the machining of Duralcan™. The best coefficient $R^2$ was obtained for $FfN_{RMS}$ - root mean square value of cutting force measured in the feed normal direction ($R^2$=0.85).

## References

[1] Hajkowski J., Popielarski P., Sika R., Prediction of HPDC Casting Properties Made of Al-Si9Cu3 Alloy, Advances In Manufacturing, Lecture Notes in Mechanical Engineering, Springer (2018) 621-631. https://doi.org/10.1007/978-3-319-68619-6_59

[2] K. Gawdzińska, L. Chybowski, W. Przetakiewicz, R. Laskowski, Application of FMEA in the Quality Estimation of Metal Matrix Composite Castings Produced by Squeeze Infiltration, Arch. Metall. Mater. 62 (2017) 2171-2182. https://doi.org/10.1515/amm-2017-0320

[3] Kawalec, M., Przestacki, D., Bartkowiak, K., Jankowiak, M., Laser assisted machining of aluminium composite reinforced by SiC particle, ICALEO - 27th Int. Congress on Applications of Lasers and Electro-Optics (2008) 895-900. https://doi.org/10.2351/1.5061278

[4] Hajkowski J., Popielarski P., Ignaszak Z., Cellular Automaton Finite Element Method Applied for Microstructure Prediction of Aluminium Casting Treated by Laser Beam, Arch. Foundry Eng. 19 (2019) 111-118.

[5] Nicholls C.J., BoswellIan B., Davies I.J., Islam M.N., Review of machining metal matrix composites, INT J ADV MANUF TECH vol.90 (2017) pp. 2429–2441. https://doi.org/10.1007/s00170-016-9558-4

[6] P. Krawiec, K. Waluś, Ł. Warguła, J. Adamiec, Wear evaluation of elements of V-belt transmission with the application of optical microscope, MATEC Web of Conf. 157 (2018) art. 01009. https://doi.org/10.1051/matecconf/201815701009

[7] M. Kujawski, P. Krawiec, Analysis of Generation Capabilities of Noncircular Cog belt Pulleys on the Example of a Gear with an Elliptical Pitch Line, J. Manuf. Sci. Eng. – Trans. ASME 133 95) (2011) art. 051006. https://doi.org/10.1115/1.4004866

[8] D. Przestacki, R. Majchrowski, L. Marciniak-Podsadna, Experimental research of surface roughness and surface texture after laser cladding, App. Surf. Sci. 388 (2016) 420-423. https://doi.org/10.1016/j.apsusc.2015.12.093

[9] K. Erdenechimeg, H. Jeong, C. Lee, A Study on the Laser-Assisted Machining of Carbon Fiber Reinforced Silicon Carbide, Materials 12 (2019) art. 2061. https://doi.org/10.3390/ma12132061

[10] D. Przestacki, A. Bartkowska, M. Kukliński, P. Kieruj, The Effects of Laser Surface Modification on the Microstructure of 1.4550 Stainless Steel, MATEC Web of Conf. 237 (2018) art. 02009. https://doi.org/10.1051/matecconf/201823702009

[11] Bartkowska A., Pertek A., Popławski M., ,Przestacki D., Miklaszewski A., Effect of laser modification of B-Ni complex layer on wear resistance and microhardness. Optics and Laser Technology 72 (2015) 116-124. https://doi.org/10.1016/j.optlastec.2015.03.024

[12] Twardowski P., Tabaszewski M., Wojciechowski S., Turning process monitoring of internal combustion engine piston's cylindrical surface, MATEC Web of Conf. 112 (2017) art.10002. https://doi.org/10.1051/matecconf/201711210002

[13] Tang J, Li W.X, Zhao,B., The Application of GA-BP Algorithm in Prediction of Tool Wear State, IOP Conference Series: Materials Science and Engineering 398 (2018) art. 012025. https://doi.org/10.1088/1757-899X/398/1/012025

[14] Ignaszak Z., Popielarski P., Hajkowski J., Codina E., Methodology of comparative validation of selected foundry simulation codes, Arch. Foundry Eng. 15 (2015) 37-44. https://doi.org/10.1515/afe-2015-0076

[15] Krawiec P., Marlewski A., Spline description of not typical gears for belt transmissions. J. Theor. Appl. Mech. 49 (2011) 355-367.

[16] Behera M. P., Dougherty T., Singamneni S., Conventional and Additive Manufacturing with Metal Matrix Composites: A Perspective, Procedia Manufacturing 30 (2019) 159-166. https://doi.org/10.1016/j.promfg.2019.02.023

[17] Krawiec P., Marlewski A., Profile design of noncircular belt pulleys, J. Theor. Appl. Mech. 54 (2016) 561-570. https://doi.org/10.15632/jtam-pl.54.2.561

[18] W. Zorawski, R. Chatys, N. Radek, J. Borowiecka-Jamrozek, Plasma-sprayed composite coatings with reduced friction coefficient. Surf. Coat. Technol. 202 (2008) 4578-4582. https://doi.org/10.1016/j.surfcoat.2008.04.026

[19] R. Dwornicka, N. Radek, M. Krawczyk, P. Osocha, J. Pobedza, The laser textured surfaces of the silicon carbide analyzed with the bootstrapped tribology model. METAL 2017 26th Int. Conf. on Metallurgy and Materials (2017), Ostrava, Tanger 1252-1257.

[20] N. Radek, A. Szczotok, A. Gadek-Moszczak, R. Dwornicka, J. Broncek, J. Pietraszek, The impact of laser processing parameters on the properties of electro-spark deposited coatings. Arch. Metall. Mater. 63 (2018) 809-816.

[21] B. Singh et al. Technical design report for the (P)over-barANDA Barrel DIRC detector. J. Phys. G-Nucl. Part. Phys. 46 (2019) art. 45001.

[22] J. Pietraszek, Fuzzy Regression Compared to Classical Experimental Design in the Case of Flywheel Assembly. In: Rutkowski L., Korytkowski M., Scherer R., Tadeusiewicz R., Zadeh L.A., Zurada J.M. (eds) Artificial Intelligence and Soft Computing ICAISC 2012. Lecture Notes in Computer Science, vol 7267. Berlin, Heidelberg: Springer, 2012, 310-317. https://doi.org/10.1007/978-3-642-29347-4_36

[23] L. Wojnar, A. Gadek-Moszczak, J. Pietraszek, On the role of histomorphometric (stereological) microstructure parameters in the prediction of vertebrae compression strength. Image Analysis and Stereology 38 (2019) 63-73. https://doi.org/10.5566/ias.2028

Terotechnology XI
Materials Research Proceedings **17** (2020) 185-190

Materials Research Forum LLC
https://doi.org/10.21741/9781644901038-28

# Verification of Correlations for Pool Boiling Heat Transfer on Horizontal Meshed Heaters

RADEK Norbert [1,a], ORMAN Łukasz J. [1,b *], PIETRASZEK Jacek [2,c] and BRONČEK Jozef [3,d]

[1]Kielce University of Technology, al. Tysiaclecia P.P. 7, 25-314 Kielce, Poland

[2]Cracow University of Technology, Al. Jana Pawła II 37, 31-864 Cracow, Poland

[3]University of Zilina, Univerzitna 1, 01026 Zilina, Slovakia

[a]norrad@tu.kielce.pl, [b]orman@tu.kielce.pl, [c]pmpietra@mech.pk.edu.pl, [d]jozef.broncek@fstroj.utc.sk

**Keywords:** Boiling, Correlations, Heat Exchangers

**Abstract.** The paper considers the problem of the accuracy of enhanced boiling heat transfer correlations. The experimental results of boiling of distilled water and ethyl alcohol have been compared with models available in literature regarding heat flux values in the nucleate boiling mode. Based on the obtained data it can be stated that the correlations are generally inaccurate in the whole range of superheats. Only a modified correlation by Xin and Chao provided comparable results to the experimental data.

## Introduction

Metal meshes are one of the passive methods of the enhancement of boiling heat transfer. They may be applied onto heating surfaces and provide the advantage of increasing heat transfer coefficients or heat flux at the same superheat (temperature difference). Due to their positive impact, meshes have been tested in various studies. Smirnov et al. [1] performed tests of boiling of water and ethanol on heaters with copper and brass mesh layers. These meshes were bonded mechanically to the base surfaces. The general finding was that heat flux was independent of the layer height and not particularly dependent on the kind of material from which it was made. Franco et al. [2] studied the performance of heaters covered with stainless steel, aluminum, copper and brass meshes under boiling of refrigerant R141b. The meshes were joined with the surface using a specially designed mounting that enabled the control of channels in the screen. It was observed that if meshes of different aperture were used together, the highest enhancement was obtained for layers consisting of finer meshes at the base on the surface and coarser meshes higher. Li et al. [3] experimentally analyzed water boiling on meshed surfaces. The authors used sintering as a technique to produce samples. The sintering temperature was 1030°C and took place in the mixture of nitrogen and hydrogen. The specimens had maximally nine meshes. It was reported that the proper contact conditions between the heat exchanger elements led to an increased heat transfer - all meshed heaters enhanced boiling in comparison with the smooth surface (without such covering). It needs to be noted that other types of microstructures are also tested in view of boiling heat transfer augmentation. For examples in [4], data on R113 pool boiling of heaters with sintered powders made of bronze were presented. The porous coating tested by the authors had the height of 2 mm and the research was conducted under ambient pressure.

Terotechnology XI
Materials Research Proceedings **17** (2020) 185-190

Materials Research Forum LLC
https://doi.org/10.21741/9781644901038-28

The use of meshes usually enhances boiling heat transfer. The level of enhancement depends on the number of factors. However, up to now the literature does not provide a model of heat transfer that would be successful in determining heat flux from meshed surfaces.

One of the easiest models of boiling heat transfer was proposed by Nishikawa et al. [4]. The authors used the assumption that the porous layer is filled with liquid, however due to a small Rayleigh number, natural convection may be disregarded and heat is transferred through conduction. Smirnov and co-workers [1] conduced experimental and theoretical study focused on boiling heat transfer on meshed surfaces. The model that the authors proposed is based on the assumption that vaporization occurs in each elementary cell of a microstructure, while the system of meshes located on the heater can be regarded as an array of microfins. Consequently, adequate formulae can be used for calculations. Xin and Chao [5] produced a model originally addressed to analyze the thermal performance of Gewa – T tunnel structures. It was later extended to consider also Thermoexcel porous coatings. In this model it was assumed that the internal tunnel within the structure is filled with vapor. Vaporization occurs within the internal surface of the covering from a thin liquid film. The thickness of this film was considered to decrease to zero with the approach to the heater surface.

The results obtained in this study seem to be very interesting for many branches of industry, both designing such heaters and using them utilitarily in their activities e.g. hydraulics of heavy-duty machines [6, 7], corrosion protection of agricultural machines [8], coupled problems of fatigue and thermomechanical loads [9, 10], especially at very high temperatures of techno-logical processes [11, 12]. The results will also influence the management and investment decision-making processes of the companies concerned [13-15], as well as the orientation of quantitative material assessments provided by image analyzes[16-18].

**Material and method**

The investigations were focused on the copper mesh layers sintered onto copper disks of 3 cm in diameter (Fig. 1). Such a system works as a heater located horizontally and supplied with heat from the lower part of the experimental set-up. A glass vessel is placed above the sample and sealed to the base. Here, the boiling process occurs, while condensation recovers the generated vapor so that the liquid level is kept constant. The details of the experimental stand have been discussed by the co-author in [5]. Two boiling liquids were used for the analyses, namely distilled water and ethyl alcohol of high purity (99.8%). The experiments were performed under ambient pressure. Cooling of the condensate was provided with the use of cold water in a glass condenser.

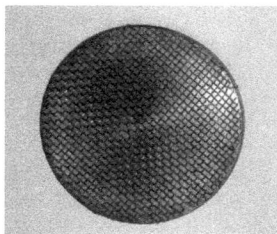

*Fig. 1. Example sample of a single mesh layer sintered onto the copper base.*

Terotechnology XI                                                        Materials Research Forum LLC
Materials Research Proceedings **17** (2020) 185-190           https://doi.org/10.21741/9781644901038-28

The thermal performance of the samples was obtained as boiling curves – representing the dependences of the heat flux values vs. wall superheat. The consecutive data points were determined with the rising heat flux values (obtained precisely with the autotransformer). The experiments were focused on the nucleate boiling mode of pool boiling heat transfer due to its highest practical applicability for the design of phase –change heat exchangers.

## Results and discussion

The application of meshed surfaces generally provides additional advantage comparing to a smooth surface without such a coating (as indicated by many literature reports). Fig. 2 presents the enhancement ratio generated for a double mesh layer (raw data adapted from [19]). It is the ratio of the heat flux ($q_{meshed}$) exchanged from the meshed surface to the heat flux from the smooth surface ($q_{smooth}$) assuming the same superheat ($\theta$). It is clearly visible that the performance of the microstructural coating is highest at small superheats (the enhancement ratio reaches the value of almost 6.5). As the heat flux is increased and the superheat rises, the thermal performance of the mesh coated samples becomes weaker but is still a few times higher than for the smooth reference surface (exactly three times at ca. 10 K superheat). It might be explained by the activation of more and more nucleation sites (locations where vapor bubbles grow and from where they depart) on the smooth surface, while they are already there on the mesh coated heaters due to the presence of metal wires on the surface. As a consequence, the smooth surface becomes more and more efficient as the temperature rises, while the microstructural coatings offer no significant additional advantage in this range of superheat.

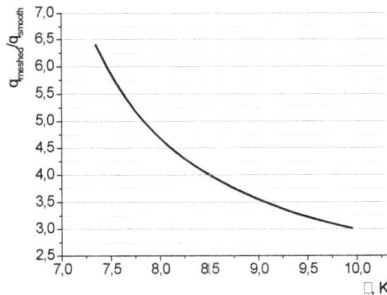

*Fig. 2. The enhancement ratio of the double mesh layer.*

Apart from experimental results obtained in the laboratory, another vital issue is a correct design of such phase – change heat exchangers. In order to do it properly, a reliable model of the phenomenon is necessary. However, despite many papers written on this subject and many research projects conducted, up to now models and correlations from literature that provide formulae for heat flux are often inaccurate. Some might provide good congruence with the experimental results, however, it is only for a certain kind of microstructural coatings or a very limited range of heat flux.

In order to verify if the selected models and correlations available in literature are precise in determination of heat flux based on the physical and chemical properties of the samples and boiling agents, an analysis was performed. Fig. 3 and 4 present the comparison of the experimental data adopted from [19] for meshed surfaces (respectively, for a single mesh of wire

Terotechnology XI
Materials Research Proceedings **17** (2020) 185-190

Materials Research Forum LLC
https://doi.org/10.21741/9781644901038-28

diameter 0.50 mm and aperture 0.75 and a double mesh of wire diameter 0.32 and aperture 1.50) with correlations from literature. The following models were used: the one developed by Smirnov et al. [1], Nishikava et al. [4] and Xin & Chao [5]. The proper performance of calculations according to the model presented by Xin and Chao was only possible after certain modifications done onto the original morel. They included the width of a single cell to be considered as the total of wire diameter and aperture and the width of the tunnel as aperture. The results of the calculations are presented for both the working liquids (a –distilled water, b – ethyl alcohol).

a)                                              b)

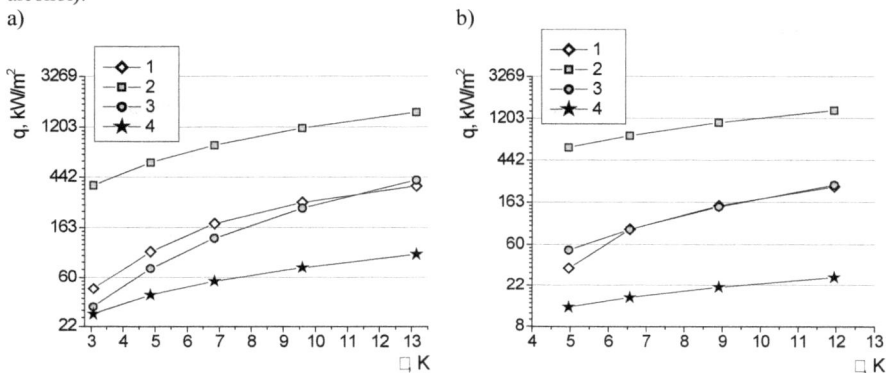

*Fig. 3. Comparison of the experimental data for the single mesh 0.75x0.50 with the correlations; 1 - experimental results, 2 - calculation results with Nishikava et al. correlation, 3 - calculation results with Xin and Chao correlation, 4 - calculation results with Smirnov et al. correlation; a – distilled water, b – ethyl alcohol*

a)                                              b)

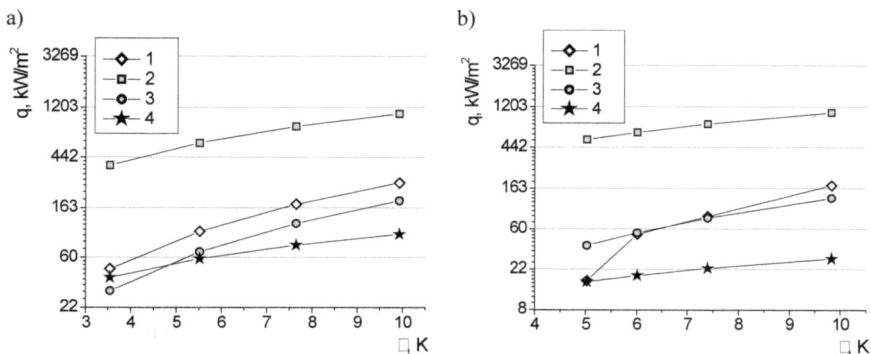

*Fig. 4. Comparison of the experimental data for the double mesh 1.50x0.32 with the correlations; 1 - experimental results, 2 - calculation results with Nishikava et al. correlation, 3 - calculation results with Xin and Chao correlation, 4 - calculation results with Smirnov et al. correlation; a – distilled water, b – ethyl alcohol*

Terotechnology XI                                                    Materials Research Forum LLC
Materials Research Proceedings **17** (2020) 185-190          https://doi.org/10.21741/9781644901038-28

The analysis of Fig. 2 and 3 reveals that the model by Xin and Chao with modifications proved to be the most accurate for the analyzed samples. The simplest model proposed by Nishikava et al. produced too high values probably due to the assumed model of heat transfer within the structural coating (and, thus, it was the least effective). The correlation by Smirnov et al. usually provided too low values, although in some cases the congruence, especially in the range of low superheats was high. The general findings of the paper are in agreement with literature of boiling on enhanced surfaces.

**Conclusions**
Sintered metal meshes are believed to be very effective in dissipating large heat fluxes with low superheat during pool boiling, although the treated surfaces also appear to be very efficient at flow boiling, as shown e.g. in experimental research [20], and in numerous computational studies, both in approaches directly based on Trefftz functions [21, 22] and inverse problems [23, 24]. However, models and correlations available in literature are often inaccurate. Their limitations often comprise a certain kind of the microstructural coating that was considered by the authors or a limited range of superheat values. Consequently, a new model is necessary to be developed and applied for the porous layers.

**References**
[1] G.F. Smirnov, A.L. Coba, B.A. Afanasiev, The heat transfer by boiling in splits, capilaries, wick structures, AIAA Paper (1978) 78-461. https://doi.org/10.2514/6.1978-461

[2] A. Franco, E.M. Latrofa, V.V. Yagov, Heat transfer enhancement in pool boiling of a refrigerant fluid with wire nets structures, Exp. Therm. Fluid Sci. 30 (2006) 263-275. https://doi.org/10.1016/j.expthermflusci.2005.07.002

[3] C. Li, G.P. Peterson, Y. Wang, Evaporation/boiling in thin capillary wicks (I) – wick thickness effects, J. Heat Transfer 128 (2006) 1312-1319. https://doi.org/10.1115/1.2349507

[4] K. Nishikawa, T. Ito, K. Tanaka, Enhanced heat transfer by nucleate boiling on a sintered metal layer, Heat transfer – Japanese Research 8 (1979) 65-81.

[5] M.-D. Xin, Y.-D. Chao, Analysis and experiment of boiling heat transfer on T-shaped finned surfaces, Chem. Eng, Comm. 50 (1987) 185-199. https://doi.org/10.1080/00986448708911825

[6] M. Domagala, H. Momeni, J. Domagala-Fabis, G. Filo, D. Kwiatkowski, Simulation of cavitation erosion in a hydraulic valve. Materials Research Proceedings 5 (2018) 1-6. https://doi.org/10.21741/9781945291814-1

[7] J. Krawczyk, A. Sobczyk, J. Stryczek, P. Walczak, Tests of new methods of manufacturing elements for water hydraulics. Materials Research Proceedings 5 (2018) 200-205.

[8] T. Lipinski, D. Karpisz, Corrosion rate of 1.4152 stainless steel in a hot nitrate acid. METAL 2019: 28th Int. Conf. on Metallurgy and Materials, Ostrava, TANGER, 2019, 1086-1091. https://doi.org/10.37904/metal.2019.911

[9] M.S. Kozien, J. Wiciak, Passive structural acoustic control of the smart plate - FEM simulation. Acta Phys. Pol. A 118 (2010) 1186-1188. https://doi.org/10.12693/APhysPolA.118.1186

[10] R. Ulewicz, P. Szataniak, F. Novy, Fatigue properties of wear resistant martensitic steel. METAL 2014: 23rd Int. Conf. on Metallurgy and Materials. Ostrava, TANGER (2014) 784-789.

[11] M. Hebda, S. Gadek, J. Kazior, Influence of the mechanical alloying process on the sintering behaviour of Astaloy CrM powder mixture with silicon carbide addition. Arch. Metall. Mater. 57 (2012) 733-743. https://doi.org/10.2478/v10172-012-0080-x

[12] D. Przestacki, M. Kuklinski, A. Bartkowska, Influence of laser heat treatment on microstructure and properties of surface layer of Waspaloy aimed for laser-assisted machining. Int. J. Adv. Manuf. Technol. 93 (2017) 3111-3123. https://doi.org/10.1007/s00170-017-0775-2

[13] J. Nowakowska-Grunt, M. Mazur, Safety management in logistic processes of the metallurgical industry. METAL 2015: 24th Int. Conf. on Metallurgy and Materials, Ostrava, TANGER, 2015, 2020-2025.

[14] J. Nowakowska-Grunt, M. Mazur, Effectiveness of logistics processes of SMES in the metal industry. METAL 2016: 25th Int. Conf. on Metallurgy and Materials, Ostrava, TANGER, 2016, 1956-1961.

[15] A. Pacana, K. Czerwinska, R. Dwornicka, Analysis of non-compliance for the cast of the industrial robot basis, METAL 2019 28th Int. Conf. on Metallurgy and Materials (2019), Ostrava, Tanger 644-650. https://doi.org/10.37904/metal.2019.869

[16] A. Gadek-Moszczak, S. Kuciel, L. Wojnar, W. Dziadur, Application of computer-aided analysis of an image for assessment of reinforced polymers structures. Polimery 51 (2006) 206-211. https://doi.org/10.14314/polimery.2006.206

[17] A. Szczotok, M. Sozanska, a comparison of grain quantitative evaluation performed with standard method of imaging with light microscopy and EBSD analysis. Prakt. Metallogr.-Pract. Metallogr. 46 (2009) 454-468. https://doi.org/10.3139/147.110043

[18] A. Gadek-Moszczak, History of stereology. Image Anal. Stereol. 36 (2017) 151-152. https://doi.org/10.5566/ias.1867

[19] L. Dąbek, A. Kapjor, Ł.J. Orman, Distilled water and ethyl alcohol boiling heat transfer on selected meshed surfaces, Mech. Ind. 20 (2019) 701. https://doi.org/10.1051/meca/2019068

[20] K. Strąk, M. Piasecka, B. Maciejewska, Spatial orientation as a factor in flow boiling heat transfer of cooling liquids in enhanced surface minichannels, Int. J. Heat Mass Transf. 117 (2018) 375-387. https://doi.org/10.1016/j.ijheatmasstransfer.2017.10.019

[21] M. Grabowski, S. Hożejowska, A. Pawińska, M.E. Poniewski, J. Wernik, Heat transfer coefficient identification in minichannel flow boiling with hybrid Picard-Trefftz method, Energies 11 (2018) 1-13. https://doi.org/10.3390/en11082057

[22] B. Maciejewska, M. Piasecka, An application of the non-continuous Trefftz method to the determination of heat transfer coefficient for flow boiling in a minichannel, Heat Mass Transf. 53 (2017) 1211-1224. https://doi.org/10.1007/s00231-016-1895-1

[23] L. Hożejowski, S. Hożejowska, Trefftz method in an inverse problem of two-phase flow boiling in a minichannel, Eng. Anal. Bound. Elem. 98 (2019) 27-34. https://doi.org/10.1016/j.enganabound.2018.10.001

[24] B. Maciejewska, M. Piasecka, Trefftz function-based thermal solution of inverse problem in unsteady-state flow boiling heat transfer in a minichannel, Int. J. Heat Mass Transf. 107 (2017) 925-933. https://doi.org/10.1016/j.ijheatmasstransfer.2016.11.003

Terotechnology XI
Materials Research Proceedings 17 (2020) 191-202

Materials Research Forum LLC
https://doi.org/10.21741/9781644901038-29

# Strength Testing and Ring Stiffness Testing of Underground Composite Pressure Pipes

KRYSIAK Piotr[1,a] *, OWCZAREK Radosław[2,b], BŁAŻEJEWSKI Wojciech[3,c]
and BŁACHUT Aleksander[3,d]

[1]Military Institute of Engineer Technology, Obornicka 136, 50-961 Wrocław, Poland

[2]Energetyka Sp. Z.o.o, ul. M. Skłodowskiej-Curie 58, 59-300 Lubin, Poland

[3]Politechnika Wrocławska, Katedra Mechaniki i Inzynierii Materiałowej, ul. Smoluchowskiego 25, 50-370 Wrocław, Poland

[a]krysiak@witi.wroc.pl*, [b]r.owczarek@interia.pl, [c]wojciech.blazejewski@pwr.edu.pl, [d]aleksander.blachut@pwr.edu.pl

Keywords: Pipe, Fibre Reinforced Composite, Strength Testing

**Abstract.** The paper presents issues related to the production and testing of filament-wound glass fibre reinforced epoxy pipes. At the beginning, a strength analysis of a pipe embedded in the ground subjected to loads from ground pressure and passing vehicles was carried out. Based on the calculations, composite pipe samples were manufactured by winding, and then tested using a universal testing machine. During the tests, the ring stiffness and strength of a pipe wall structure were studied, and the latter was characterised by a variety of materials in its radial cross-section (various reinforcing fibres such as glass, basalt, carbon).

## Introduction

The research area studied in this work concerns the behaviour of a pipe (culvert) embedded in the ground at the appropriate depth. Therefore, considerations taken into account concern two objects: the composite pipe (material, technology, properties) and the soil (type, properties).

Polymer matrix composites which include, among others, glass-, carbon,- basalt fibre- and organic fibre- (Kevlar) reinforced polymer matrix composites, form the largest group among the fibre composites used. A significant class of reinforced plastic products includes pipes and tanks produced as a result of filament winding. In many technical fields, they effectively replace products manufactured out of traditional construction materials. However, in order to obtain optimal strength properties of such a product manufactured out of polymers, the winding process should be designed in such a way that the load is transferred via the reinforcing fibres. As these materials behave in a particular way under load, it is important to know the mechanics behind strain and the methodology for verifying basic properties of these materials [2, 6].

When designing structures embedded in the ground, the main issue concerns the determination of the size and distribution of loads acting on their external surface. These difficulties result from the random character of the factors affecting the operation of the structures buried in the ground. An important element here is the analysis of all these factors and matching the right parameters of the manufactured product to the load [4].

### Determining the pipe embedment depth

Determining the load exerted onto a pipe by the overburden requires the determination of a pipe burial depth. This depth is also crucial when determining live (transport) loads, as these loads functionally decrease with an increased burial depth. It should be underlined here that the trench

Materials Research Forum LLC
https://doi.org/10.21741/9781644901038-29

width and materials used for filling should also be specified. Most of these parameters are governed either by the national- or industry standards:

- bedding for pipelines can be made of crushed stone, sandstone, gravel or sand. The bedding material used should meet the requirements of the PN-EN 13043: 2004 standard;
- soil compaction index at the pipeline foundation level should be $I_s \geq 0.97$ according to Proctor impact compaction test. In case of lower compaction values, additional soil improvement should be done via cement or lime soil stabilization;
- the burial depth of pipelines should be such that the thickness of the soil layer above the pipe equals at least 0.9 m;
- backfill in the pipe zone (area immediately surrounding the pipe) should only consist of easily compacted soil, e.g. sand, sand or gravel. The height of the pipe zone should reach from the bottom of the trench to a level of 0.3 m above the top of the pipe. The width of the pipe zone should be equal to the width of the trench (at least four times the pipe diameter);
- the minimum distance on each side of the pipe from the trench/pipe shoring should be 0.3 ÷ 0.5m, depending on the external diameter of the pipe, while in the case of mechanical soil compaction, the distance from the pipe to the trench shoring should be at least 0.4 ÷ 0, 5m, regardless of the diameter of the trench [4].

Based on these guidelines, a bedding of a pipe in a trench was designed. Dimensions dependent on the diameter of the pipe were determined for the outer diameter of the pipe $d_e = 150mm$. Fig. 1 presents the designed pipe placed in the trench.

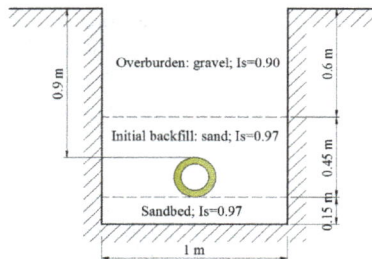

*Fig. 1. Pipe embedded in a trench with compaction zones marked*

**Guidelines for the pipe-soil system**

Along with the progress of civilization, the development of sewer networks and road networks, as well as the issues of sewer pipes, transmission pipes and culverts became of significant interest for the engineers. The emergence of new materials and technologies meant that composite materials were being used more and more often. These materials are characterized by high strength and very good resistance to structural degradation in environments, such as soil. Therefore, pipes made of composite materials, especially continuous fibre (glass, carbon) reinforced composites in a polymer (polyester resin, epoxy resin) matrix are most often used as lines and pipelines for gas transmission. Strength calculations of the structure of such pipes usually focus on determining the appropriate wall thickness due to the internal pressure of the gas within the pipeline.

In the case of pipelines laid in the ground, the dominant load on the pipe may include external loads caused by the impact of the ground, of ground water and overburden load. Such load pattern occurs when there is no internal pressure in the pipe, e.g. during pipeline maintenance.

The correct calculation of pipe stress and loads requires qualifying the pipe to the appropriate group of appropriate elastic properties.

The cables are divided into three groups, which result from the PN-EN 805: 2002 and PN-EN 476: 2011 standards:

1) rigid pipes – those for which the relative deformation of the wall is ~ 0%. This includes pipes made of traditional materials, e.g. concrete, stoneware, cast iron. These pipes are an independent static system and do not cooperate with the soil medium;

2) semi-rigid pipes – those for which minor relative wall deformations (~ 0.5%) are allowed. Apart from these materials being capable of withstanding high mechanical stress, soil compaction parameters are important as well. This group includes pipes made of glass fibre reinforced epoxy resins (duroplastics): GRP-EP, GRP-UP;

3) flexible pipes – for which relative deformation (< 5%) is allowed. These pipes "cooperate" with the soil and together they form a static system. These include pipes made of PVC-U, PE and PP, among others.

On the basis of the guidelines above, filament fibre composite pipes analysed in this work can be classified as belonging to the semi-rigid pipe group.

**Determination on the pressure**
When analysing the load distribution impacting both the plastic- and pipes made of traditional materials, it should be noted that it is not the same. Traditional (rigid) pipes are practically non-deformable, therefore the stress arising from the load concentrates in the upper and lower parts of the pipe. The resultant bending stresses in the walls are unfavourable from the perspective of the durability of the pipe.

When the plastic pipe is under load, the stress in the upper and lower part of the pipe decreases, while the lateral stress increases. The deforming pipe exerts pressure on the soil and causes passive earth pressure, which in turn reduces the bending stress within the pipe wall. The force with which the soil around the pipe is able to resist the pressure of the pipe depends on the size of the vertical load and the type of soil, as well as its compaction [5].

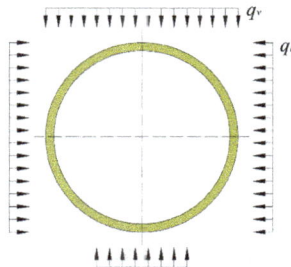

*Fig. 2. Diagram of loads impacting the pipe embedded in the trench*

Fig. 2. shows the distribution of pressures impacting the pipe embedded in the soil, assuming that the pipe is rigid or semi-rigid (relatively inelastic, e.g. a polyethylene pipe).

According to the diagram, the following load components act on the pipe:
1) vertical earth pressure $q_v$:

$$q_v = p_v + p_t,\qquad(3.1)$$

where:
$p_v$ – dead load (soil load),
$p_t$ – live load (traffic load);

Taking the theory of embankment for calculations, the formula for dead load assumes the following form:

$$p_v = \frac{5}{3}(\gamma \cdot H),\qquad(3.2)$$

where:
$\gamma$ – unit weight of bedding material and initial backfill (for sand $\gamma$=19.5kN/m$^3$);
$H$ – height of soil cover over top of pipe (acc. to Fig. 1.1, H=0.9 m),

therefore:

$$p_v = \frac{5}{3} \cdot 19.5 \cdot 0.9 = 29 kPa\qquad(3.3)$$

Depending on the embedment depth, live load is calculated based on PN-85/S-10030 standard. Assuming that the embedment depth is 0.9 m, live load would equal the following:

$$p_t = 52 kPa\qquad(3.4)$$

Therefore the total vertical earth pressure will equal:

$$q_v = 29 + 52 = 81 kPa\qquad(3.5)$$

2) horizontal earth pressure $q_h$:

$$q_h = q_v \cdot K_0,\qquad(3.6)$$

where:
$K_0$ – coefficient of active earth pressure at rest.
horizontal earth pressure is calculated with $K_0$ equal 0.5,
Therefore:

$$q_h = 74 \cdot 0.5 = 37 kPa.\qquad(3.7)$$

**Manufacture of test samples**
The test samples were manufactured out of three types of composite by using continuous filaments in the form of roving, wound on a suitable core. The exact properties of glass fibres [8], basalt fibres [9] and carbon fibres [10] are given in Table 1.

Epolam 5015 epoxy resin and AXSON Epolam 2016 hardener were used in the composite matrix. The properties of the cured resin are given in Table 2 [11].

The test samples were manufactured by winding the fibre strands around the core. A specially prepared core was mounted to the winder handle, on which fibre strands were being wound. The core had been constructed in such a way that it was possible to fabricate several samples as part of a single manufacturing process. The core was fitted with sliding steel rings which formed a plane of resistance for the manufactured samples. The fibre was unwound from a roving spool, and then supersaturated in a tub with liquid resin and then goes through a supersaturation system.

*Table 1. Physical and mechanical parameters of fibres used*

| Property | ER 3005 (Krosglass) fibreglass | BCF 13-1200-KV12 (Basfibre) Basalt fibre | UTS 5631 12K (TohoTenax) Carbon fibres |
|---|---|---|---|
| Flexural modulus [GPa] | 73 | 85 | 240 |
| Tensile strength [MPa] | 3400 | 2900 | 4800 |
| Poisson ratio [–] | 0.21 | 0.26 | 0.285 |
| Elongation at break [%] | 3.5 | 3.1 | 1.8 |
| Density [g/cm$^3$] | 2.55 | 2.75 | 1.79 |
| Linear density [tex] | 1200±7% | 1200 | 800 |
| Monofilament diameter [μm] | 10÷15 | 13 | 6.9 |

*Table 2. Properties of Epolam 5015 (Axson) resin*

| Property | Value |
|---|---|
| Flexural modulus [GPa] | 2.9 |
| Tensile strength [MPa] | 73 |
| Elongation at break [%] | 7 |
| Poisson ratio [–] | 0.35 |
| Hardness [Shore D15] | 84 |
| Mixing ratio [by weight] | 32 |
| Density at 25°C [g/cm3] | 1.12÷1.16 |
| Pot life (on 500 g ) at 25°C [min.] | 360÷450 |
| Glass transition temperature [°C] | 82 |
| Brookfield viscosity at 25°C [mPa.s] | 400÷500 |

Five rings were manufactured from each composite for the research purposes. The geometrical parameters of the samples taken are shown in Table 3, while Fig. 3 presents the samples prepared for testing.

*Table 3. Geometric parameters of samples adopted for testing*

| No. | Sample material (composite) | Average wall thickness e [mm] | Sample width [mm] |
|---|---|---|---|
| 1. | Epoxy/glass fibre (ES) | 3.91 | 25 |
| 2. | Epoxy/basalt (EB) | 3.90 | 25 |
| 3. | Epoxy/carbon (EW) | 4.46 | 25 |

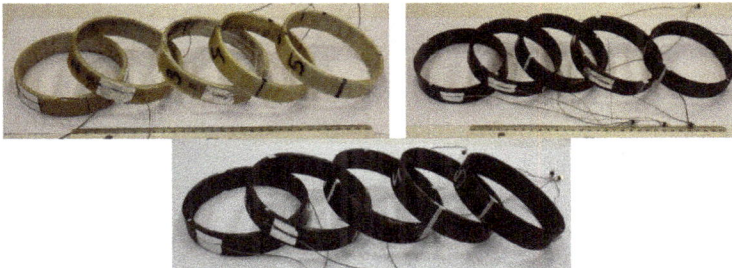

*Fig. 3. Manufactured samples*

Terotechnology XI

Materials Research Proceedings **17** (2020) 191-202

Materials Research Forum LLC

https://doi.org/10.21741/9781644901038-29

The examination of the actual strength properties of the manufactured structures was conducted and included investigation of the fibre volume ratio in the entire composite, as the strength of the composite is mainly determined by the reinforcing fibres contained in it.

The determination of fibre volume ratio was done by analysing the content of the surface occupied by the fibres in relation to the entire surface of the structure in microscopic images, assuming that the fibres have a circular cross-section, they are packed in the matrix in a hexagonal lattice, and that the matrix and fibre reinforcement exhibit isotropic properties. The determination of fibre content was carried out via image analysis.

Fig. 4. shows examples of microscopic images of the composites analysed, while Table 4 provides a list of fibre volume ratios in individual structures.

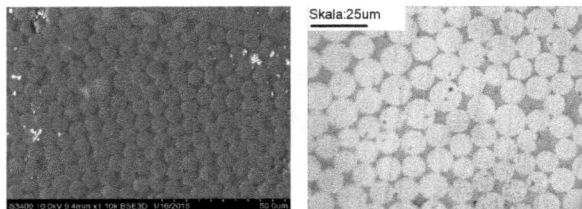

*Fig. 4. Sample photos of the EW composite microstructure taken with magnification of 1100 (left) and ES composite taken with magnification of 1000 (right)*

*Table 4. Volume ratio of the reinforcing fibres within the manufactured composite*

| Composite type | Fibre volume ratio [%] | Resin volume ratio [%] | Void volume ratio [%] |
|---|---|---|---|
| Epoxy/glass fibre composite | 65.5 | 23.6 | 10.9 |
| Epoxy/basalt composite | 63.9 | 22.3 | 13.8 |
| Epoxy/carbon composite | 61.9 | 25.8 | 12.3 |

*Table 5. Effective strength properties of composites*

| Properties | Epoxy/glass fibre composite | Epoxy/basalt composite | Epoxy/carbon composite |
|---|---|---|---|
| $E_1$ [GPa] | 48.8 | 55.4 | 149.7 |
| $E_2 = E_3$ [GPa] | 12.1 | 11.8 | 12.3 |
| $G_{12}$ [GPa] | 4.5 | 4.3 | 4.4 |
| $G_{23}$ [GPa] | 4.3 | 4.2 | 4.3 |
| $v_{12}$ [−] | 0.25 | 0.29 | 0.31 |
| $v_{23}$ [−] | 0.39 | 0.40 | 0.42 |

Then, knowing the strength properties of fibres and resin (Table 1 and Table 2), as well as the fibre volume ratio in the composite (Table 4), the effective strength properties of the composites were determined using the homogenization method. These properties are listed in Table 5. Eshelby's inclusion [3] was used for calculations, assuming that:
− the material consists of a matrix and reinforcing fibres;
− materials are homogeneous and linear elastic

Terotechnology XI                                                          Materials Research Forum LLC
Materials Research Proceedings **17** (2020) 191-202              https://doi.org/10.21741/9781644901038-29

− the fibres have a circular cross-section and are evenly distributed;
− there are no voids or discontinuities in the composite.

## Strength testing

Structural elements produced out of composite materials are subjected to rigorous strength and qualification tests. This approach is dictated on the one hand by the complexity of the material structure and, on the other hand, by high load values during operation. For determining strain and stress distributions in isotropic materials, analytical and numerical computational methods have been developed and are still developing rapidly. They are also used to elaborate models of anisotropic and orthotropic materials, however, reliable determination of the mechanical properties needed for these models is problematic and costly. Therefore, approval of any structures made of composited has to be preceded by thorough experimental testing, in accordance with the relevant directives and standards. Rigidity testing of flexible and rigid pipes analysed in this paper can also be carried out via analytical methods (according to the Scandinavian- and German methods, respectively), while semi-rigid pipes are tested experimentally according to DIN 53769.

Experimental tests are mainly aimed at determining the strength of the structures produced by experimental testing of pipe ring stiffness and ring deformations. Tests of the manufactured pipe samples were carried out on the basis of DIN 53769 standard, according to the diagram shown in Fig. 5.

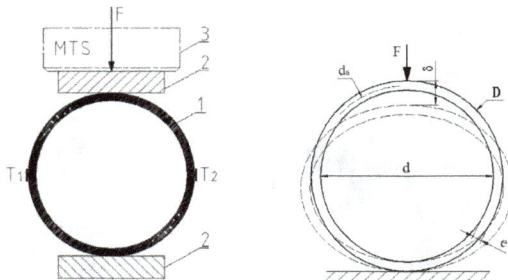

*Fig. 5. Diagram of the rig for testing ring stiffness (left) and diagram of sample deformation after loading (right); 1 – tested sample, 2 – support elements, 3 – universal testing machine crosshead, T1 and T2 – strain gauges*

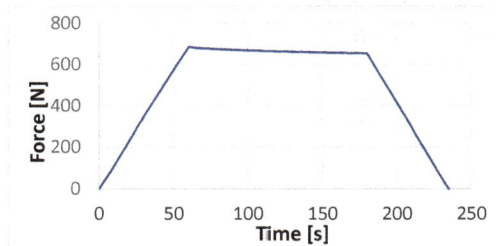

*Fig. 6. An example of how to test a ring sample according to DIN 53769 standard*

The test consists in selecting annular samples from a batch and subjecting them to such a linear load which, within one minute, would cause pipe deformation equal to 3% of its diameter ($\delta/D$ = 3%). After 2 minutes, during which a deformation is being maintained, the bending force and the deflection value are measured. The procedure is repeated twice. An example of the testing process is shown in Fig. 6.

**Experimental determination of pipe ring stiffness.** Test samples were made in the form of rings with an internal diameter of 113 mm and thicknesses given in Table 6, where average thicknesses were read from five samples from each material. Based on the known outside diameter, the required ring deflection was calculated during the tests.

*Table 6. Geometric parameters of test samples*

| No. | Type of sample material (composite) | Average wall thickness e [mm] | Ugięcie próbki $\delta$=0,03D [mm] |
|---|---|---|---|
| 1. | ES | 3.91 | 3.62 |
| 2. | EB | 3.90 | 3.62 |
| 3. | EW | 4.46 | 3.66 |

During the experiment, rings were placed successively between the universal testing machine supports and appropriate crosshead movement was directed. Fig. 7. presents a photograph from the EB composite ring compression test.

*Fig. 7. An example of how to EB composite sample test is tested. Strain gauge is visible*

The movement of the universal testing machine crosshead caused compression of the rings, while due to the rigidity of the rings which prevented its deformation of the support elements a reaction was set off, which was measured at each load test. Two measurements were made for each sample. Table 7 presents the test results for all the materials.

*Table 7. Results from ring stiffness tests for ES, EB and EW samples*

| | Measurement No. | ES composite | EB composite | EW composite |
|---|---|---|---|---|
| Reaction value F [N] | 1 | 754.7 | 580.4 | 2049.8 |
| | 1.1 | 703.4 | 592.1 | 2081.3 |
| | 2 | 756.6 | 577.8 | 1772.0 |
| | 2.1 | 792.6 | 604.2 | 1588.8 |
| | 3 | 555.9 | 569.4 | 1791.4 |
| | 3.1 | 583.2 | 593.4 | 1821.7 |
| | 4 | 703.4 | 633.9 | 1557.2 |
| | 4.1 | 731.8 | 625.1 | 1414.9 |
| | 5 | 735.2 | 680.5 | 1747.9 |
| | 5.1 | 754.2 | 686.0 | 1760.4 |
| | Average | **707.1** | **614.3** | **1758.5** |
| | Standard deviation | **77.4** | **41.6** | **206.0** |
| | Standard deviation [%] | **10.9** | **6.8** | **11.7** |

Ring stiffness is defined as its resistance to peripheral deflection as a result of dividing the force acting on the sample by the length of the tested sample and the deflection [7].

$$S = \frac{F \cdot f}{l \cdot \delta} \tag{5.1}$$

where:
F – force [N],
l – sample length [m];
f – deflection coefficient of the pipe deformed as the result of its ovalization, as determined from the following formula:

$$f = 10^{-5}\left(1860 + 2500\frac{\delta}{d_s}\right) \tag{5.2}$$

From the data obtained and summarized in Table 5.2, and on the basis of dependence 5.1, ring stiffness of the manufactured pipes was determined and summarized in Table 8.

*Table 8. Experimentally determined ring stiffness of manufactured samples*

| No. | Pipe material | Reaction force F [N] | External diam. of the pipe D [m] | Pipe length l [m] | Ring stiffness S [N/m²] |
|---|---|---|---|---|---|
| 1. | ES | 707.1 | 120.82 | 0.025 | 6 194 |
| 2. | EB | 614.3 | 120.80 | 0.025 | 5 381 |
| 3. | EW | 1758.5 | 121.92 | 0.025 | 15 329 |

The analysis of the obtained results reveals that for samples manufactured of differing materials but with similar ring wall thickness, different reaction values were obtained, which is obviously related to the strength properties of the materials from which the samples were fabricated.

**Experimental determination of pipe ring stiffness.** During the ring stiffness test, deformation measurements were also taken for each ring at two points on opposite sides of the ring (Fig. 6). The graph (Fig. 8) shows an example distribution of strain values for a sample from ES composite at a full load cycle.

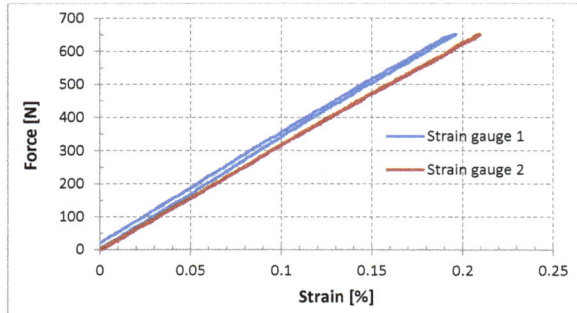

*Fig. 8. Distribution of ring deformation values for a ring made out of ES composite*

The measured maximum values of ring deformations from the measurements taken are summarized in Table 9. The mean value and standard deviation were also determined, which provides the opportunity to compare the results obtained for different materials quickly.

*Table 9. Ring deformation results for a rings made out of ES, EB and EW composites*

| | Measurement No. | Composite ES | | Composite EB | | Composite EW | |
|---|---|---|---|---|---|---|---|
| | | T1 | T2 | T1 | T2 | T1 | T2 |
| Deformation ε [‰] | 1.0 | 1.82 | 1.61 | 1.93 | 1.97 | 2.27 | 2.64 |
| | 1.1 | 1.88 | 1.68 | 1.98 | 2.01 | 2.31 | 2.68 |
| | 2.0 | 2.12 | 1.63 | 1.86 | 1.89 | 2.04 | 2.10 |
| | 2.1 | 2.25 | 1.75 | 1.98 | 2.01 | 1.96 | 1.99 |
| | 3.0 | 1.74 | 1.51 | 1.92 | 1.99 | 2.06 | 2.26 |
| | 3.1 | 1.85 | 1.61 | 2.01 | 2.08 | 2.10 | 2.30 |
| | 4.0 | 1.96 | 2.09 | 1.96 | 2.14 | 1.84 | 1.63 |
| | 4.1 | 2.07 | 2.22 | 1.97 | 2.13 | 1.89 | 1.56 |
| | 5.0 | 2.18 | 2.12 | 2.08 | 2,23 | 2,04 | 1,75 |
| | 5.1 | 2.22 | 2.16 | 2.09 | 2,25 | 2,07 | 1,78 |
| | Average | 1.92 | | 2.02 | | 2.06 | |
| | Standard deviation | 0.24 (12.7%) | | 0.11 (5.2%) | | 0.29 (14.2%) | |

Comparing the results of the tests carried out with the assumptions regarding the design of this type of objects, which determine the maximum peripheral deformation to be at the level of 0.5% (5 ‰), the results presented in Table 5.4 indicate values almost 2.5 times smaller. The analysis of the obtained values points to the fact that the smallest dispersion of the results is characteristic of basalt fibres. It can be noted that composites made from carbon fibres, which exceed the strength properties of both the glass- and basalt fibres by almost three times, exhibited a similar level of ring deformation.

**Application of the test results**

The pipe embedded in the ground works under conditions of internal load, due to gas pressure (e.g. CNG), and external load, resulting from soil pressure and road traffic. An extremely unfavourable situation occurs when the pipe is temporarily exposed to only one of the loads listed (e.g. when testing the pressure of the pipe on the surface or when laying the pipeline without introducing operating pressure or during maintenance works). For some comparison, Table 10 presents maximum allowable loads relative to the weight and price for 1 m of pipe from materials analysed in the paper.

*Table 10. Calculation results for manufactured pipes*

|  | Composite ES | Composite EB | Composite EW |
|---|---|---|---|
| Internal diameter d [mm] | 113 | 113 | 113 |
| Wall thickness e [mm] | 3.91 | 3.90 | 4.46 |
| Tensile strength $R_m$ [MPa]* | 2244 | 1879 | 2990 |
| Safety factor n [–]** | 3.65 | 3.65 | 2.35 |
| Allowable stress $R_m/n$ [MPa] | 615 | 514 | 1272 |
| Young modulus $E_1$ and $E_2$ [GPa] | 48.8 12.1 | 55.4 11.8 | 149.7 12.3 |
| Density [kg/m3] | 1944 | 2016 | 1407 |
| Maximum internal pressure $p_0$ [bar]*** | 440 | 368 | 910 |
| Traffic load per axis [kN] | ~150 | ~130 | ~370 |
| Min. embedment depth [m] | 0.5 | 0.5 | 0.5 |
| Mass/1m [kg] | 2.85 | 2.95 | 2.36 |
| Price/1m [zł] | ~40 | ~50 | ~134 |

*determined basing on mixture theory;
**item [1];
***calculated basing on Lamé's equation

**Summary**

1. For all composites, the packing of fibres in the matrix is at a high level; the values of 66% for ES, 64% for EB and 62% for ES are almost limit values (e.g. the highest tensile strength for ES oscillates at about 72% fibre content in the composite).
2. Due to the high fibre content in the composites and their unidirectional arrangement in the matrix, high strength parameters of the structures produced were obtained.
3. Tests of the ring stiffness of the obtained structures prove that circumferential winding positively the said value positively. Typical composite pipes exhibit ring stiffness of $5 \div 10$ kN /$m^2$. The results within the range of of $5.4 \div 15.3$ kN/$m^2$ were obtained.
4. Comparing the results of the strength tests carried out with the assumptions regarding the design of semi-rigid pipes, which determine the maximum ring deformation at the level of 0.5%, the values were almost 2.5 times smaller.

Terotechnology XI                                                                                          Materials Research Forum LLC
Materials Research Proceedings **17** (2020) 191-202                          https://doi.org/10.21741/9781644901038-29

## References

[1] Bełzowski A., Zasady doboru współczynników bezpieczeństwa konstrukcji z materiałów kompozytowych, Kompozyty, Tom 4, nr 12, str. 396–403, Wydawnictwo Politechniki Częstochowskiej, 2004.

[2] Boczkowska A., Kapuściński J., Lindemann Z., Perzyk-Witemberg D., Wojciechowski S., Kompozyty, Wydanie II zmienione, Oficyna Wydawnicza Politechniki Warszawskiej, Warszawa, 2003.

[3] Eshelby J.D., The Determination of the Elastic Field of an Ellipsoidal Inclusion and Related Problems, Proceedings of the Royal Society, vol. 241, pp. 376–396, 1957. https://doi.org/10.1098/rspa.1957.0133

[4] Madryas C., Kolonko A., Wysocki L., Konstrukcje przewodów kanalizacyjnych, Oficyna Wydawnicza Politechniki Wrocławskiej, Wrocław, 2002.

[5] Marzejon K., Różnice w projektowaniu sieci z tworzyw sztucznych w porównaniu z sieciami z materiałów tradycyjnych. http://www.docplayer.pl

[6] Walczak K., Wpływ parametrów technologicznych przetwórstwa na własności rur ciśnieniowych poliestrowo-szklanych, Rozprawa doktorska, Politechnika Śląska, Gliwice, 1975.

[7] Information on www.hobas.pl.

[8] Information on www.krosglass.pl.

[9] Information on www.basfibre.com.

[10] Information on www.tohotenax-eu.com.

[11] Information on www.fatol.nl.

Terotechnology XI
Materials Research Proceedings 17 (2020) 203-210

Materials Research Forum LLC
https://doi.org/10.21741/9781644901038-30

# Analysis of the Diagnostic Process of Castings used in Automotive

CZERWIŃSKA Karolina[1,a*], PACANA Andrzej[2,b] and ULEWICZ Robert[3,c]

[1] Rzeszów University of Technology, Aleja Powstańców Warszawy 12, 35-959 Rzeszów, Poland

[2] Rzeszów University of Technology, Aleja Powstańców Warszawy 12, 35-959 Rzeszów, Poland

[3] Czestochowa University of Technology, J.H. Dabrowskiego 69, 42-201 Czestochowa, Poland

[a*] k.czerwinska@prz.edu.pl, [b] app@prz.edu.pl, [c] robert.ulewicz@pcz.pl

**Keywords:** Eddy Current Testing, Penetration Testing, Ishikawa Diagram, Piston Combustion Chamber, Quality Control

**Abstract.** The durability of aluminum pistons determines the resistance of the combustion chamber of a piston (bottom) to pressure and temperature. For this reason, the condition of a combustion chamber surface must be monitored scrupulously and any incompatibilities corrected if necessary. The paper uses the eddy current method and the penetration test method for surface quality control of the combustion chamber of diesel engine pistons used in light vehicles. The aim of the study was to determine the sources of the most acute non-compliance detected by the study, using traditional quality management methods. Ultimately, the aim of the analysis was to reduce the number of non-compliant products or to eliminate them completely.

## Introduction

Internal combustion engines are the primary source of vehicle drive. Despite intensive work related to the improvement of alternative sources (electric motors, fuel cells), none of them is at a level of development that would enable them to compete with combustion engines (referring to aspects of their operation such as versatility, ease of use taking into account the universality of fuel/energy, traction characteristics, operating costs, spare parts, service network or vehicle users, habits) [1, 2]. Piston with piston ring package and cylinder is one of the most important kinematic nodes of combustion engine. The piston-crankshaft system of a combustion engine is a unit that is exposed to extreme heat and mechanical loads. It is estimated that the resistance, as a result of friction of pistons and piston rings against cylinder smoothness, reaches from 50÷65% of all mechanical losses occurring in the engine [3, 4]. Aspects such as reduction of operating and production costs, requirement for reliability, constant improvement of travel standards, as well as care for the environment affect the need to work on new solutions in the area of construction, production and operation of combustion engines [5, 6]. Both the cooperation of various components of the piston-cylinder group are key to the operation of the engine efficiency and its durability [7, 8], quality of engine components, which determines the need to eliminate non-compliance products subsurface and surface [9, 10]. As part of the diagnostic process in foundry engineering, a trend can be observed in the implementation of comprehensive testing methods [11, 12], in which castings are controlled using several non-destructive testing methods. Quality control with the use of eddy current, penetration, ultrasonic, radiographic or magnetic methods is applied at various stages of the production process.

Continuous progress in the development of research facilities enables high-precision measurements to be made with more information at the same time [13, 14] to determine the

Terotechnology XI                                                    Materials Research Forum LLC
Materials Research Proceedings 17 (2020) 203-210          https://doi.org/10.21741/9781644901038-30

technical condition of equipment and, on this basis, to enable action to be taken to improve its durability, reliability and operational efficiency [14].

The approach described below may be applied in analogous issues of quality control and failures detection e.g. logistics in SMES [15], flow elements in biotechnological installations [16, 17] or heavy-duty actuators [18, 19]. It may also be interesting in the case of corrosion protection [20] and the application of protective coatings [21-23] and their laser modification [24], where the physicochemical processes are highly unstable. Also in the analysis of microstructures, carried out with the use of image analysis techniques [25, 26], such an approach can be very useful.

## Methodology of research

In order to diagnose the condition of the casting surface, two non-destructive testing methods were used, i. e. the eddy current and penetration method. Both methods used are non-destructive tests - a group of test methods that provide information on the properties of the material of an object without affecting its surface or structural properties [27]. The priority objective of non-destructive testing is to detect and assess non-conformity supremacy of material. The use of non-destructive studies mainly justifies safety considerations and the economic aspect of the occurrence of unforeseen failures [28, 29].

The centrifugal current method is a surface method. Electromagnetic induction consisting of inducing current in a closed electrical circuit as a result of the action of a variable magnetic field is a fundamental phenomenon that is used in wirocurrent studies. Inducted in the test material, centrifugal currents produce their own magnetic field, which is directed in accordance with Lenz rule, to the opposite to the inciting field of [30]. The intensity of the magnetic field produced by the operation of centrifugal currents depends on the electromagnetic properties of the product area checked (magnetic permeability relative and electrical conductivity appropriate). All changes in the analysed material, such as e. g: change of structure, change of hardness, discontinuity, affect the value of electromagnetic parameters, and thus the value of eddy current intensity and induced magnetic field. Diagnostics of the values of electromagnetic field changes and amplitude as well as phase shift of voltage and intensity create the possibility to assess the condition of the examined product area [31, 32].

Peetric studies use the phenomenon of capillary, that is, penetration of liquids into capillaries. An important role here is played by the moisture of the material, the surface voltage of the liquid and the width of the slot (the continuity of the material). When the penetrant penetrates the slot, its excess is removed on the surface of the object and the residue of the penetrant which is located in the slot is pulled to the surface by means of a developer. Most often, colour (red) or fluorescent penetrants are used for penetration tests [33]. Penetration tests enable detection of incompatibilities emerging to the surface such as narrow-slit cracks, open surface defects, flat and other surface defects, no melt, through defects (leaks), non-metallic surface inclusions, folds, pores, porosity and rolling. By means of this method it is possible to detect differently-oriented open surface discontinuities, i. e. discontinuities not drawn and not filled with contaminants [34-36].

Terotechnology XI
Materials Research Proceedings **17** (2020) 203-210

Materials Research Forum LLC
https://doi.org/10.21741/9781644901038-30

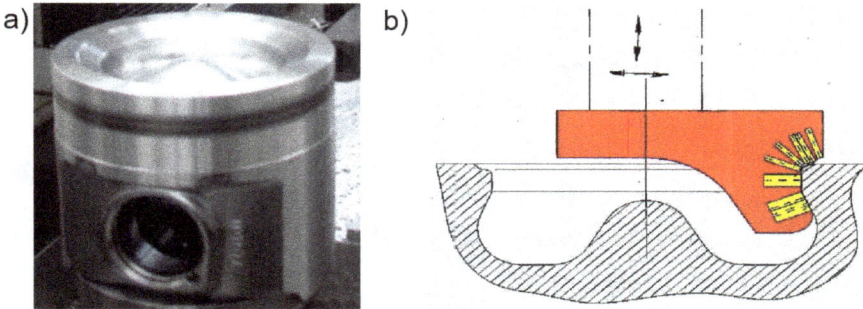

*Figure 1.* Subject of research - (a) model of piston used in diesel passenger cars,
(b) area subject to research – combustion chamber

## Analysis

The aim of the research was to diagnose the condition of a combustion chamber surface in a piston used in a diesel engine in between-operational quality control, by means of eddy current and penetration tests. The research was to determine the reasons for the occurrence of incompatibility of castings and to propose remedial actions that could ultimately contribute to the reduction of the number of non-compliant castings in the combustion chamber surface in the piston used in the diesel engine.

The conducted research concerned batches of products made in the 2nd and 3rd quarter of 2019 in one of the production companies located in the southern part of Poland. Quality control was performed in accordance with the internal procedure of the company according to each production order.

In order to assess the possibility of detecting internal inconsistencies in the material of the product, experimental tests were conducted. A piston (Fig.1a.) designed for a Toyota diesel engine for use in passenger cars was the subject of the study. The piston combustion chamber (Fig.1b).

Pistons are cast from B2 alloy (designation functioning in the company), which is a eutectic aluminum and silicon alloy designed for the production of petrol and diesel pistons used in light vehicles. B2 alloy has no international or national equivalents.

## Results

The obtained results of tests carried out with the use of eddy current method in the combustion chamber of the diesel engine piston are presented in Figure 2.

The result of the tests showed the presence of an unacceptable material discontinuity in the combustion chamber of the piston. As a result of the in-depth inspection and confirmation of the presence of non-compliances, a penetration survey of the pig bottom in the area was carried out to determine whether any non-compliances existed. The result of the penetration test is shown in Fig. 3.

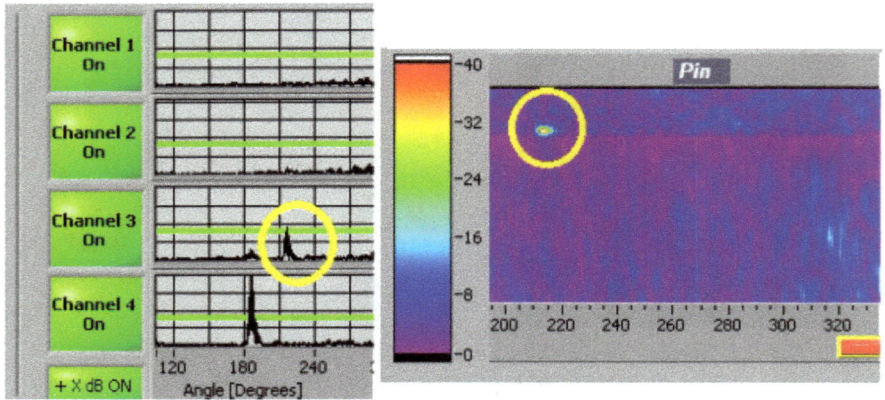

***Figure 2.*** *The result of the eddy current testing of the combustion chamber of the piston with the indication of the detected, unacceptable surface discontinuity*

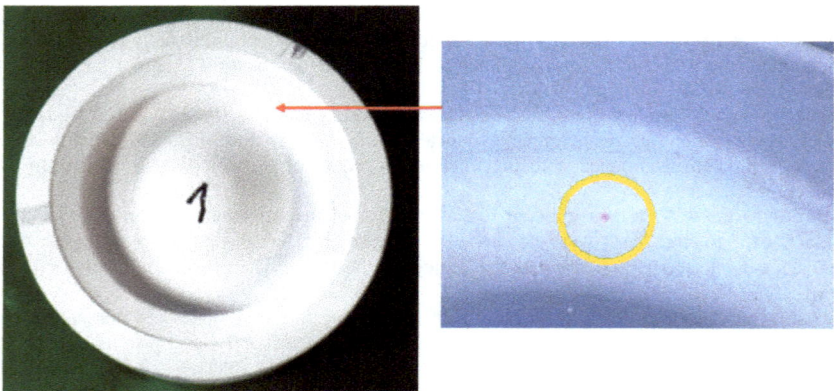

***Figure 3.*** *Piston combustion chamber research result - view after color luminescence*

Penetration tests confirmed the presence of incompatibilities in the combustion chamber. Metallographic surveys were carried out to deepen the analysis from the area of non-compliance (Fig. 4).

The occurrence of observed discontinuity - near-surface material discontinuity (fracture) of the chamber base results in a qualitative disqualification of the piston. A brainstorming session and an Ishikawa diagram were the methods used in order to identify the root cause of the discrepancy. The working group consisted of the following employees: quality control manager, quality control employee and non-destructive testing specialist. The brainstorming session was used to analyze the cause of non-compliance in the product. Potential causes of the analyzed (most severe for the company) non-compliance were identified on the Ishikawa diagram in the 6M system (Fig. 5).

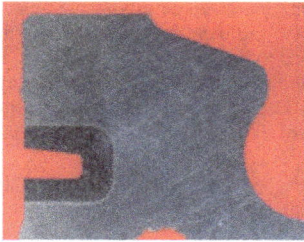

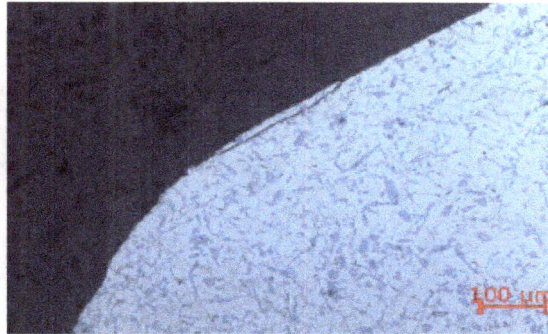

**Figure 4.** *The result of metallographic research - a near-surface material*

Figure 5 presents factors which influence the occurrence of one of the most important piston inconsistencies for the company – a near-surface material discontinuity in the combustion chamber. The most important factor influencing the occurrence of incompatibilities in the discussed series of products was distinguished in the scope of the method. In this group, the inadequate removal of castings from the mould was the most important.

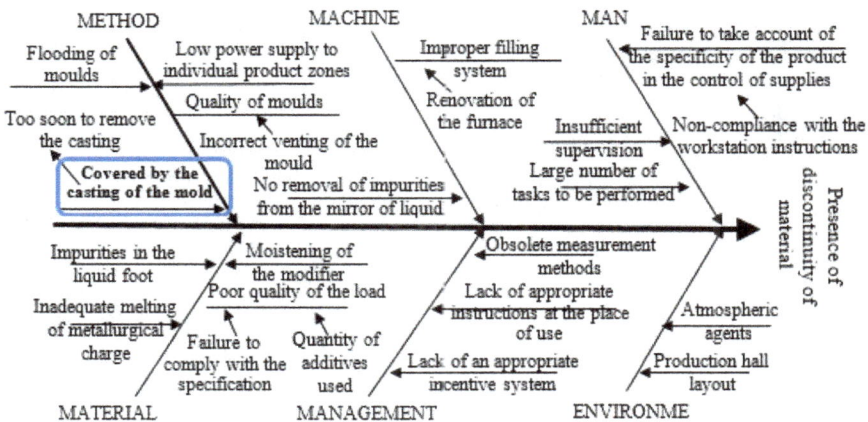

**Figure 5.** *Ishikawa Diagram of Causes of Surface Material Discontinuity in the Piston Combustion Chamber*

Terotechnology XI                                                                                  Materials Research Forum LLC
Materials Research Proceedings 17 (2020) 203-210                        https://doi.org/10.21741/9781644901038-30

**Conclusion**
In this work, diagnostic tests of the surface of the combustion chamber of a diesel engine piston were carried out using the eddy current method and penetration test. The aim of the test was to control the quality of a batch of products, check the usefulness of the control and diagnostic test in the production process and analyze the identified inconsistencies.

The non-destructive eddy current method was used to locate the discontinuity in the area of the combustion chamber – a near-surface material discontinuity, and then, in order to verify the indication, a luminescent test was performed, which confirmed the presence of discontinuities. Due to the fact that the discontinuity indication eliminates the piston, metallographic surveys were carried out on the area where the discrepancy occurred. During the brainstorming session and the Ishikawa diagram, it was diagnosed that the key cause of the discrepancy was inadequate (too fast) removal of castings from the mould.

The applied non-destructive testing methods combined with the quality management method largely complement each other. The proposed combination may be a component of methods supporting quality management processes.

**References**
[1]    A. Lisec, K.Lisec, M. Obrecht. Cost and safety aspects of using electric and hybrid vehicles in local food supply chain, Production Engineering Archives 25 (2019) 35-38. https://doi.org/10.30657/pea.2019.25.06

[2]    A. Pacana, K. Czerwińska, L. Bednarova, Discrepancies analysis of casts of diesel engine piston, Metalurgija 57 (2018) 324-326.

[3]    Motor Service, Pierścienie tłokowe do silników spalinowych, MS Motor Service International GmbH, T.M.S., 2013.

[4]    R. Lewkowicz, P. Piątkowski, T. Kiedrowski, R. Ściegienka, Badania wpływu rozruchu i jakości olejów silnikowych na zużycie pierścieni tłokowych, Autobusy: technika, eksploatacja, systemy transportowe, INW SPATIUM 12 (2016) 1127-1132.

[5]    A. Baberg, M. Freidhager, H. Mergel, K. Schmidt, Aspekte der Kolbenmaterialwahl bei Dieselmotoren, MTZ, 73 (12) (2012) 964–969. https://doi.org/10.1007/s35146-012-0526-8

[6]    E. Tillová, M. Chalupová, L. Kucharíková, Quality control of cylinder head casting, Production Engineering Archives 14 (2017) 3-6. https://doi.org/10.30657/pea.2017.14.01

[7]    M. Blümm, A. Baberg, F. Dörnenburg, D. Leitzmann, Innovative Schaftbeschichtungen für Otto- und Dieselmotorkolben, MTZ 77 (2) (2016) 54-59. https://doi.org/10.1007/s35146-015-0170-1

[8]    E. Wróblewski, A. Iskra, M. Babiak, Minimalizacja zużycia elementów grupy tłokowo-cylindrowej, Autobusy: technika, eksploatacja, systemy transportowe, INW SPATIUM 18 (2017) 1137-1141.

[9]    E.K. Vukelja, I. Duplančič, B. Lela, Continuous roll casting of aluminum alloys– casting parameters analysis, Metalurgija 49 (2) (2010) 115-118.

[10]  R. Ulewicz, F. Novy, The influence of the surface condition on the fatigue properties of structural steel, Journal of the Balkan Tribological Association 22 (2) (2015) 1147-1155.

[11]  D. Malindzak, A. Pacana,  H. Pacaiova, An effective model for the quality of logistics and improvement of environmental protection in a cement plant, Przemysł Chemiczny 96 (9) (2017) 1958-1962.

[12]  A. Pacana, A. Radon-Cholewa, J. Pacana, A  Woźny, The study of stickiness of packaging film by Shainin method, Przemysl Chemiczny 94 (8) (2015) 1334-1336.

[13]  W. Orłowicz, Ultradźwiękowa kontrola jakości odlewów z żeliwa, Archiwum Odlewnictwa 1 (1/2) (2001) 211-226.

[14]  P. Zientek, Metody badań nieniszczących wybranych elementów konstrukcji turbozespołu małej mocy, Napędy i sterowanie 19 (3) (2017) 114-119.

[15]  J. Nowakowska-Grunt, M. Mazur, Effectiveness of logistics processes of SMES in the metal industry. METAL 2016: 25th Int. Conf. on Metallurgy and Materials, Ostrava, TANGER, 2016, 1956-1961.

[16]  E. Skrzypczak-Pietraszek, K. Reiss, P. Zmudzki, J. Pietraszek, J. Enhanced accumulation of harpagide and 8-O-acetyl-harpagide in Melittis melissophyllum L. agitated shoot cultures analyzed by UPLC-MS/MS. PLOS One. 13 (2018) art. e0202556. https://doi.org/10.1371/journal.pone.0202556

[17]  E. Skrzypczak-Pietraszek, K. Piska, J. Pietraszek, Enhanced production of the pharmaceutically important polyphenolic compounds in Vitex agnus castus L. shoot cultures by precursor feeding strategy. Engineering in Life Sciences 18 (2018) 287-297. https://doi.org/10.1002/elsc.201800003

[18]  E. Baussan et al. Neutrino super beam based on a superconducting proton linac. Physical Review Special Topics-Accelerators and Beams 17 (2014) art. 031001. https://doi.org/10.1103/PhysRevSTAB.17.031001

[19]  M. Domagala, H. Momeni, J. Domagala-Fabis, G. Filo, D. Kwiatkowski, Simulation of cavitation erosion in a hydraulic valve. Materials Research Proceedings 5 (2018) 1-6. https://doi.org/10.21741/9781945291814-1

[20]  D. Klimecka-Tatar, G. Pawlowska, K. Radomska, The effect OF Nd12Fe77Co5B6 powder electroless biencapsulation method on atmospheric corrosion of polymer bonded magnetic material. METAL 2014: 23rd Int. Conf. on Metallurgy and Materials. Ostrava, TANGER (2014) 985-990.

[21]  R. Dwornicka, N. Radek, M. Krawczyk, P. Osocha, J. Pobedza, The laser textured surfaces of the silicon carbide analyzed with the bootstrapped tribology model. METAL 2017 26th Int. Conf. on Metallurgy and Materials (2017), Ostrava, Tanger 1252-1257.

[22]  N. Radek, A. Szczotok, A. Gadek-Moszczak, R. Dwornicka, J. Broncek, J. Pietraszek, The impact of laser processing parameters on the properties of electro-spark deposited coatings. Arch. Metall. Mater. 63 (2018) 809-816.

[23]  A. Szczotok, N. Radek, R. Dwornicka, Effect of the induction hardening on microstructures of the selected steels. METAL 2018: 27th Int. Conf. on Metallurgy and Materials, Ostrava, TANGER, 2018, 1264-1269.

Terotechnology XI                                                    Materials Research Forum LLC
Materials Research Proceedings 17 (2020) 203-210        https://doi.org/10.21741/9781644901038-30

[24]  N. Radek, K. Bartkowiak, Laser treatment of electro-spark coatings deposited in the carbon steel substrate with using nanostructured WC-Cu electrodes. Physics Procedia. 39 (2012) 295-301. https://doi.org/10.1016/j.phpro.2012.10.041

[25]  A. Gadek-Moszczak, N. Radek, S. Wronski, J. Tarasiuk, Application the 3D Image Analysis Techniques for Assessment the Quality of Material Surface Layer Before and After Laser Treatment. Advanced Materials Research-Switz. 874 (2014) 133-138. https://doi.org/10.4028/www.scientific.net/AMR.874.133

[26]  L. Wojnar, A. Gadek-Moszczak, J. Pietraszek, On the role of histomorphometric (stereological) microstructure parameters in the prediction of vertebrae compression strength. Image Analysis and Stereology 38 (2019) 63-73. https://doi.org/10.5566/ias.2028

[27]  J. Pietraszek, Fuzzy Regression Compared to Classical Experimental Design in the Case of Flywheel Assembly. In: Rutkowski L., Korytkowski M., Scherer R., Tadeusiewicz R., Zadeh L.A., Zurada J.M. (eds) Artificial Intelligence and Soft Computing ICAISC 2012. Lecture Notes in Computer Science, vol 7267. Berlin, Heidelberg: Springer, 2012, 310-317. https://doi.org/10.1007/978-3-642-29347-4_36

[28]  R. Sikora, T. Chady, Badania nieniszczące metodami elektromagnetycznymi, Przegląd Elektrotechniczny 92 (9) (2016) 1-7. https://doi.org/10.15199/48.2016.09.01

[29]  A. Lewińska-Romicka, Badania nieniszczące. Podstawy defektoskopii. Warszawa, WNT, 2001.

[30]  M. Kochel, D. Pasieka, 2017. Badania nieniszczące konstrukcji spawanych. Szybkobieżne Pojazdy Gąsienicowe 1 (43) (2017) 5-13.

[31]  E. Wyslocka, R. Ulewicz, Magnets: History, The Current State And The Future, METAL 2015 24th Int. Conf. on Metallurgy and Materials, Tanger, Ostrava, 2015, 1680-1686.

[32]  M. Ikonić, B. Barišić, D. Blažević, Identification and Quantification of Raw Materials During Designing of Cast Producing Process, Metalurgija 46 (3) (2007) 179-184.

[33]  L.S. Tian, Y.C. Guo., J.P. Li, J.L. Wang, H.B. Duan, F. Xia, M.X. Liang, Elevated re-aging of a piston aluminum alloy and effect on the microstructure and mechanical properties. Materials Science and Engineering A 738 (2018) 375-379. https://doi.org/10.1016/j.msea.2018.09.078

[34]  J. Czurchryj, H. Papkala, A. Winiowski, Niezgodności w złączach spajanych, Gliwice, Instytut Spawalnictwa, 2005.

[35]  Radek, N., Kurp, P., Pietraszek, J., Laser forming of steel tubes. Technical Transactions 116 (2019) 223-229. https://doi.org/10.4467/2353737XCT.19.015.10055

[36]  Lovejoy D., Penetrant testing. A practical guide. New York, Chapman & Hall, 1991.

Materials Research Forum LLC
https://doi.org/10.21741/9781644901038-31

# Influence of Irradiance Level on the Toxic Gases Emission During the Combustion of Materials used in Floor Constructions in Rolling Stock

MILCZAREK Danuta[1,a*] and DUSZYNSKA-ZAWADA Patrycja[1, b]

[1]Instytut Kolejnictwa (The Railway Research Institute), Materials & Structure Laboratory, 50, Chłopicki Street, 04-275 Warsaw, Poland

[*a]dmilczarek@ikolej.pl, [b]pduszynska-zawada@ikolej.pl

**Keywords:** Fire Tests of Railway Materials, Fire Safety, Fire Behavior, Test Methods, Toxicity, FTIR, Rail Vehicles, Heat Radiation Intensity

**Abstract.** This article presents the results of studies on the influence of irradiance level on the thermal decomposition of a non-metal materials used in floor constructions in rolling stock. This paper contains an analysis of emitted products of combustion and their evolution rate during the potential fire of railway vehicles. The irradiance level of the most intense individual gas emissions was determined, which allowed estimating the safe evacuation time of passengers and staff.

## Introduction

Construction of modern rolling stock vehicles is associated with the need to apply the latest technical and material solutions. This must be connected with necessity to ensure fire safety. The main purpose of fire protection in the event of fire on board in a rail vehicle is to allow passengers and crew to evacuate to the area of maximum safety. On the other hand, the time when passengers are stranded in the vehicle should be as short as possible. One of the worst effects of fire is toxic gases release, which poses a lethal threat to passengers and impedes or even makes the evacuation impossible. Floor composites are an important group of materials (in terms of weight and volume) included in the construction of railway vehicles.. Thus, it is very important to select them properly in the scope of requirements related to fire protection of rolling stock. Therefore, materials that meet the requirements of PN-EN 45545-2+A1:2015 [9] are used for the construction of railway vehicles. In addition to fire parameters characterized by resistance to external sources of fire such as:

- flame propagation (length and speed of flame front travel over a specimen of the material after initiating its combustion) [1],
- heat release rate (amount of heat released in a unit of time during sample combustion) [1-3],
- smoke emission (optical density of air in the environment of a burning specimen) [4],

they must also meet the requirements for toxic gases emissions.

Floor composites consist of a floor substrate (including thermal insulation) and floor covering elements (together with fixing elements and an adhesives used in final use conditions). The most common materials used for floor constructions are composites based on, e.g.:

- flame-retardant polyester resins [5, 6], which are characterized by good mechanical properties and technological processing [8], most often installed in toilets,
- fireproof impregnated plywood with a rubber-cork spacer,
- aluminum cork systems

to which the following floor coverings are glued:

- carpets,
- elastics based on plasticized polyvinyl chloride with the addition of anti-pyrenes,
- elastics based on rubber.

**Materials and methods – Toxic gas emission test**

Floor composites made on the components mentioned above were used for the tests. Their detailed material composition is presented in Table 1. The tests were carried out in accordance with the methodologies described below.

*Table 1. Marks and composition of tests specimens of prepared floor composites.*

| Specimen | Composition of floor composites |
|---|---|
| A36/17 | – covering based on PCV, GRABO, thickness - 2.5 mm,<br>– adhesive for coverings Macroplast,<br>– plywood with a rubber-cork spacer, thickness 18 mm, |
| A44/16 | – carpet, Lantal, thickness - 3 mm,<br>– adhesive for carpets,<br>– aluminum-cork laminate,  thickness -18 mm |
| A117/17 | – rubber covering, Noraplan Mobile Stone, thickness - 2 mm,<br>– adhesive for coverings Macroplast,<br>– plywood with spacer, thickness - 18 mm, |
| A27/19 | – rubber covering, Noraplan Mobile Stone, thickness - 2 mm,<br>– adhesive for coverings Macroplast,<br>– glass - polyester laminate GRRALITHE, thickness - 3 mm, |

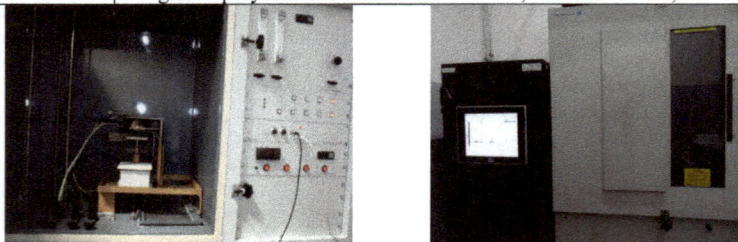

*Fig. 1 Smoke chamber (on the left) and FTIR analyzer conjugated with it*
*for determination of gas toxicity according to PN-EN 45545-2:A1:2015 [1]*
*(source: Materials & Structure Laboratory Instytut Kolejnictwa)*

The commonly used method for determination of toxic gases emitted during a fire uses the Fourier Transform Infrared Spectroscopy FTIR technique [8]. However, the method does not take into account the irradiant level on the thermal decomposition of products during a potential fire. Still, it exerts the main influence on the resulting of combustion products and the rate of emission. For testing, the FTIR analyzer was used in accordance with PN-EN 45545-2+A1:2015 Annex C [9], ISO 19702:2015 [10] and PN-EN 17084:2019 [11] conjugated with a smoke chamber in accordance with the EN ISO 5659-2:2017 [12]. The test method relies on measuring the amount of smoke (by measuring of optical density) formed during the combustion of a specimen subjected to a specific level of thermal radiation, and then determination of the amount and type of toxic gases released.

The evaluation of toxic products that arise in a closed smoke chamber is made by analyzing the concentration of the following 8 gases: carbon (IV) oxide $CO_2$, carbon (II) oxide CO, hydrogen bromide HBr, hydrogen chloride HCl, hydrogen cyanide HCN, hydrogen fluoride HF, nitrogen (IV) oxide $NO_2$, nitrogen (II) oxide NO and sulfur (IV) oxide $SO_2$. The gases are taken with a sampling probe in the 4th and 8th minute of the test during the measurement of the optical density of smoke in the smoke chamber according to PN-EN ISO 5659-2 [12]. The smoke chamber is shown in Fig. 1 (on the left) for determination of optical density and FTIR analyzer conjugated with it (on the right) for determination of the concentration of toxic gases.

**Results — Toxicity testing of floor composites**
The tests results for determination of emission of toxic gases emitted during the combustion of floor composites are summarized in Tables 2-5. Values recorded below the limit of detection are marked as n.o.

***Table 2.*** *Emission of toxic gases during the combustion of the material A36/17*

| Gas | Time at sampling time | Concentration, mg/m$^3$ | | | | |
|---|---|---|---|---|---|---|
| | | $10 \text{ kW/m}^2$ | $20 \text{ kW/m}^2$ | $30 \text{ kW/m}^2$ | $40 \text{ kW/m}^2$ | $50 \text{ kW/m}^2$ |
| $CO_2$ | 4 min. | 756.89 | 845.82 | 903.2 | 1112.4 | 1284.6 |
| | 8 min. | 910.5 | 1246.9 | 1542.6 | 2145.7 | 2525.3 |
| CO | 4 min. | 32.6 | 54.7 | 86.5 | 101.1 | 106.5 |
| | 8 min. | 71.9 | 93.6 | 132.4 | 153.4 | 179.2 |
| $SO_2$ | 4 min. | n.o | 19.6 | 35.6 | 54.3 | 69.2 |
| | 8 min. | 21.3 | 48.8 | 56.4 | 77.8 | 95.7 |
| HCL | 4 min. | 35.6 | 53.6 | 106.9 | 121.4.4 | 146.7. |
| | 8 min. | 102.5 | 135.6 | 184.6 | 206.8 | 225.2 |

***Table 3.*** *Emission of toxic gases during the combustion of the material A44/16*

| Gas | Time at sampling time | Concentration, mg/m$^3$ | | | | |
|---|---|---|---|---|---|---|
| | | $10 \text{ kW/m}^2$ | $20 \text{ kW/m}^2$ | $30 \text{ kW/m}^2$ | $40 \text{ kW/m}^2$ | $50 \text{ kW/m}^2$ |
| $CO_2$ | 4 min. | 716.5 | 1054.2 | 1625.6 | 2154.7 | 2864.3 |
| | 8 min. | 1856.8 | 2256.4 | 3354.4 | 4556.3 | 5325.7 |
| CO | 4 min. | 14.4 | 23 5 | 31.3 | 38.5 | 41.3 |
| | 8 min. | 26.4 | 38 2 | 58.4 | 76.5 | 89.5 |
| $SO_2$ | 4 min. | 19.6 | 35 1 | 54.3 | 72.3 | 88.4 |
| | 8 min. | 36.4 | 51 2 | 98.3 | 121.3 | 143.2 |
| $NO_x$ | 4 min. | n.o. | 11.5. | 16.4 | 22.3 | 32.5 |
| | 8 min. | n.o. | 22 4 | 49.7 | 61.2 | 76.8 |

***Table 4.** Emission of toxic gases during the combustion of the material A117/17*

| Gas | Time at sampling time | Concentration, mg/m$^3$ | | | | |
|---|---|---|---|---|---|---|
| | | 10 kW/m$^2$ | 20 kW/m$^2$ | 30 kW/m$^2$ | 40 kW/m$^2$ | 50 kW/m$^2$ |
| CO$_2$ | 4 min. | 916.5 | 1854.2 | 2869.4 | 3546.2 | 4125.8 |
| | 8 min. | 1546.7 | 2157.3 | 4894.2 | 7256.8 | 8952.9 |
| CO | 4 min. | 19.6 | 32.6 | 43.8 | 55.1 | 63.2 |
| | 8 min. | 31.9 | 43.6 | 82.4 | 115.8 | 136.5 |
| SO$_2$ | 4 min. | 12.1 | 31.4 | 61.4 | 85.6 | 103.8 |
| | 8 min. | 34.9 | 58.6 | 114.6 | 154.6 | 174.9 |
| HCN | 4 min. | n.o. | n.o. | n.o. | n.o. | 24.3 |
| | 8 min. | n.o. | n.o. | n.o. | n.o. | 31.5 |
| NO$_x$ | 4 min. | n.o. | n.o. | 11.2 | 18.6 | 24.7 |
| | 8 min. | 16.5 | 24.7 | 44.8 | 72.7 | 82.6 |

***Table 5.** Emission of toxic gases during the combustion of the material A27/19*

| Gas | Time at sampling time | Concentration, mg/m$^3$ | | | | |
|---|---|---|---|---|---|---|
| | | 10 kW/m$^2$ | 20 kW/m$^2$ | 30 kW/m$^2$ | 40 kW/m$^2$ | 50 kW/m$^2$ |
| CO$_2$ | 4 min. | 1689.4 | 2861.7 | 3629.8 | 4026.4 | 4634.3 |
| | 8 min. | 3517.4 | 6186.8 | 8181.4 | 8752.6 | 9052.1 |
| CO | 4 min. | 21.6 | 43.7 | 59.3 | 62.4 | 70.1 |
| | 8 min. | 44.5 | 80.6 | 115.4 | 127.6 | 138.8 |
| SO$_2$ | 4 min. | 10.8 | 33.7 | 67.6 | 89.9 | 106.5 |
| | 8 min. | 42.1 | 66.4 | 130.6 | 171.1 | 177.6 |
| HCN | 4 min. | n.o. | n.o. | n.o. | 21.4 | 52.6 |
| | 8 min. | n.o. | n.o. | 32.0 | 43.1 | 76.8 |
| NO$_x$ | 4 min. | n.o. | n.o. | 15.7 | 22.6 | 27.4 |
| | 8 min. | 14.6 | 33.8 | 39.1 | 71.4 | 85.1 |

**Analysis – The influence of irradiance level on the emission of toxic gases depending of the composition of floor composites**

The floor composite with PVC lining A36/17 (plasticized polyvinyl chloride) is characterized by a relatively low level of CO$_2$ emission, which comes mainly from a layer of wood material in the compact (plywood with a spacer). The high level of emitted HCL is caused by the thermal decomposition of PVC from the surface layer of the composite. The process of PVC dechlorination begins at low temperatures, therefore this gas was determined at the irradiance level of 10kW/m$^2$. After exceeding the initial temperature of 180°C on the specimen surface, which corresponds to the irradiance level of 30 kW/m$^2$ (according to Fig. 2), increasing HCL emission in 4$^{th}$ min. of the test was observed. The same process, only intensified, took place in the 8$^{th}$ minute of the test. From this irradiance level, an increase in CO emission was also noted. The graph of toxic gases concentrations as a function of irradiance level is presented in Fig. 2.

Terotechnology XI                                              Materials Research Forum LLC
Materials Research Proceedings **17** (2020) 211-217          https://doi.org/10.21741/9781644901038-31

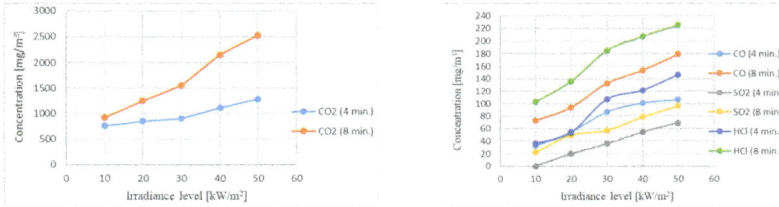

*Fig. 2 Graph of toxic gases concentrations as a function of irradiance level for the floor composite A36/17.*

The floor composite with rubber covering A44/16 (polyamide fleece) has a medium level of $CO_2$ emission related to the thermal decomposition of polyamide from the covering and cellulose from the base and a low level of CO emission. The amount of $CO_2$ increased after exceeding 180°C of the initial temperature on the surface which is a response to the irradiance level of 30 $kW/m^2$. In addition to these gases, nitrogen oxides were also detected which were emitted in quantities allowing their determination when the irradiance level equaled 20 $kW/m^2$. They came from the thermal decomposition of the polyamide as the top layer of the composite.

The graph of toxic gases concentrations as a function of irradiance level is presented in Fig. 3.

The floor composite with rubber covering A117/17 and A27/19 (cross-linked NBR rubber) shows a high level of $CO_2$ emission, especially in the 8th minute of the test for a higher irradiance level on the specimen surface (40 $kW/m^2$ and 50 $kW/m^2$). It is caused by the influence of a layer of wood material in the compact (plywood, pine scantlings).

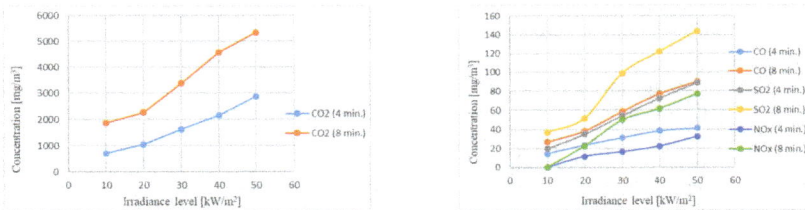

*Fig. 3 Graph of toxic gases concentrations as a function of irradiance level for a floor composite A44/16*

A large amount of $SO_2$ evolved due to the oxidation of sulfide bonds to sulfenic and tiosulfoxylic acids, and then to $SO_2$ to a small extent to $SO_3$ in the boundary layer of the burning elastomer cross-linked with sulfur. The toxicity of emitted gases of the tested composite is also significantly affected by the release of amounts of toxic nitrogen compounds, especially in the 8th minute of the test, at the irradiance level exceeding 30 $kW/m^2$. Particularly toxic HCN was also determined for the highest irradiance level. A similar distribution of toxic gas emissions was observed during the thermal decomposition of a floor composite consisting of the same rubber but glued onto a glass-polyester laminate.

The graph of toxic gases concentrations as a function of irradiance level is presented in Fig. 4 and 5.

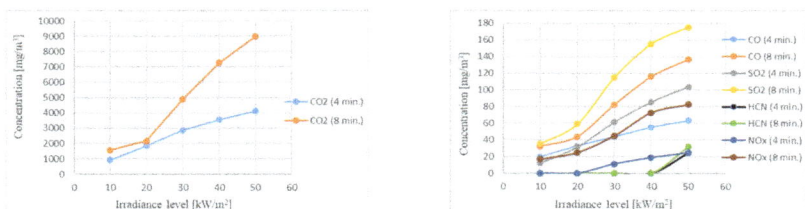

*Fig. 4 Graph of toxic gases concentrations as a function of irradiance level for a floor packet no. A117/17*

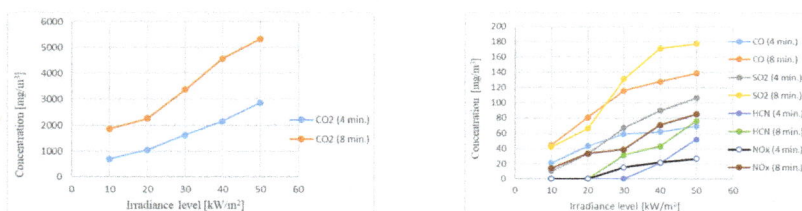

*Fig. 5 Dependencies of toxic gases concentrations as a function of heat radiation intensity for a floor packet no A27/19*

## Summary

The performed tests allowed for determining the influence of irradiant level on the amount and type of gases emitted during combustion. The results of experiments carried out in laboratory conditions were used to identify particularly hazardous substances, which even in small quantities pose a threat to the health of passengers and crew. These substances include, in particular, nitrogen oxides, hydrogen cyanide, sulfur oxide (IV) and carbon oxide (II). In contrast, large amounts of toxic and highly corrosive HCL were recorded when testing a compact floor with PVC coverings. It was shown during the tests that after exceeding the irradiant level of 30 kW/m$^2$, there was an intensive increase in toxic gases emissions. Obtained concentration values increased significantly after exceeding the first four minutes of testing. This confirms that the safe evacuation time in the event of a fire should be as short as possible, not exceeding three minutes according to TSI LOC&PAS [13], taking into account the large amount of non-metallic materials used to build the rolling stock and the associated emissions of toxic compounds for humans.

## References

[1] J. Radziszewska-Wolińska, D. Milczarek, Fire Tests of Non-Metallic Materials for Walls and Ceiling in Rolling Stock, Material Research Proceedings 5 (2018) 90-95. https://doi.org/10.21741/9781945291814-16

[2] J. Radziszewska-Wolińska, A. Kaźmierczak, Fire Properties of Upholstery and Fire Resistance of the Complete Passenger Seat, Material Research Proceedings 5 (2018) 31-36. https://doi.org/10.21741/9781945291814-6

[3] J. Radziszewska-Wolińska, Selection of Components of Upholstery Systems, Material Research Proceedings 5 (2018) 142-147. https://doi.org/10.21741/9781945291814-25

[4] J. Radziszewska-Wolińska, I. Tarka,The Influence of Reinforcing Layers and Varnish Coatings on the Smoke Properties of Laminates Based on Selected Vinyl Ester and Polyester Resins Material Research Proceedings 5 (2018) 210-215. https://doi.org/10.21741/9781945291814-37

[5] D. Riegert, Sposoby modyfikowania właściwości palnych tworzyw sztucznych, Bezpieczeństwo i Technika Pożarnicza 30 (2) (2013) 51-57.

[6] W. Zatorski, K. Sałasińska, Combustibility studies of unsaturated polyester resins modified by nanoparticles, Polimery 11-12 (2016) 815-823. https://doi.org/10.14314/polimery.2016.815

[7] J. Radziszewska-Wolińska, D. Milczarek, Uniepalnienie materiałów niemetalowych a ich właściwości funkcjonalne, TTS Technika Transportu Szynowego 11-12 (2012) 56-59

[8] Z. Kęcki, Podstawy spektrofotometrii molekularnej, PWN, Warszawa, 1998.

[9] PN-EN 45545-2+A1:2015 Railway applications – Fire protection on railway vehicles – Part 2: Requirements for fire behaviour of materials and components

[10]ISO 19702:2015 Toxicity testing of fire effluents. Guidance for analysis of gases and vapours in fire effluents using FTIR gas analysis

[11]PN EN 17084:2019 Railway applications. Fire protection on railway vehicels. Toxicity test of materials and components

[12]PN-EN ISO 5659-2:2017-08 Plastics - Smoke generation - Part 2: Determination of optical density by a single-chamber test

[13]COMMISSION REGULATION (EU) No 1302/2014 of 18 November 2014 concerning a technical specification for interoperability relating to the 'rolling stock — locomotives and passenger rolling stock' subsystem of the rail system in the European Union

Terotechnology XI
Materials Research Proceedings 17 (2020) 218-225

Materials Research Forum LLC
https://doi.org/10.21741/9781644901038-32

# Fire Properties of Railway Rubber Products

RADZISZEWSKA-WOLIŃSKA Jolanta[1, a *] and ŚWIETLIK Aneta[1, b]

[1] Instytut Kolejnictwa (The Railway Research Institute), Materials & Structure Laboratory, 50, Chlopicki Street, 04-275 Warsaw, Poland

[*a]jradziszewska-wolinska@ikolej.pl, [b]aswietlik@ikolej.pl

Keywords: Non-Metallic Materials, Rail Vehicles, EN 45545-2, Fire Safety

**Abstract.** The article discusses the use of rubber elements in rail vehicles and their desired functional properties. Then, the requirements for materials in the field of fire safety as well as fire test results for individual groups of rubber products are presented. Next, the directions and results of modifications of elastomer mixtures for railway elements are discussed.

## Introduction

Rubber is an elastomer made of aliphatic polymer chains, crosslinked in the vulcanization process of natural rubber (obtained from the resin of the *Hevea brasiliensis* tree) or synthetic (polybutadiene and other polyolefins), or mixtures thereof. It was invented and used for the first time in South America by the Maya and Aztecs, and it was brought to Europe in the 15th century by Christopher Columbus. The rubber production method was patented in 1843, while after processing in 1946 the technology for cold vulcanizing of rubber with sulfur chloride, the production of rubber products on an industrial scale was developed. This material has gained great popularity, especially due to its high flexibility, which is characterized by resistance to deformation and vibration damping, as well as mechanical strength. In addition, rubber is characterized by high chemical resistance and good dielectric properties. However, rubber in the strict sense is not resistant to high temperatures and burns giving off black, pungent smoke. At the same time, the properties of particular mixtures strictly depend on the type and proportion of individual components and can be very diverse [19]. Thanks to the above, these materials have been used in many industrial sectors, including railways.

### Rubber products in rail vehicles

Rubber in rolling stock began to be introduced as a replacement for leather and fabrics in the second half of the 20th century. In currently manufactured vehicles, elements made of various types of rubber compounds, depending on the type of rolling stock, constitute about 7-10% of the mass of all non-metallic materials used for its construction and equipment.

The most popular group are longitudinal seals. These are window seals, door connectors, loop handles, panel joints and technical cabinet seals, for which, above all, high mechanical strength and flexibility are required, and for elements used outdoors, resistance to environmental conditions such as rain, snow and large temperature differences from -30°C to + 40°C. EPDM (ethylene propylene) and the increasingly popular silicone rubber and its mixtures are the most commonly used rubber compounds for their production. These mixtures are characterized by high resistance to atmospheric conditions.

Flexible rubber and metal-rubber assemblies which can be found in the chassis are another group of railway products. These include, among others: suspension elements of the bogies, shock absorbers, bumpers, steering pins, air bags for pneumatic suspension joints, tyres as well as various types of rings, washers and sleeves. These products should be characterized above all by very good thermal and mechanical resistance (low permanent deformation under compression) and resistance to oils. For its production nitrile (NBR), natural (NR), styrene butadiene (SBR) mixtures are used.

Rubber is also used for the production of a wide range of hoses used inside and outside vehicles. For example, they are:

- brake hoses, often made of two different external and internal mixtures,
- pressure hoses in compressed air systems, among others for systems for removing excess gravel from tracks, pneumatic control of pantographs and contactors (silicone mixes),
- hoses in cooling systems are usually suction and pressure hoses made of silicone rubber.

All hoses should be highly flexible in a wide range of temperatures. In addition, resistance to pressure and transported media (fuel, oils, hydraulic fluids, water) is necessary.

Membranes of intercommunication gangways constitute an important mass group in the passenger vehicle. These elements should be characterized by good mechanical properties (resistance to loads, stretching and occurrence of vibrations), resistance to changing environmental conditions and high stability of parameters, i.e. resistance to ageing. Earlier they were produced from rollers made of thick EPDM rubber plates. In contrast, new vehicles use lighter skeletal structures covered with rubberized fabrics using silicone compounds.

Another group of rubber products are insulation materials used inside the body shell as insulation of walls (internal vertical surfaces), ceilings (internal horizontal surfaces facing down) and floors (internal horizontal surfaces facing up). In this case, properties such as flexibility, high resistance to water vapor diffusion, low thermal conductivity and low specific weight are necessary. Products made of foam rubber based on synthetic rubber, e.g. Armaflex, are used.

As the last group, cable insulation should be mentioned. They must above all meet the requirements in the field of electrical parameters (electric permeability, dielectric strength), but also have to be resistant to mechanical factors, environmental conditions andoils. However, currently this type of insulation is being abandoned, because, among others, cross-linked polyolefins, plastic and EVA (ethylene vinyl acetate copolymer) are more and more popular.

**Fire properties requirements**

In order to ensure the required level of fire safety for rail vehicles, non-metallic materials including rubber products should also, in addition to the functional properties described above, meet the requirements for fire parameters. These requirements are contained in the TSI (Technical Specifications for Interoperability) [6] and harmonized with European standards. The series of standards EN 45545 [7- 13] discuss all aspects that should be taken into account when designing rail vehicles to minimize the risk of fire, and in the event thereof - minimizing the spread of fire and smoke to the inside and outside of the vehicle. Part 2 of the above mentioned standards [8] contains requirements for materials. Table 1 below presents the groups of requirements (R) assigned to the product groups described earlier, while Table 2 below shows the required values of the applicable parameters. At the same time, these requirements vary depending on the level of hazard (HL) for vehicles, resulting from its operational category and the railway infrastructure in which it moves. The most rigorous requirements are assigned to the HL3 level, i.e. for vehicles that drive on underground sections, tunnels and/or elevated structures, where the available evacuation time is very short.

It should be noted that according to the previously applicable national standards (including PN standards), many rubber products that did not have direct contact with passengers did not require testing for fire performance. Only EN 45545-2 [8] introduced the obligation to test all elements located in the chassis, cabinets and technical compartments. Unfortunately, it is not always possible to meet applicable requirements,especially for rubber. Therefore, the standard

[8] (in p. 4.7) allows for the use of a material that does not meet the requirements in terms of fire protection due to its other properties necessary for the proper functioning of the vehicle. This provision applies to:

- the situation when, prior to the conclusion of the contract, there is no product available on the market in the given group that meets the requirements specified in p. 4.1 of the standard [8],
- use of a limited amount of material and requires a risk analysis.

*Table 1. Groups of R requirements for rubber products according to EN 45545-2 [8]*

| Product No | Product Name | Set of requirement |
|---|---|---|
| IN16 | Interior seals | R22 |
| EX12 | Exterior seals | R23 |
| M1 | Flexible metal/rubber units | R9 |
| M2 | Hoses - Interior | R22 |
| M3 | Hoses - Exterior | R23 |
| EX9 | Air bags for pneumatic suspension | R9 |
| EX11 | Tyres | R9 |
| EX7 | Exterior surfaces of gangways | R7 |
| IN1A | Interior vertical surfaces | R1 |
| IN1B | Interior horizontal downward- facing surfaces | R1 |
| IN1C | Interior horizontal upwards- facing surfaces | R22 |
| EL2 | Cable containment (linear product) – circular crossection | Depending on the application and circuit: R22, R23, R6 or R9 |
| | Cable containment (linear product) – rectangular crossection | Depending on the application and circuit: R1, R6, R7, R17, R9, R22 or R23. |

*Table 2. Sets of requirements for rubber products according to EN 45545-2 [8]*

| Set of requirement | Test method | Parameter Unit | HL1 | HL2 | HL3 |
|---|---|---|---|---|---|
| R1 | ISO 5658-2 | CFE [kW/m$^2$] | $\geq 20$ | $\geq 20$ | $\geq 20$ |
| | ISO 5660-1, 50 [kW/m$^2$] | MARHE [kW/m$^2$] | - | $\leq 90$ | $\leq 60$ |
| | EN ISO 5659-2, 50 [kW/m$^2$] | D$_{s4}$ [-] | $\leq 600$ | $\leq 300$ | $\leq 150$ |
| | | VOF$_4$ [min] | $\leq 1200$ | $\leq 600$ | $\leq 300$ |
| | EN ISO 5659-2, 50 [kW/m$^2$] | CIT $_G$ [-] | $\leq 1.2$ | $\leq 0.9$ | $\leq 0.75$ |
| R6 | ISO 5660-1, 50 [kW/m$^2$] | MARHE [kW/m$^2$] | $\leq 90$ | $\leq 90$ | $\leq 60$ |
| | EN ISO 5659-2, 50 [kW/m$^2$] | D$_{s4}$ [-] | $\leq 600$ | $\leq 300$ | $\leq 150$ |
| | | VOF$_4$ [min] | $\leq 1200$ | $\leq 600$ | $\leq 300$ |
| | EN ISO 5659-2, 50 [kW/m$^2$] | CIT $_G$ [-] | $\leq 1.2$ | $\leq 0.9$ | $\leq 0.75$ |
| R7 | ISO 5658-2 | CFE [kW/m$^2$] | $\geq 20$ | $\geq 20$ | $\geq 20$ |
| | EN ISO 5659-2, 50 [kW/m$^2$] | MARHE [kW/m$^2$] | - | $\leq 90$ | $\leq 60$ |
| | EN ISO 5659-2, 50 [kW/m$^2$] | D$_{smax}$ [-] | - | $\leq 600$ | $\leq 300$ |
| | | CIT G [-] | - | $\leq 1.8$ | $\leq 1.5$ |
| R9 | ISO 5660-1, 25 [kW/m$^2$] | MARHE [kW/m$^2$] | $\leq 90$ | $\leq 90$ | $\leq 60$ |
| | EN ISO 5659-2, 25 [kW/m$^2$] | D$_{smax}$ [-] | - | $\leq 600$ | $\leq 300$ |
| | | CIT G [-] | - | $\leq 1.8$ | $\leq 1.5$ |
| R17 | ISO 5658-2 | CFE [kW/m$^2$] | $\geq 13$ | $\geq 13$ | $\geq 13$ |
| | ISO 5660-1, 50 [kW/m$^2$] | MARHE [kW/m$^2$] | - | $\leq 90$ | $\leq 60$ |
| | EN ISO 5659-2, 50 [kW/m$^2$] | D$_{smax}$ [-] | - | $\leq 600$ | $\leq 300$ |
| | | CIT $_G$ [-] | - | $\leq 1.8$ | $\leq 1.5$ |
| R22 | EN ISO 4589-2 | Zawartość tlenu [%] | $\geq 28$ | $\geq 28$ | $\geq 32$ |
| | EN ISO 5659-2, 25 [kW/m$^2$] | D$_{smax}$ [-] | $\leq 600$ | $\leq 300$ | $\leq 150$ |
| | NF X 70-100-1 oraz 2 600 [°C] | CIT $_{NLP}$ [-] | $\leq 1.2$ | $\leq 0.9$ | $\leq 0.75$ |
| R23 | EN ISO 4589-2 | Oxygen Index [%] | $\geq 28$ | $\geq 28$ | $\geq 32$ |
| | EN ISO 5659-2, 25 [kW/m$^2$] | D$_{smax}$ [-] | - | $\leq 600$ | $\leq 300$ |
| | NF X 70-100-1 and 2 600 [°C] | CIT $_{NLP}$ [-] | - | $\leq 1.8$ | $\leq 1.5$ |

## Results of laboratory fire tests

Due to the implementation of EN 45545-2 standard [8], manufacturers of rubber components and rail vehicles began testing their products. The results of laboratory tests carried out at the Instytut Kolejnictwa for rubber materials intended for particular elements are presented in Table 3 below and in charts (Fig. 1-4). At the same time, due to the fact that smoke properties are the most difficult to meet for rubber, tests were started as a rule from this parameter and no other parameters were determined when negative smoke results were obtained. In the case of gaskets, as part of control tests, their scope was limited.

*Table 3. Laboratory test results for selected rubber mixtures carried out in 2015-2019*

| Application | Material | Sample No | CFE, kW/m² ISO 5658-2 | MARHE, kW/m² ISO 5660-1 | $D_{s4}$, EN ISO 5659-2 | $VOF_4$, min EN ISO 5659-2 | $D_{smax}$, EN ISO 5659-2 | $CIT_G$, PN-EN 45545-2, Annex C | CIT NLP NF X 70-100-1 oraz 2 | OI, % EN ISO 4589-2 |
|---|---|---|---|---|---|---|---|---|---|---|
| | | | | | Requirements R22, R23 | | | | | |
| internal and external seals, loop handles and profiled cords | silicone mix | A54/15 | nb | nb | 8.5 | 6.96 | 68.77 | nb | nb | >48.6% |
| | | A33.1/17 | nb | nb | nb | nb | 160.2 | nb | nb | 34.4 |
| | | A33.2/17 | nb | nb | nb | nb | 187.7 | nb | nb | 34.4 |
| | | A34.1/17 | nb | nb | nb | nb | 273.5 | nb | nb | 29.7 |
| | | A34.2/17 | nb | nb | nb | nb | 188.1 | nb | nb | 29.7 |
| | | A35.1/17 | nb | nb | nb | nb | 255.9 | nb | nb | 31.7 |
| | | A35.2/17 | nb | nb | nb | nb | 192.9 | nb | nb | 31.7 |
| | | A86/17 | nb | nb | nb | nb | 257 | nb | nb | 30.5 |
| | | A87/17 | nb | nb | nb | nb | 162.6 | nb | nb | 34.6 |
| | | A126/17 | nb | nb | nb | nb | 162.2 | nb | nb | 36.1 |
| | | A127/17 | nb | nb | nb | nb | 176.1 | nb | nb | 37.3 |
| | | A198/17 | nb | nb | nb | nb | 177.1 | nb | nb | 32.6 |
| | | A199/17 | nb | nb | nb | nb | 230.9 | nb | nb | 31.2 |
| | | A136/17 | nb | nb | nb | nb | 124.9 | nb | nb | 37.0 |
| | | A1/15 | nb | nb | nb | nb | 92.81 | nb | nb | >48.6% |
| | | A2/15 | nb | nb | nb | nb | 92.8 | nb | nb | |
| | rubber EPDM | A159/18 | nb | nb | 288.9 | 528.9 | 295.4 | nb | nb | 26.0 |
| | | A160/18 | nb | nb | 214.3 | 338.8 | 257.1 | nb | nb | 29.9 |
| | | A203/19 | nb | nb | nb | nb | nb | nb | nb | 32.5 |
| | | A204/19 | nb | nb | nb | nb | nb | nb | nb | 33.1 |
| | | A205/19 | nb | nb | nb | nb | nb | nb | nb | 33.6 |
| | | | | | Requirements R1, R7 | | | | | |
| intercommunication gangways | silicone mix on textile | A156/18 | 22.6 | 85.1 | 150.4 | 343.4 | 176.9 | 0.01 | nd | nd |
| | | A5/19 | 28.6 | 87.4 | 155.5 | 190.7 | 413.9 | 0.01 | nd | nd |
| | | A6/19 | 25.3 | 102.6 | 119.8 | 184.2 | 423.2 | 0.01 | nd | nd |
| | | A7/19 | 26.6 | 84.7 | 148.5 | 253.4 | 271.0 | 0.02 | nd | nd |
| | | A8/19 | 26.9 | 85.6 | 185.4 | 347.3 | 361.8 | 0.02 | nd | nd |
| | rubber mix | A40/16 | 16.7 | 82.0 | nb | nb | nb | nb | nd | nd |
| | | A98/16 | 13.3 | 84.3 | nb | nb | nb | nb | nd | nd |
| | | A173/16 | 17.8 | nb | nb | nb | nb | nb | nd | nd |
| | | A21/17 | nb | 80.6 | 484.2 | 1037.3 | 1151.5 | 2.2 | nd | nd |
| | | A144/17 | nb | nb | 512.9 | 1145.3 | 1320.0 | 1.2 | nd | nd |
| | | A47/18 | nb | nb | 499.9 | 931.1 | 741.2 | nb | nd | nd |
| | | A115/19 | 18.1 | 85.0 | 491.5 | 1027.5 | 1100.1 | 2.3 | nd | nd |
| | | | | | Requirements R22, R23 | | | | | |
| brake hoses | two layer of rubber mix | A148/19 | nd | nd | nd | nd | 251 | nd | nb | 47.9 |
| | | A149/19 | nd | nd | nd | nd | 227.1 | nd | nb | nb |

nd - not referred, nb – not tested

As it results from the presented tables and charts, gasket products made of silicone and most of silicone mixtures meet the requirements for all parameters at the HL1 and HL2 hazard levels.

The tests carried out for the materials of the gangways have shown that EPDM rubber mixtures do not meet any of the acceptable criteria. However, for silicone mixtures used in skeletal constructions, the requirements for HL1 and HL2 have been met.

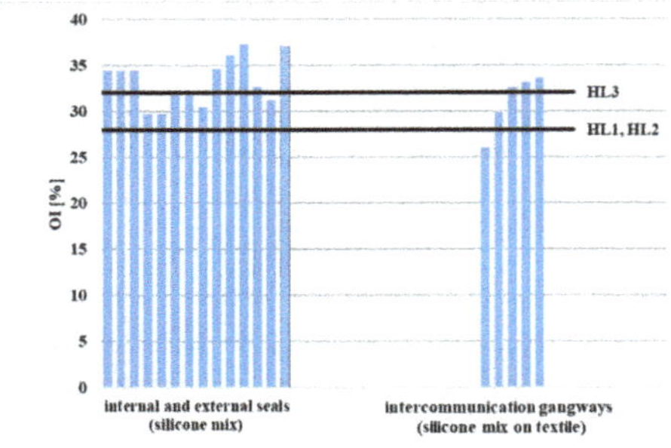

*Fig. 1. Oxygen Index (OI) values for rubber mixtures for various products*

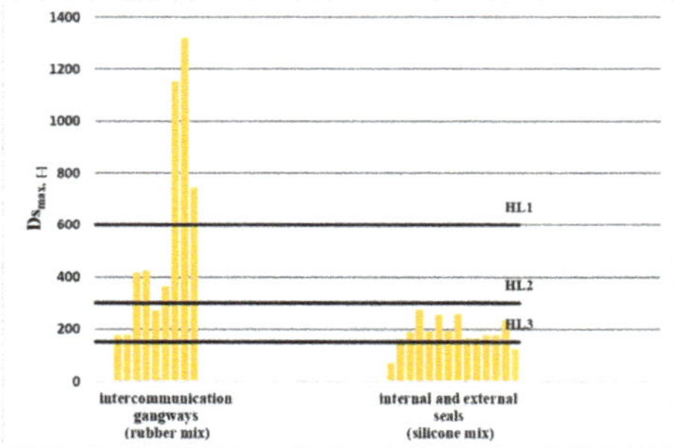

*Fig. 2. Maximum optical density (D_{smax}) values } for rubber mixtures for various products*

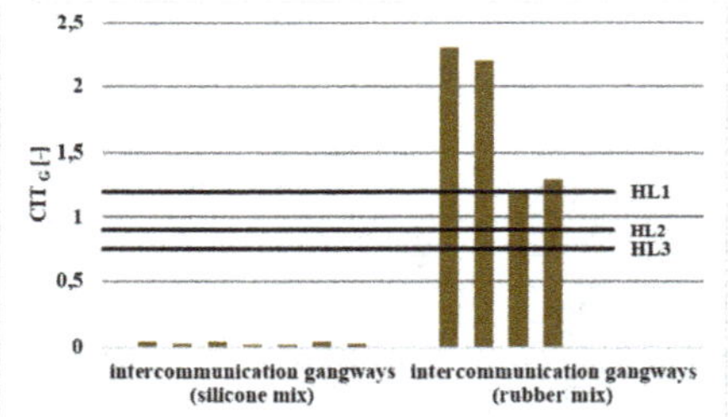

*Fig. 3. Conventional Index of Toxity (CIT_{G}) values for various rubber mixtures*

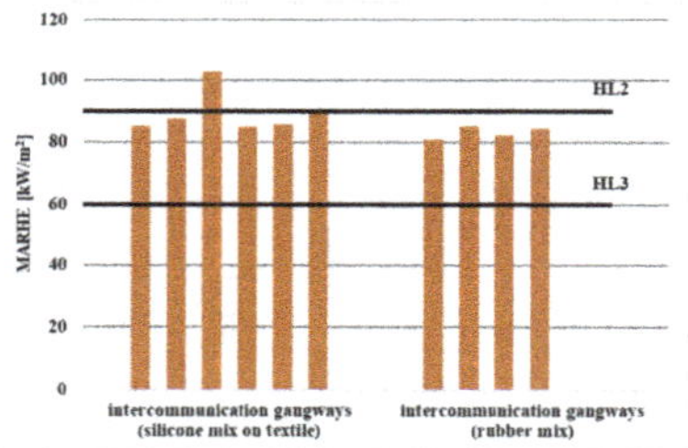

*Fig. 4 MARHE values for various rubber mixtures*

Exemplary photographs (Figs. 5. 6) show that some rubbers burn intensively. The sample within the OI test exceeded the criterion of time and length of the burnt part.

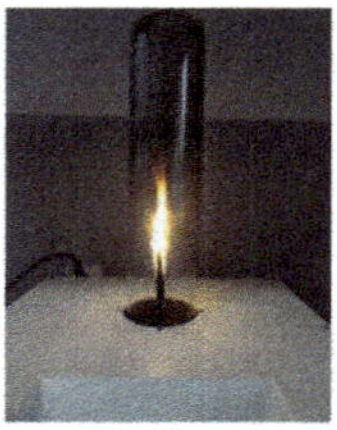

*Fig. 5. Oxygen Index test for EPDM.*

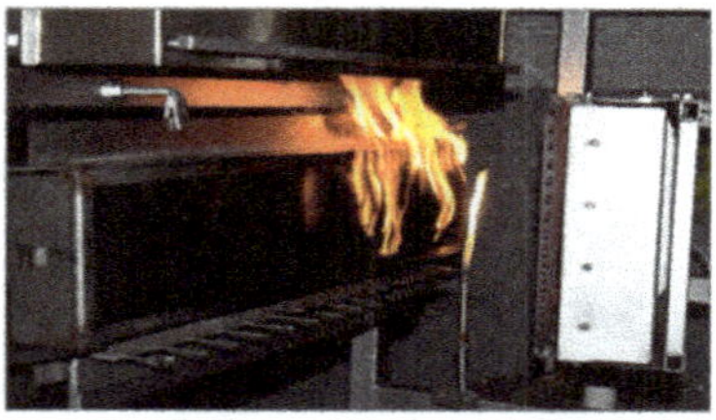

*Fig.6. Test according to ISO 5658-2 (determination of CFE) for EPDM.*

**Modification directions for rubber mixtures**

As it was shown earlier, silicone rubber is the least susceptible to burning. However, it is not possible to use it for all rail vehicle applications. On the other hand, flame retardant rubber is not an easy task to be done because it usually causes a significant reduction in impact strength. tensile strength and elongation at break [1]. The methods used include [1- 3. 20]:

- introduction of an inorganic filler (among others: magnesium hydroxide $Mg(OH)_2$, aluminum hydroxide $Al(OH)_3$, zinc hydroxyzinate $ZnSn(OH)_6$ and hydrated zinc borate $2ZnO \cdot 3B_2O_3 \cdot 3.5H_2O$).
- applying a synergistic effect by simultaneously introducing $Al(OH)_3$, graphite and paraffin.
- introduction of phthalocyanides.
- introduction of nanoparticles (e.g. zinates and borates of zinc. silicates).

However (despite the provision in p. 4.7 of the standard [8]), many manufacturers have undertaken research and development work aimed at modifying rubber mixtures in order to meet, or at least improve fire properties while maintaining very important functional features.

According to publication [4], the Austrian manufacturer has already managed to develop a technology for producing rubber that meets the requirements of R9 for the purpose of flexible rubber and metal-rubber assemblies found in the chassis (M1. EX9. EX11), while maintaining the required mechanical parameters. BATEGU rubber mixes BATEGU® 9559 and BATEGU®9713 for elements with anti-vibration properties (with the required hardness in the range of 40-60 Shore A) meet the fire requirements of R9 at the HL2 level. However, in the hardness range of 65-83 they meet the requirements of R9 at the HL3 level.

On the other hand, modification works for mixtures used for coating fabrics in skeleton solutions showed that materials from one manufacturer marked A156/18, A5/19, A6/19, A7/19, A8/19 (Table 3) met the requirements of HL1 and HL2 threats. Therefore, the producers of this type of product face the challenge to create such a material that will meet the requirements at HL3 hazard level.

However, not all of the work undertaken brought the expected effect, as in the case of rollers for intercommunication gangways. The test results for samples A40/16, A98/16, A173/16, A21/17, A144/17, A47/18 and A115/19 presented in Table 3. refer to subsequent product modifications of the same manufacturer, for which a slight improvement of parameters was obtained. However, the most difficult to meet are parameter CFE (the best result is 18.1 $kW/m^2$ with a minimum requirement $\geq 20$ kW / $m^2$) and CIT (the best result is the minimum required. i.e. 1.2).

**Summary**

The introduction from January 1. 2018 of the requirements of EN 45545-2 [8] to the mandatory application proved to be a great challenge for rubber products used in rail vehicles. Rubber is used for elements that require specific mechanical properties that are essential for the proper functioning of rolling stock. In contrast, the flammability of natural rubber found in large components (rollers for inter-wagon passages) is a serious fire hazard in a train due to the rapid spread of flame and the intense release of black obstructing visibility and very toxic smoke. Particularly dangerous conditions occur when passing through a tunnel [1]. Therefore, undertaking further research and development work aimed at developing new rubber compounds that meet the requirements in the field of fire properties while maintaining very important mechanical properties should be considered extremely important.

**References**

[1]  J. M. Radziszewska-Wolińska. D. Milczarek Uniepalnienie materiałów niemetalowych a ich właściwości funkcjonalne. TTS 11-12/2012. pages 56-59 (in Polish)

[2]  I. Gajlewicz. M. Lenartowicz. Nowe kierunki uniepalniania tworzyw polimerowych. PRZETWÓRSTWO TWORZYW 3 (May-June) 2014. pages 216-223 (in Polish)

[3]  D. M. Bieliński. M. Kmiotek. Uniepalnianie i poprawa stabilności termicznej elastomerów metodami inżynierii matreriałowej. Inżynieria materiałowa Nr 1/2009. pages 1-4 (in Polish)

[4]  D. Grefen. Fire Protection of Flexible Metal/Rubber Components Including Elements in Bogies. PowerPoint presentation on 3th International Conference MODERN TRENDS OF FIRE PROTECTION IN ROLLING STOCK. Warsaw. 18th of May 2016

[5]  J.M. Radziszewska-Wolińska - Development of Requirements for Fire Protection of Rolling Stock In Poland and its Comparison with EN 45545. PROBLEMY KOLEJNICTWA-RAILWAY REPORTS. vol. 57. issue 160. ISSN 0552-2145. Warsaw. 2013. pages 109 – 119.

[6]  Commission Regulation (EU) No 1302/2014 of 18 November 2014 concerning a technical specification for interoperability relating to the 'rolling stock - locomotives and passenger rolling stock' subsystem of the rail system in the European Union

[7]  EN 45545-1 Railway applications – Fire protection on raiway vehicles – Part 1: General

[8]  EN 45545-2 Railway applications – Fire protection on raiway vehicles – Part 2: Requirements for fire behaviour of materials and components

[9]  EN 45545-3 Railway applications – Fire protection on raiway vehicles – Part 3: Fire resistance requirements for fire barriers

[10] EN 45545-4 Railway applications – Fire protection on raiway vehicles – Part 4: Fire safety requirements for railway rolling stock design

[11] EN 45545-5 Railway applications – Fire protection on raiway vehicles – Part 5: Fire safety requirements for electrical equipment including that of trolley buses. track guided buses and magnetic leviation vehicles

[12] EN 45545-6 Railway applications – Fire protection on raiway vehicles – Part 6: Fire control and management systems

[13] EN 45545-7 Railway applications – Fire protection on raiway vehicles – Part 7: Fire safety requirements for flammable liquid and flammable gas installations

[14] ISO 5658-2 Flame spread laterally on vertically mounted products

[15] ISO 5660-1 Reaction-to-fire tests - Heat release. smoke production and mass loss rate - Part 1: Heat release rate (cone calorimeter method) and smoke production rate (dynamic measurement)

[16] EN ISO 5659-2 Plastics – Smoke generation – Part 2: Determination of optical density by a single-chamber test

Terotechnology XI                                         Materials Research Forum LLC
Materials Research Proceedings 17 (2020) 218-225            https://doi.org/10.21741/9781644901038-32

[17] EN ISO 4589-2 Plastics – Determination of burning behaviour by oxygen index – Part 2: Ambient temperature test

[18] NFX 70 100-1 Fire tests - analysis of gaseous effluents - Part 1: Methods for analysing gases stemming from thermal degradation. Part 2: Tubular furnace thermal degradation method

[19] https://sciaga.pl/tekst/58424-59-historia_i_zastosowanie_gumy_przemysl (access 22.11.2019)

[20] https://encyklopedia.pwn.pl/haslo/uniepalnianie-materialow-polimerowych;3991346.html (in Polish) (access 22.11.2019)

Terotechnology XI
Materials Research Proceedings **17** (2020) 226-232

Materials Research Forum LLC
https://doi.org/10.21741/9781644901038-33

# Fire and Smoke Properties of Electric Cables and Wires in Case of Different Geometric Structure and Composition

TARKA Izabela[1,a*], PIERGIES Jakub[1, b*] and ŁYSZCZ Marta[1, c*]

[1] Instytut Kolejnictwa (The Railway Research Institute), Materials & Structure Laboratory, 50, Chlopic ki Street, 04-275 Warsaw, Poland

[*a] itarka@ikolej.pl , [*b] jpiergies@ikolej.pl , [*c] mlyszcz@ikolej.pl

**Keywords:** Fire Safety, Fire Behavior, Test Methods, Heat Release Rate, Rail Vehicles, Heat Radiation Intensity

**Abstract.** The purpose of this work was to assess the influence of multi-core cables construction on their fire properties determined when the bunch of cables was burned, and determination of their flammability class. Seven-core cables with a 1.5 mm$^2$ single core cross-section were selected for the tests. These cables consist of two types of insulation materials, four types of filling materials and three types of coating materials.

**Introduction**

Modern trends and safety in rolling stock seek to limit potential losses through the use of such materials and design solutions that inhibit the spread of fire and retain functional abilities of a system during fire. The cables used in rolling stock must provide a low level of total fire risk, which includes minimizing the probability of emergence and spread of fire as well as minimizing its effects on people and equipment. Fire properties are taken into account here (ignition point, self-extinguishing, smoke density, emission of toxic gases etc.) that must meet specific requirements to ensure fire hazard on a minimum level. Insulation materials and sheathed cables, like other materials, components construction, finishing or equipment elements in rolling stock should be characterized by the following features [1]:

• low flammability,
• low ability to spread fire,
• low smoke emission,
• low emission of toxic and irritating gases.

Damage caused directly by fire is not usually the main cause of people's death. The emission of dense smoke containing irritating and toxic gases proves to be much more dangerous.

Electric cables have two roles in a fire protection aspect. Firstly, they are part of fire protection systems and assist in escape and rescue. Secondly, they may be the cause of fires, increase the propagation of fire and contribute to greater damage [2].

Halogens (fluorine, chlorine, bromine) whose presence significantly decreases flammability the total hazard level is increasing. In the fire zone, where the flame is supported by other burning materials, halogen-containing plastics decompose, emitting toxic gases. Halogen formed at that time, together with hydrogen compounds (hydrogen chloride, hydrogen fluoride, hydrogen bromide) have a very high irritating and toxic effect on humans and cause damage to electronic equipment, especially after contact with fire extinguishing water (corrosion of electronic equipment and constructing structures). Plastic containing halogen also emits a lot of thick smoke under the influence of fire, which significantly limits visibility, making it difficult to

Terotechnology XI                                                    Materials Research Forum LLC
Materials Research Proceedings **17** (2020) 226-232            https://doi.org/10.21741/9781644901038-33

evacuate people from a burning train. Halogen containing plastics also pose a threat to the natural environment by difficulties in their decomposition [1].

Furthermore, by putting pressure on the electrical industry, the consumer lobby and environmental organizations prompted to look for halogen-free alternatives in manufacture technology of flame retardant products. As a result of introduction of halogen-free materials in the cable industry, it is possible to simultaneously ensure a high level of fire safety, health care, and environmental protection. Halogen-free materials are already offered for all typical applications for thermoplastic and thermosetting plastics. Low content of halogens in cable insulations are characterized by low toxicity, low corrosivity and low smoke emission while maintaining other fire properties like flame spread and flammability. That is why halogen-free electric cables are more often used in technical and transport infrastructure elements. Currently, halogen-free cables in Europe are required for rail transport in accordance with the requirements of EN-45545-2:2015 [5]. This type of cables are also recommended for use in buildings with large clusters of people or where, as a result of fire and the emission of corrosive gases, material losses and casualties can occur e.g. in hospitals, airports, department stores, skyscrapers, hotels, theaters, cinemas or schools. For these applications, electric cables in higher flammability classes are used in accordance with the requirements of the Regulation of the European Parliament and of the Council (EU) no. 305/2011 (CPR for Construction Products Regulation) [6]. This applies especially to escape routes in buildings, tunnels and fire protection installations. Therefore, it is very important to refine the construction of cables so that their plastic components such as insulation, filling, coating, insulating or shielding tapes are not only halogen-free, but also meet the flammability classes B2ca-s1, d0, a1 according to EN-13501-1: 2018 [7].

## Experimental

The purpose of this work is to assess the influence of multi-core cable construction on their fire properties determined when a bunch of cables was burned, and thus on their flammability class. Seven-core cables with a 1.5 mm2 single core cross-section were selected for testing. This type of cable uses a filler between insulated conductors as a reinforcing element. During the tests, the influence of plastics used as filler (four types) in two thicknesses was checked. The most exposed parts of the cable are the outer sheath and filler material. Therefore, their type and amount have the greatest influence on the rate of flame spread and the amount of heat released. Thus, selected cables for testing contain only one type of insulation - cross-linked flame retardant polyethylene. To confirm the necessity for flame retardant polyolefin insulation, one type of polyethylene insulation cable without flame retardants was also tested and the obtained fire properties were compared.

To sum up, two types of insulation materials, four types of filler material and three types of coating material were used in the tested cables. Samples for testing were prepared by the Polish manufacturer Technokabel S.A. and delivered to the Railway Research Institute. Nine different seven-core halogen-free cables were selected for testing. The tested cables consisted of the elements shown in Table 1 and were arranged in different configurations. An oxygen index characterizing ignition ability was also determined for each plastic element. Three types of EVA copolymer flame-retardant with aluminum and magnesium compounds were used as the filler in the tested cables [3]. Table 1 presents information on the types of materials used, whereas Table 2 - the composition of tested cables.

In order to determine fire properties mentioned above such as flammability, flame spread, smoke and toxic gases emission, the tests are carried out according to Polish and European standards.

*Table 1 Composition of the tested cables.*

| Element of cable | Type | Density [g/cm3] | Oxygen Index [%] |
|---|---|---|---|
| insulation | XLPE  Cross-linked polyethylene non fire retardant (a) | 0.92 | 17 |
| | XLHFFR -Cross linked halogen-free, crosslinked with silanes flame-retardant compound (b) | 1.42 | 31 |
| filler thickness | TPE-O - flame retardant based on Olefin Thermoplastic Elastomer (c) | 1.94 | 45 |
| | flame retardant EVA based on thermoplastic elastomer (d) | 1.82 | 45 |
| | flame retardant EVA based on thermoplastic elastomer (e) | 1.8 | 52 |
| | flame retardant EVA based on thermoplastic elastomer (g) | 1.78 | 60 |
| sheet | HFFR -Halogen Free Flame retardant based on polypropylene (h) | 1.54 | 37 |
| | HFFR - Halogen Free Flame retardant (i) | 1.55 | 35 |
| | HFFR -Halogen Free Flame retardant (j) | 1.5 | 36 |

*Table 2 Composition and geometric of tested cables.*

| Specimen number | Symbol | Ø [mm] | Insulation materials | Filler thickness | Filler | Coating material |
|---|---|---|---|---|---|---|
| 1 | Cable N2XH-J B2ca 0,6/1kV 7x1,5mm2 | 11.9 | XLHFFR (b) | 1.00 | EVA (g) | HFFR (h) |
| 2 | Cable N2XH-J B2ca 0,6/1kV 7x1,5mm2 | 11.9 | | 1.00 | | HFFR (i) |
| 3 | Cable N2XH-J B2ca 0,6/1kV 7x1,5mm2 | 11.8 | | 1.00 | | HFFR (j) |
| 4 | Cable N2XH-J B2ca 0,6/1kV 7x1,5mm2 | 12 | | 1.00 | EVA (e) | HFFR (h) |
| 5 | Cable N2XH-J 0,6/1 kV 7x1,5 mm2 (2) | 10.7 | | 0.30 | EVA (g) | |
| 6 | Cable N2XH-J 0,6/1 kV 7x1,5 mm2 (1) | 10.6 | | 0.30 | EVA (d) | |
| 7 | Cable N2XH-J 0,6/1 kV 7x1,5 RE | 10.5 | | 0.30 | | |
| 8 | Cable N2XH-J 0,6/1 kV 7x1,5 RE (3) | 12 | | 1.00 | TPE-O (c) | |
| 9 | Cable N2XH-J 0,6/1 kV 7x1,5 RE | 11.2 | XLPE (a) | 0.30 | | |

The test performed for bunch of cables and wires in large scale was conducted according to EN 50399 [8]. A bunch of cables or wires were mounted vertically in a test chamber and exposed to propane burner with the nominal heat of power 20.5kW or 30kW.

This test method simulates fire of wires in cables mounted on building or vehicle and it presented the whole range of fire properties. In this test, the following parameters were determined:

- related to heat release:

      heat release rate (HRR)

      total heat release rate (THR)

      fire growth rate index (FIGRA)

- related to smoke production:

      smoke production rate (SPR)

      total smoke production (TSP)

- related to physical properties:

      damage length

      droplets and flaming particles.

The heat release rate and derivatives such as THR and FIGRA are determined by the measurement of oxygen consumption derived from the oxygen atmospheric concentration and oxygen concentration after burning with the flow rate in the combustion product stream. The smoke production rate and total smoke production is calculated from the measurement of the obscuration of a laser light beam by the combustion product stream. The damage length is measured after the test with measuring scale and droplets and flaming particles are recorded by an operator during the whole test. The results obtained on this apparatus due to the registration of so many fire parameters can be used not only for classification tests but also used to develop fire simulation models. Such studies were carried out as part of the TRANSFEU project. [4] The test chamber is presented in Figure 1 while the results of the test are presented in Table 3.

*Fig. 1 Test chamber*

*Fig. 2 Rack with oxygen analyzer*

The same insulation material and the same filling material but two different coatings with similar OI values were used for specimens 1 and 2. There were no significant differences in the obtained

Materials Research Forum LLC
https://doi.org/10.21741/9781644901038-33

parameters. Both cables constructed in this way are characterized by good fire properties - low heat release and low TSP smoke. During the test of specimen 2, dropping of burning droplets occurred.

*Table 3 Results of the tests*

| Cable number | Ø [mm] | Damage length [m] | Peak HRR(30) [kW] | THR (1200) [MJ] | FIGRA [W/s] | TSP [m²] | Class | Smoke production | Flaming droplets |
|---|---|---|---|---|---|---|---|---|---|
| 1 | 11.9 | 0.72 | 26.9 | 13.2 | 53.7 | 8.2 | B2ca | s1 | d0 |
| 2 | 11.9 | 0,9 | 22.9 | 12.3 | 49.8 | 12.8 | B2ca | s1 | d2 |
| 3 | 11.8 | 0.7 | 18.1 | 9.3 | 45.2 | 11.1 | B2ca | s1 | d2 |
| 4 | 12 | 0.76 | 23.5 | 15.7 | 82.1 | 16.3 | Cca | s1 | d2 |
| 5 | 10.7 | 1.07 | 23.6 s | 12.3 | 43.9 | 22.6 | B2ca | s1 | d0 |
| 6 | 10.6 | 1.32 | 35.0 | 18.9 | 69.4 | 57.4 | Cca | s2 | d0 |
| 7 | 10.5 | 2 | 45.2 | 25.9 | 83.8 | 50.1 | Cca | s2 | d0 |
| 8 | 12 | 1.4 | 44.3 | 20.8 | 83.3 | 62.1 | Cca | s2 | d2 |
| 9 | 1.2 | 3.3 | 332.5 | 89.4 | 428.4 | 135.3 | Dca | s2 | d2 |

Specimens 3 and 4 used the same insulation material and the second type of filler material and two different coatings with similar OI values. During the test, differences in HRR values and the amount of smoke emission were observed. The tested specimen 4 obtained worse parameters due to the application of a polypropylene-based coating (h).

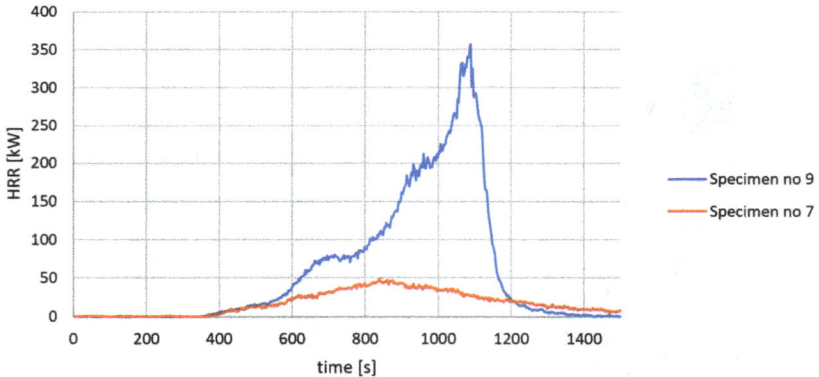

*Fig 3. Graph of HRR as a function of time for specimen no 7 and 9.*

Specimens 1, 4 and 8 used the same insulation and coating material, but three different fillers with different OI values and 1 mm thickness. Difference in HRR values and large difference in

Terotechnology XI
Materials Research Proceedings **17** (2020) 226-232

Materials Research Forum LLC
https://doi.org/10.21741/9781644901038-33

the amount of smoke emission were obtained. This relationship was as follows: the lower the filler oxygen index, the worse the fire and smoke parameters.

For specimens 5, 6 and 7, the same insulation and coating material were used, as well as three different fillings (different OI) in this case for the thickness of 0.3 mm. Different HRR values were observed; for specimen 5 the lowest value associated with a very high oxygen index. A significant difference in the amount of smoke emission was also registered - only specimen 5 meets B2ca class requirements.

For specimens 7 and 9 the same shell material and the same material and filling thickness were used but two different insulations: one made of flame-retardant PE (cable 7) and the other of PE without flame retardants (cable 9). A very large difference in the value of HRR and the amount of smoke emitted was observed. In the case of cable 9, because of the rapid increase in HRR, the test was stopped due to danger that could cause damage of the apparatus. The difference in HRR values is presented in Fig. 3

Specimens 1, 5, 7 and 8 with the same composition were compared in pairs, occurring in two variants of filling thickness: 0.3 mm and 1 mm. For the filling with symbol (a) there was a decrease in the value of recorded parameters (HRR, THR, FIGRA) for smaller thickness. However, no such relationship was observed for filling (c).

**Summary**

The tests showed that the fire properties of electric cables determined in accordance with EN 50399: 2011 depend on the type of material used for their construction and the flame retardants contained therein. The greatest influence here is the degree of flame retardancy of the insulation, as well as filler and coating materials, which is associated with the value of the oxygen index. However, no significant effect of the thickness of the filler used was demonstrated. Of the tested electric cable variants, the highest class B2ca-s1, d0 is achieved only for specimens 1 and 5 built on the same components (thinner layer of filler - 0.3 mm) in the following configuration: insulation (b) + filler (g) + coating ( h). Due to the specifics of large-scale tests and the influence thereon of factors not related to the composition of the specimens, the performed tests should be extended by a larger number of further tests, taking into account also other types of flame retardant insulation. It is important to refine the increasing number of flame retardant halogen-free cable components, taking into account favorable technological and price parameters. This will increase the fire safety of critical infrastructure components using electric wires.

**References**

[1] Kable i przewody elektryczne przeznaczone do taboru szynowego Damian MAJCHRZYK, Izabela TARKA, Prace Instytutu Kolejnictwa – Zeszyt 149 (2016), 14-21

[2] The effects causing the burning of plastic coatings of fire-resistant cables and its consequences ZsuzsannaKerekes, ÁgostonRestás, Eva Lublóy Journal of Thermal Analysis and Calorimetry, Springer 2019

[3] Lei Ye, Baojun Qu, Flammability characteristics and flame retardant mechanism of phosphate-intercalated hydro, ScienceDirect, Polymer Degradation and Stability 93 (2008) 918-924. https://doi.org/10.1016/j.polymdegradstab.2008.02.002

[4] S. Brzozowski, J. M. Radziszewska- Wolińska, Modelowanie spalania kabli metodą FDS, Problemy Kolejnictwa – Zeszyt 159 (2013)

Terotechnology XI                                                    Materials Research Forum LLC
Materials Research Proceedings 17 (2020) 226-232          https://doi.org/10.21741/9781644901038-33

[5]  PN-EN 45545-2+A1:2015 Railway applications – Fire protection on railways vehicles –
Part2: Requirements for fire behavior of materials and components.

[6]  REGULATION (EU) No 305/2011 OF THE EUROPEAN PARLIAMENT AND OF THE
COUNCIL of 9 March 2011 laying down harmonised conditions for the marketing of
construction products and repealing Council Directive 89/106/EEC (Text with EEA relevance)

[7]  EN-13501-1: 2018 Fire classification of construction products and building elements –
Part 1: Classification using data from reaction to fire tests

[8]  EN 50399:2011 Common test methods for cables under fire conditions – heat release and
smoke production measurement on cables during flame spread test – Test apparatus, procedures,
results

Materials Research Forum LLC
https://doi.org/10.21741/9781644901038-34

# Effectiveness of Anti-Graffiti Coatings used in Rolling Stock

GARBACZ Marcin[1, a*] and KOWALIK Paweł[1, b]

[1] Instytut Kolejnictwa (The Railway Research Institute), Materials & Structure Laboratory, 50, Chlopicki Street, 04-275 Warsaw, Poland

[*a] mgarbacz@ikolej.pl, [b] pkowalik@ikolej.pl

**Keywords:** Rolling Stock, Coatings, Graffiti Removal

**Abstract** The article presents the problem of graffiti on protective coatings on the wagon bodies of rail vehicles. It presents a way of dealing with this phenomenon by adding an anti-graffiti clear coat, allowing easy cleaning of graffiti without damaging the paint coat. The anti-graffiti clear coats available on the market and the agents used to wash them are briefly characterized. The research methodology and research results on the anti-graffiti protection effectiveness of various paint coats used in the railway industry, tested between the years 2010 and 2018 in the Railway Research Institute in the Materials & Structure Laboratory, are also presented.

## Introduction

Graffiti vandalism is a serious aesthetic and technical problem for owners and services responsible for the maintenance of rolling stock. Despite the fact that graffiti is not such a common problem as at the turn of the century, it still appears quite often on the rolling stock infrastructure, mainly causing material damage and large financial losses resulting from cleaning and, in the worst case, repainting. To effectively counteract unwanted graffiti painting, it is necessary to quickly remove the effects of the graffiti artists' work to discourage them from similar practices in the future, as they primarily care about their popularity, often signing their paintings with a pseudo-artistic nick and posting the effects of their work on the Internet. The best solution for this purpose is to protect the surface with a suitable quick and easy anti-graffiti remover. The currently applied paint coating types for new rolling stock have appropriate anti-graffiti protection in the form of a clear coat, which additionally protects the coatings against harmful UV radiation and other external factors, while old rolling stock often cannot be cleaned, because graffiti is removed with the factory paint during its cleaning and the only solution to the problem is to re-paint the train. For many years now, Poland has been bound by the Normative Document DN 001/08/A2/2016 [1], which sets out the requirements for paint coatings intended for use in rolling stock. This document requires that such paint coatings have an anti-graffiti clear coat, which enables the easy removal of graffiti from a paint coating with appropriate protection.

## Railway anti-graffiti clear coats used in railway applications

The most common anti-graffiti clear coats used in rolling stock to protect paint coats include polyurethane clear coats, which are classified as permanent systems without the need to renew the protective coating after washing (repeated washing possible) [2]. They are characterized by high resistance to washing agents used for cleaning rail vehicles and for graffiti removal. Such a clear coat is usually applied onto the base-coat or topcoat of the paint. When writing about an anti-graffiti clear coat, we refer to a product that allows undesirable subsequent paint coatings to be removed without damaging the original paint coat under reduced adhesion. Any substance present on the paint surface affects its structure, causing irreversible changes in the coating to a greater or lesser degree, until it is completely damaged and the non-resistant base-coat is exposed [3]. Other less common safety features used in rolling stock include so-called lost systems, which have a shorter protection time. The protection consists of applying a protective coating which separates the protected base-coat from paints. Once the coating is covered with graffiti, it is washed away with paints. Such coats usually consist of waxes, polyacrylates, microcrystalline waxes or paraffin emulsions. In this case, it is necessary to reapply the protective coating after the removal of graffiti (on the cleaned spot) as well as to apply the remover periodically on the surfaces subject to current maintenance, i.e. wagon bodies. The advantage of such systems is that there is no need to use any cleaning agents or removers (graffiti removal by means of a pressure washer with hot water), which generates lower graffiti removal costs [2,5].

**Means and methods used to remove graffiti from railway infrastructure**
For the removal of unwanted graffiti in the railway industry, mechanical cleaning with hot water under pressure or special chemicals are applied, or both cleaning methods are combined. Various chemical agents are used to remove graffiti from paint coats, but they are usually in the form of liquids or gels. They contain organic substances, mainly aliphatic and aromatic hydrocarbons and tensides. The type of remover used should be adequate to the type of anti-graffiti protection applied and is usually determined by the manufacturer of a given paint or an entire painting system [3]. They are applied with a soft brush or sprayed on a large surface over graffiti and left for about 5 minutes, after which they are washed off by abrasion or stripping. Finally, the cleaned wagon is washed with soapy water and, at the end of the application, the wagons are rinsed with hot water. It should be noted here that, even though there is a satisfactory visual effect, invasive chemical graffiti removal methods affect the cleaned surface. Their repeated use may result in permanent damage. Chemical graffiti removal facilitates the repair and painting of the surface (in case the cleaning effect is not satisfactory). Painting over graffiti is very difficult and, in most cases, requires several coats of paint on it.

**Means used for graffiti painting**
The range of materials used by graphic artists is very wide, but the most common graffiti painting means used in the railway industry are spray paints and permanent markers. The main components of spray paints or graffiti markers are the following: pigment (a material that ensures color), binder (a film-forming transparent material in which pigment particles are dispersed and harden and bind the pigment on the painted surface) and a solvent that allows the flow of the pigment mixture/binder, where, in the case of markers, fast evaporating solvents are used to fix the writing instantly. Pigments can generally be divided into two groups:
- inorganic pigments, most of which occur in nature as minerals, such as titanium oxides TiO, $TiO_2$, ZnO zinc oxide (white colors), lead chromate $PbCrO_4$, zinc chromate $ZnCrO_4$ (yellow colors), iron oxide $Fe_2O_3$ or lead oxide $Pb_3O_4$ (red colors). Blue and green pigments include ultramarine, Paris blue and chromium oxide ($Cr2O3$). The metallic colors in gold, silver and brown graffiti spray paints are obtained by adding aluminum, zinc, bronze, stainless steel or pearlescent pigments.
- organic pigments that are currently used to a much greater extent than their inorganic equivalents. Used in spray paints, they usually include substances, such as soot (the most common black pigment), phthalocyanine dyes, mainly blue and green, various azo compounds (e.g. toluidine red, lithium-bar red, yellow benzimidazolone) which cover about 70% of all organic pigments, carbonyl dyes as their salts and metal complexes [4,6].

Binding agents are divided into natural (plant or animal) agents, agents prepared by chemical modifications of natural materials – semi-synthetic agents and synthetic polymers, which are the most commonly applied and will be briefly characterized. Synthetic agents (synthetic resins) can be divided into three main types: alkyd, acrylic and polyvinyl acetate [4,6].

Alkyd agents (used as one of the first) are obtained by a reaction between polyalcohols and dicarboxylic acid. Most spray paints contain alkyd resins as binding agents. Such paints dry faster, are more durable and more difficult to remove than oil paints [4,6].

Acrylic agents are obtained by polymerization of one or more monomers, which include mainly esters, acrylic acid and methacrylic acid. Acrylic polymer resins can be thermoplastic (the paint film is formed by evaporation of the solvent without chemical action) or thermosetting (the paint coat is hardened by heat or in reaction with another chemical substance to form a cross-linked structure) [4,6].

Polyvinyl agents (PVA) are obtained by polymerization of vinyl acetate. PVA emulsions require the addition of plasticizers and emulsifiers, such as dibutyl phthalate or external plasticization by copolymerization [4,6].

Other binders used in spray paints are polyurethanes, polyesters, cellulose, cellulose nitrate, epoxy resins, polycyclohexanone and chlorinated rubber (a mixture of natural styrene/butadiene rubber and chlorine). Paint containing chlorinated rubber is highly resistant to water and chemical products [4,6].

Solvents used in the production of spray paints and markers can be divided into the following: hydrocarbon (aliphatic, naphthenic and aromatic), oxygen solvents (ketones, esters, glycol esters and alcohols) and water (components of the continuous phase of most emulsion paints), where hydrocarbon solvents are most commonly applied [4,6].

In addition to pigments, binders and solvents, paints also contain additives. Their amount is very small and no more than 5%. These include plasticizers, dispersants (increasing the plasticity or fluidity of spray paints), surfactants, wetting agents (dispersing pigments), thickeners, pH buffers, and anti-foaming agents (changing the surface tension of paints) [4,6].

**Research methodology:**
The anti-graffiti clear coat protection effectiveness was tested in the Laboratory of Materials & Structure on appropriately prepared test panels of entire painting systems (with the same functional parameters of coatings used on wagon bodies). The tests were conducted on the basis of the ASTM D6578/D6578M-13 (2018) [7] standard, in accordance with the modified procedure. The modification of the standard included the adoption of an effectiveness assessment scale of anti-graffiti protection and the fact that only agents commonly used by graffiti artists, i.e. markers and sprays, were used for painting. The evaluation was performed using the A method, i.e. the assessment with corrected vision and the washing substances included specialist agents for removing graffiti based on organic solvents available on the market in the form of liquids, gels or tissues soaked in the appropriate substance.

The test procedure consisted of painting the coating on surfaces bounded by a suitable template on at least 3 samples evaluated against an extra, unpainted sample. The graffiti was left on the coating surface at room temperature for 24 hours and then it was removed by means of specified removers, where it was evaluated visually in natural daylight. The painting and washing process included 10 complete cycles. The test result is the specified efficiency of action of the anti-graffiti clear coat in the form of a number of cycles in relation to the point at which, after removal of a given graffiti painting on the coating surface, the change occurs for the first time (evaluation relative to gloss loss and color change).

Terotechnology XI                                                              Materials Research Forum LLC
Materials Research Proceedings 17 (2020) 233-239                     https://doi.org/10.21741/9781644901038-34

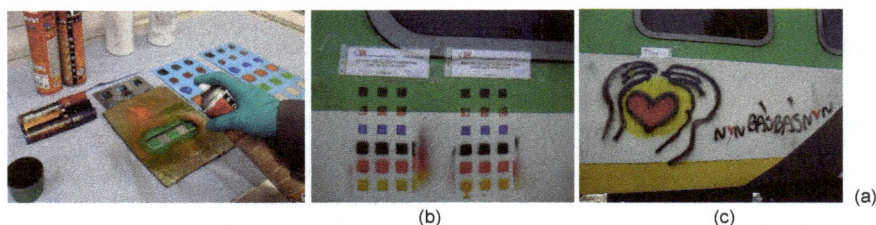

*Figure 1. Example of graffiti made in a laboratory and on wagon bodies according to ASTM D6578/D6578M-13 (a and b), and example of graffiti found on rolling stock (c).*

Table 1 summarizes the test results for several coating systems with anti-graffiti topcoats, which were painted and washed according to the procedure described above and which passed the remaining tests included in DN 001/08/A2/2016 [1]. The expanded measurement uncertainty (P=95%, k-2) is ± 2 painting cycles and has been determined from internal and external inter-laboratory comparisons. Figure 1 shows the painting method in the laboratory and on wagon bodies according to ASTM D6578/D6578M-13 (2018) [7].

**Analysis of the results and conclusions:**
The above results show that the vast majority of the examined painting systems with the anti-graffiti clear coat were characterized by a high protection effectiveness against painted graffiti. At least five coats of paint applied to the tested graffiti coatings and washed with graffiti removers recommended by the manufacturers of these coatings did not cause any visual changes, such as the loss of gloss or shadow formation and damage, such as blistering or peeling (they were imperceptible to the human eye). Thus, the majority of anti-graffiti clear coats currently available on the market meet the requirements for paint coats in rail vehicles contained in DN 001/08/A2/2016 [1]. According to this document, it is essential for the anti-graffiti clear coat to ensure that no changes are made to the paint coat surface after five cycles of application and washing of graffiti. In addition, in accordance with the document, painting systems should meet a number of other criteria, including high chemical and physical resistance, such as corrosion resistance, aging resistance and high resistance to chemicals due to frequent contact with cleaning agents on the railways.

The combined data in Table 1 also shows that there were differences in the ease of graffiti removal, expressed as the number of cycles without changes to the coating surface, depending on the type of material used to create the graffiti characters. It was harder to remove markers than spray paints. For Coat 14, three tested markers (Pentel black, blue and red) caused changes in the coating color after the first cycle, while for three tested spray paints, no changes were observed on the paint coat after performing 10 full cycles of painting and washing, which indicates that the coating reacts strongly with the applied markers. For Coats 2 and 12 (the same anti-graffiti clear coat, different marking agents) for markers of different manufacturers, completely different test results were obtained, whereas for applied spray paints of different manufacturers, no changes were observed after 10 cycles of painting and washing. The greatest difficulties in removing paint from Mołotow markers were observed for the red marker, which was the only alcohol-based marker. In the case of Pentel markers, the obtained results of resistance to painting are much weaker, but the manufacturer does not fully specify the composition of its markers, so it is not possible to determine the exact cause of the observed phenomenon. If the solvent is an alcohol, it can be a better transporting carrier for the dye and binder, which can penetrate deeper into the coating pores and bind more firmly to the coating than the acrylic markers used, for example. Differences in the results for markers and spray paints, in addition to composition, can also be explained by their different application methods. The markers are directly applied on the coating surface. The operator may exert different pressure on the coating, which may affect the ease with which the paint can be removed from the coating. This is not the case with spray paints as they are first thoroughly mixed by shaking the paint container and then sprayed onto the coating at a distance of approx. 25 cm. It can therefore be assumed that the adhesion of spray paints is lower than that of markers. This makes it easier to remove spray paints without leaving any changes on the paint surface. In addition, the coating is chemically and mechanically damaged due to repeated contact between the writing tip and the coating in the case of markers, as well as due to washing with chemical agents as it loses its performance and protective properties, thus the thickness of the anti-gravity protection coating is also important.

During many years of research, it was also observed that the type of non-woven fabric used for removing anti-graffiti agents and graffiti influences the abrasion of the top anti-graffiti clear coat and consequently worsens its protective properties. The contact time of chemical cleaners should also be as short as possible and in accordance with the declarations of the manufacturers of dishwashing agents, as it also has a negative impact on the protective coating of paint, often causing irreversible changes in the coating (permanent discoloration or peeling of the coating).

It is extremely important to note that the paint systems tested, including the top graffiti coating, were not exposed to harmful weather conditions, so the actual effectiveness of coating protection in natural conditions may be much weaker, especially with the passage of time. The protection effectiveness will be reduced under the influence of external aging factors, such as light exposure to UV radiation, environmental conditions, such as changing weather conditions, air pollution or specialized chemical cleaning agents. Literature references show that the aging of samples significantly affects the results, which should also be taken into account when designing laboratory tests to check the quality of graffiti protection for a given paint / paint system [2,5].

**Summary**

Based on the obtained data and many years of experience in laboratory research, it can be concluded that durable anti-graffiti clear coat systems seem to be one of the best possible solutions for the protection of railway wagons against vandalism. Anti-graffiti clear coat systems should be designed in symbiosis with the remover to investigate in detail the interaction between the coating, graffiti agent and solvent. This should be done to maintain the functional and decorative properties of the coating systems applied for as long and as efficiently as possible to reduce subsequent operating costs.

*Table 1. Comparison of the results of anti-graffiti paint resistance tests using various removers according to the methodology described in ASTM D6578/D6578M-13, performed in 2010-2018 in the Material & Structure Laboratory.*

| red spray paint (DECO COLOR / MOTIP DUPLI / MASTON) | yellow spray paint (DECO COLOR / MOTIP DUPLI / MASTON) | black spray paint (DECO COLOR / MOTIP DUPLI / MASTON) | red marker (PENTEL N850) | white marker (PENTEL MMP20) | blue marker (PENTEL N850, N860) | black marker (PENTEL N50, N850) | silver spray paint (Molotow) | green spray paint (Molotow) | spray orange (Molotow) | red marker (alcoholic, Molotow) | blue marker (acrylic, Molotow) | black marker (acrylic, Molotow) | Graffiti remover | Type of tested anti-graffiti clear coat | No |
|---|---|---|---|---|---|---|---|---|---|---|---|---|---|---|---|
| - | - | - | - | - | - | - | 10 | 10 | 10 | 5 | 10 | 10 | AGS 221 - gel | Teknodur 295-900 | 1 |
| - | - | - | - | - | - | - | 10 | 10 | 10 | 9 | 10 | 10 | AGS-221 - gel | XPC60036 | 2 |
| - | - | - | - | - | - | - | 10 | 10 | 10 | 8 | 10 | 10 | AGS-5 | XPC60036 | 2 |
| - | - | - | - | - | - | - | 10 | 10 | 10 | 10 | 10 | 10 | AGS-221 - gel | BO-100 AGR | 3 |
| - | - | - | - | - | - | - | 10 | 10 | 10 | 0 | 10 | 10 | AGS-221 - gel | XPC60012 | 4 |
| - | - | - | - | - | - | - | 5 | 5 | 5 | 5 | 5 | 5 | AGS 221 gel | Mipa CAG 90 2K-Antigraffiti-Klarlack | 5 |
| - | - | - | - | - | - | - | 5 | 5 | 5 | 5 | 5 | 5 | RHOBA CLEAN-CRAFT 150 | CC710 | 6 |
| - | - | - | - | - | - | - | 10 | 10 | 10 | 10 | 10 | 10 | AGS-5 | PCVZ | 7 |
| - | - | - | - | - | - | - | 10 | 10 | 10 | 10 | 10 | 10 | AGS-221 - gel | 607.0815 | 8 |
| - | - | - | - | - | - | - | 10 | 10 | 10 | 10 | 10 | 10 | Cleaner CPC 028 | 607.0815 | 8 |
| - | - | - | - | - | - | - | 10 | 10 | 10 | 10 | 10 | 10 | Cleaner CPC 029 | 607.0815 | 8 |
| - | - | - | - | - | - | - | 10 | 10 | 10 | - | 10 | 10 | Nitro remover from the coating | 923.HS 90 | 9 |
| - | - | - | - | - | - | - | 5* | 5* | 5* | 5* | 5* | 5* | AGS-221 - gel | NOVAKRYL 575 | 10 |
| 10 | 10 | 10 | 0 | - | 0 | 0 | - | - | - | - | - | - | AGS-221 - gel | DURETHANE XPC60011 | 11 |
| 10 | 10 | 10 | 3 | - | 1 | 3 | - | - | - | - | - | - | AGS-221 - gel | DURETHANE XPC60036 | 12 |
| 10 | 10 | 10 | - | 10 | 10 | 10 | - | - | - | - | - | - | AGS 5 SR | Clear 2K HS 110.0111.Y | 13 |
| 10 | 10 | 10 | 0 | - | 0 | 0 | - | - | - | - | - | - | 047, coating manufacturer's | 923.5026 | 14 |
| - | 10 | 10 | 5 | - | 5 | 5 | - | - | - | - | - | - | AGS-221 - gel | ALEXIT Topcoat 460-5A. | 15 |
| 10 | 10 | 10 | - | 10 | 10 | 10 | - | - | - | - | - | - | RHOBA CLEAN-CRAFT 150 | CC700 Clear Coat Anti-graffiti | 16 |
| 3 | 3 | 3 | - | 4 | 4 | 4 | - | - | - | - | - | - | tikkurila maalipesu (20%) | Temadur Clear | 17 |

Marking agent — The number of painting/washing cycles at which the anti-graffiti clear coat did not show any changes

*Only a cycle of five paintings/washings has been carried out.

## References

[1] Normative Document DN 001/08/A2/2016 "Wyroby lakierowe stosowane w pasażerskim taborze szynowym – w lokomotywach, wagonach i zespołach trakcyjnych." (Painting products for passenger rolling stock – locomotives, wagons and train units)

[2] S. Rossi, M. Fedel, S. Petrolli, F. Deflorian, Characterization of the Anti-Graffiti Properties of Powder Organic Coatings Applied in Train Field, Coatings — Open Access Journal 2017, 7/67. https://doi.org/10.3390/coatings7050067

[3] Ł. Pasieczyński, N. Radek, Problemy Kolejnictwa – Zeszyt 170 (marzec 2016) Badanie wybranych właściwości systemu powłokowego "antygraffiti" dla pojazdów szynowych (Examination of selected properties of anti-graffiti coating system for railway vehicles)

[4] R. Lambourne, T.A. Strivens. Paints and surface coatings. Theory and practice. Cambridge: Woodhead Publishing; 1999. https://doi.org/10.1533/9781855737006

[5] S. Rossi, M.Fedel, S.Petrolli, F.Deflorian, Behaviour of different removers on permanent anti-graffiti organic coatings, Journal of Building Engineering 5 (2016) 104–113. https://doi.org/10.1016/j.jobe.2015.12.004

[6] P. Sanmartín, F. Cappitelli, R. Mitchell, Current methods of graffiti removal: A review, Construction and Building Materials 71 (2014) 363–374. https://doi.org/10.1016/j.conbuildmat.2014.08.093

[7] ASTM D6578/D6578M-13 (2018) Standard Practice for Determination of Graffiti Resistance

Terotechnology XI
Materials Research Proceedings **17** (2020) 240-245

Materials Research Forum LLC
https://doi.org/10.21741/9781644901038-35

# The Laboratory Tests of the Embedded Block System with High Vertical Elasticity

ANTOLIK Łukasz[1,a,*] and SIWIEC Jakub[1,b]

[1]Instytut Kolejnictwa (The Railway Research Institute), Materials & Structure Laboratory, 50, Chłopicki Street, 04-275 Warsaw, Poland

[a]lantolik@ikolej.pl, [b]jsiwiec@ikolej.pl

**Keywords:** Fastening System, Embedded Block System, Slab Track

**Abstract.** The resilient system that fastens rail to the ground is a very responsible element of the railroad track construction. This article discusses the function characteristics of an embedded block system (EBS) with high vertical flexibility, a range of the product application, methods of mechanical tests carried out in the laboratory and criteria of the tests result approval.

## Introduction

Since the early beginning of the railroad transport system, there has been a parallel problem of the fixing rail to the ground technique. The world markets are mostly dominated by solutions such as types W, Fastclip, Nabla, SB, RN etc. [1]. Some areas use domestic solutions prepared for the specific exploitation conditions e.g. tram railway [10]. The multitude of design solutions and their impact on the production cost, easiness of maintenance, failure rate, atmospheric susceptibility and truck loading are so large that every market can use ready-made solutions.

Each type of the system that fastens rail to the ground is characterised by a number of advantages and disadvantages. Unfortunately, the weakness of these solutions is that they are based on placing rail pads between the rail and support. It is undoubtedly the responsible and weakest part in any fastening system, and additionally it takes only 1% of the height of the railroad construction chain. The classical fastening systems may not be used in urban areas or engineering objects, where relatively high parameters relating to the vibration absorbing properties of the track and substructures are required. In order to increase the level of vibration dumping, course lined with anti-vibration mats, rails embedded in the anti-vibration coat or embedded block system (EBS) are used. All the above-mentioned solutions fix rail by adhesion or mechanical technique and do not use any ballast.

Methods of the rail vertical elasticity in relation to the rigid bed using an EBS placed in an elastic layer made of polyurethane resin are discussed in the further part of the article.

This solution is characterised by a relatively high vertical flexibility and a high degree of vibration dumping, while maintaining a high operating resistance.

## Execution of tests

For approval purposes, some reference shall be made to the common European requirements, which the fastening systems are subjected to. This unification of functional parameters aims at designing solutions that are interchangeable in all countries belonging to the European Economic Area (EEA). The essential requirements are set out in Regulation 1299/2014 of Technical Specification for Interoperability for the Infrastructure subsystem [2]. This refers to the requirements of EN 13481-5+A1:2017 [3] regarding the fastening systems used in the slab track. Testing a new solution in the laboratory conditions is therefore associated with compliance with strictly defined requirements. Not only is the quality of the performed tests important but also

Terotechnology XI                                                          Materials Research Forum LLC
Materials Research Proceedings **17** (2020) 240-245                    https://doi.org/10.21741/9781644901038-35

their sequence due to their possible influence on the results of the tests. The embedded block system described in this article has been subjected to mechanical tests for category C, at loads corresponding to the maximum load of a single axle of the 260 kN and the minimum radius of curve r=150 m. Table 1 shows loads used in the fatigue tests of the fastening systems to the slab track depending on the track category.

*Table 1. Test loads and positions [source: EN 13481-5+A1:2017]*

| $k_{LFA}{}^c$ | < 50 MN/m | | | ≥ 50 < 75 MN/m | | | ≥ 75 < 100 MN/m | | | ≥ 100 MN/m | | |
|---|---|---|---|---|---|---|---|---|---|---|---|---|
| Category | α ° | $X^d$ mm | $P_v/\cos\alpha$ $kN^{a,b}$ | α ° | $X^d$ mm | $P_v/\cos\alpha$ $kN^{a,b}$ | α ° | $X^d$ mm | $P_v/\cos\alpha$ $kN^{a,b}$ | α ° | $X^d$ mm | $P_v/\cos\alpha$ $kN^{a,b}$ |
| A | 45 | 100 | 50 | 45 | 100 | 55 | 38.6 | 50 | 65 | 38.6 | 50 | 80 |
| B | 38.6 | 100 | 55 | 38.6 | 100 | 60 | 38.6 | 50 | 70 | 38.6 | 50 | 85 |
| C | 33 | 25 | 60 | 33 | 25 | 65 | 33 | 25 | 75 | 33 | 25 | 95 |
| D | 26 | 15 | 60 | 26 | 15 | 65 | 26 | 15 | 75 | 26 | 15 | 95 |

[a] The test loads apply only to rail sections included in EN 13674-1 (excluding 49E4) and EN 13674-4+A1
[b] The test loads take into account the possible use of slab track with higher cant deficiencies than ballasted track
[c] $k_{LFA}$ = Low frequency dynamic stiffness measured at 5 Hz according to EN 13146-9+A1 and Table 2 above
[d] For embedded rail and web supported rail, the rail section shall be unmodified (i.e. X = 0)

The tested embedded block system was first checked for main static parameters according to the specified order, i.e. the tighten force was first applied to the concrete block support according to EN 13146-7:2012 [4]. The aim of this test was to check clamping force of the rail by the fastening system. The test was performed by applying the vertical tensile force with the rising rate of 10 kN/min. until the rail pad was able to be removed and the rail foot retuned to its original position.

An example of the test diagram is presented in Fig. 1.

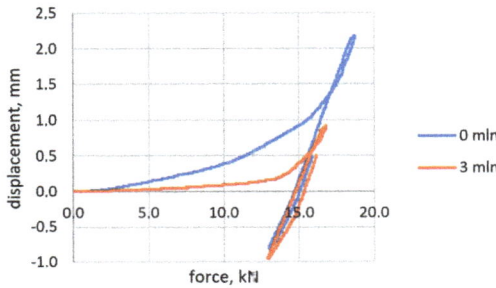

*Fig. 1. Comparison of the clamping force before and after the fatigue test. [Author: J. Siwiec]*

The subsequent test was to check a longitudinal resistance according to EN 13146-1+A1:2014 [5]. The aim of this test was to determine the characteristics of the longitudinal movement of the rail relative to the block support.

In order to do this, a horizontal load with a rising rate of 10 kN/min was applied to the rail until the rail slipped relative to the block support. The plastic and elastic displacement of the rail in relation to the concrete support was determined. The results of this test are particularly important from the point of view of the superstructure design, taking into account its susceptibility to temperature influences and resistance to longitudinal loads e.g. due to train emergency braking. An example of the test result is presented in Fig. 2.

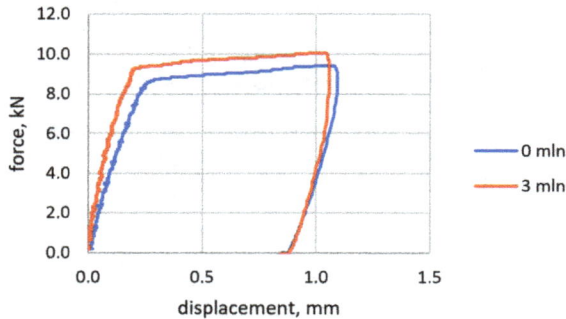

*Fig. 2. Comparison of the longitudinal rail restraint before and after the fatigue test. [Author: J. Siwiec]*

The last static test was to check the vertical stiffness, i.e. susceptibility of the concrete under static and dynamic loads in the vertical plane according to EN 13146-9+A1:2011 [6]. The construction assumption of the test was to obtain a density of polyurethane resin which would guarantee high flexibility and high operating resistance. During the test, a 3–fold linear incremental load of up to 64 kN was applied. Additionally, the support's response to the dynamic load of 5 Hz frequency was checked and the stiffness coefficient was calculated, which equals 18.2% for the new product.

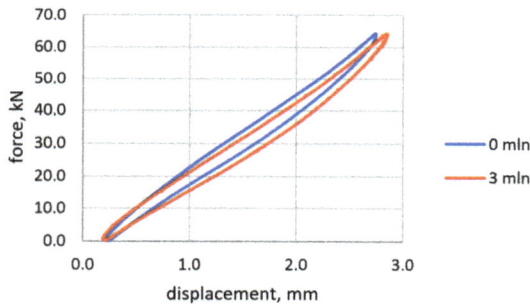

*Fig. 3. Comparison of the vertical stiffness of the embedded block system before and after the fatigue test. [Author: L. Antolik]*

The fatigue test scheme of the EBS was carried out in the number of 3 million cycles according to Table 1, cat. C for the fastening system with vertical stiffness below 50 MN/m. Figure 4 shows the object prepared for the fatigue test. The nature of the test and the maximum resultant force of $PV/\cos\alpha = 60$ kN are equivalent to maximum loads of the EBS during train running in a curve, assuming that the individual supports take the load according to the scheme shown in Fig. 5. This is equivalent to an uninterruptible load of 18 Tg transferred by support, which according to estimates gives trains the mass of about 90 Tg.

*Fig. 4. Embedded block system during the fatigue test. [Author: L. Antolik]*

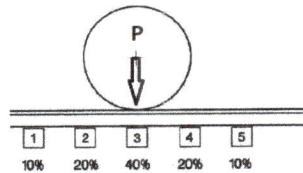

*Fig. 5. General idea of the load distribution for each support. [7]*

During the cyclic test, the displacement of the rail head and foot at 6 points was measured. The measurements were made for the first 1000 cycles and for the last 1000 cycles of 3 million load cycles. The results of the measurements in points 1÷4 in Fig. 6 are related to the vertical stiffness achieved by the block support. Particularly important measurement points are those marked with number 5 and 6 in Fig. 6. The values of plastic and elastic displacement measured at these locations lead to determination of the stability of the track gauge.

According to the Id-14 [8] maintenance instruction, the track gauge is measured at virtual points located 14 mm below the rolling surface of the rail head. Limit values of the track gauge in-service are shown in Table 2.

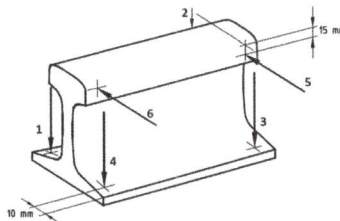

*Fig. 6. Displacement measurement places. [Author: L. Antolik]*

After the fatigue cycle, the vertical stiffness of the embedded block system, the longitudinal resistance of the rail to the block support and the rail clamping force to the support were re-examined in a fixed order. A visual inspection of the support components for mechanical damage was also carried out. It was stated that none of the elements was damaged. The polyurethane

Terotechnology XI                                                    Materials Research Forum LLC
Materials Research Proceedings **17** (2020) 240-245          https://doi.org/10.21741/9781644901038-35

resin did not show any cracks. Re-fixing or delamination that could disqualify the element were also not observed.

*Table 2. Track gauge deviations in-service [9].*

| Velocity [km/h] | Alert limit *AL* | | Intervention Limit *IL* | | Immediate Action Limit *IAL* | | Limit values acc. to TSI | |
|---|---|---|---|---|---|---|---|---|
| | min. | maks. | min. | maks. | min. | maks. | min. | maks. |
| | | | | [mm] | | | | |
| V ≤ 80 | −7 | 25 | −9 | 30 | −11 | 35 | −9 | 35 |
| 80 < V ≤ 120 | −7 | 25 | −9 | 30 | −11 | 35 | −9 | 35 |
| 120 < V ≤ 160 | −6 | 25 | −8 | 30 | −10 | 35 | −8 | 35 |
| 160 < V ≤ 230 | −4 | 20 | −5 | 23 | −7 | 28 | −7 | 28 |
| 230 < V ≤ 300 | −3 | 20 | −4 | 23 | −5 | 28 | −5 | 28 |

*Table 3. Results of the fatigue tests.*

| No. | Type of test | Results obtained after 3 mln cycles of repeated loadings | | Requirements acc. EN 13481-5+A1:2017 |
|---|---|---|---|---|
| 1 | Determination of the vertical stiffness | Sample 1 | approx. 6% | change ≤ 25 % |
| | | Sample 2 | approx. 5% | |
| 2 | Determination of the longitudinal restraint | Sample 1 | approx. 7% | $F_{min} \geq 7$ kN change ≤ 20 % |
| | | Sample 2 | approx. 17% | |
| 3 | Determination of the clamping force | Sample 1 | approx. 2% | change ≤ 20 % |
| | | Sample 2 | approx. 0% | |

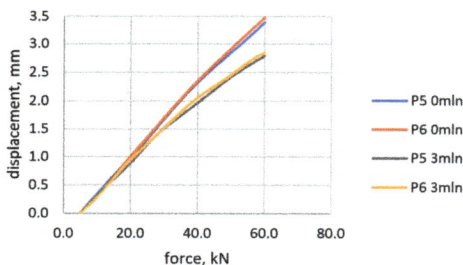

*Fig. 7. Effect of fatigue tests for measured lateral elastic displacement. [Author: L. Antolik]*

Table 3 presents the results of the fatigue tests. They are satisfactory from the point of view of the application possibilities of the tested embedded block system in the operation in the C category track according to EN 13481-5+A1:2017 [3] and at the operating speed of up to 250 km/h. At the same time, Fig 7 chart, which shows measured elastic side displacement of the rails enables to establish that within the maximum forces during laboratory tests, which should never occur in operations, the rail spacing may increase by approx. 3.5 mm due to the elastic displacement of the support, whereas after 3 million cycles this value decreases to approx. 2.8

Terotechnology XI
Materials Research Proceedings **17** (2020) 240-245

Materials Research Forum LLC
https://doi.org/10.21741/9781644901038-35

mm taking into account the plastic displacement of the support in relation to the seat, which was not measured during the experiment.

## Conclusions

The application of a specific solution for fastening the rail to the support system depends on the results expected by the infrastructure manager. Investment costs have also significant impact on the selected solution. In urban areas or engineering applications, where high damping properties are required, a slab track in combination with an embedded block system with relatively high elastic and vibration absorption is an effective solution.

The lifetime of each engineering object lasts longer if less vibration from the passing trains is absorbed. This also applies to densely populated areas, where seismic vibrations cause discomfort and impair the quality of life. Polish railway infrastructure in agglomerations use mainly wooden sleepers as a vibrations absorber. At the time of global debate on ecology, it is possible to apply a ready-made complementary solution where the parameter of high flexibility and attenuation by the rail fastening system is desired.

## References

[1] Chudyba Ł.: Fastening systems to concrete sleepers – comparison of the operating characteristics of fastening systems SB and W14, Przegląd Komunikacyjny no. 11, Warszawa 2017, pp. 27 – 32. https://doi.org/10.35117/A_ENG_17_11_05

[2] European Commission: Commission Regulation No. 1299/2014: on the technical specifications for interoperability relating to the 'infrastructure' subsystem of the rail system in the European Union, Brussels 2014

[3] EN 13481-5+A1:2017: Railway applications – Track – Performance requirements for fastening systems – Part 5: Fastening systems for slab track with rail on the surface or rail embedded in a channel.

[4] EN 13146-7:2012: Railway applications - Track - Test methods for fastening systems – Part 7: Determination of clamping force

[5] EN 13146-1+A1:2014 Railway applications – Track – Test methods for fastening systems – Part 1: Determination of longitudinal rail restraint

[6] EN 13146-9+A1:2011: Railway applications - Track - Test methods for fastening systems - Part 9: Determination of stiffness.

[7] Ł. Antolik: *Wpływ przekładki podszynowej na pracę systemu przytwierdzenia typu SB*, Problemy Kolejnictwa nr 177, Instytut Kolejnictwa, Warszawa 2017, pp. 7–12. https://doi.org/10.36137/1771p

[8] Id-14 (D-75) Instrukcja o dokonywaniu pomiarów, badań i oceny stanu torów, PKP Polskie Linie Kolejowe, Warszawa 2005

[9] H. Bałuch: Odchyłki dopuszczalne torów według normy europejskiej i wynikające stąd problemy, Technika Transportu Szynowego nr 6, Warszawa 2009, pp. 53 – 58

[10] Lakušić S., Haladin I. Ahac. M. *The Effect of Rail Fastening System Modifications on Tram Traffic Noise and Vibration.* Hindawi Publishing Corporation Shock and Vibration Volume 2016, Article ID 4671302, 15 pages. https://doi.org/10.1155/2016/4671302

Materials Research Forum LLC
https://doi.org/10.21741/9781644901038-36

# Investigations of the Head Check Defects in Rails

KOWALCZYK Dariusz[1, a*], ANTOLIK Łukasz[1,b], MIKŁASZEWICZ Ireneusz[1,c] and CHALIMONIUK Marek[2, d]

[1]Instytut Kolejnictwa (The Railway Research Institute), Materials & Structure Laboratory, 50, Chłopicki Street, 04-275 Warsaw, Poland

[2] Instytut Techniczny Wojsk Lotniczych, 6, Księcia Bolesława Street, 01-494 Warsaw, Poland

*[a] dkowalczyk@ikolej.pl, [b]lantolik@ikolej.pl, [c]imiklaszewicz@ikolej.pl ,
[d]marek.chalimoniuk@itwl.pl

**Abstract** The article presents the method and results of computed tomography (CT) tests of head-check defects occurring commonly in operation. These are one of the most common defects in rails that can lead to rail rupture. Based on previously performed CT and microscopic observations, rail models with typical defects were created and ultrasound beam propagation simulation was performed to increase the detectability of such defects.

**Keywords:** Rail, Ultrasound Examination, Head Checking, FEM Analysis

**Introduction**

Squats and checks, as the most common rail defects in operating conditions, are a serious problem in many rail systems around the world. Despite the work related to the modernization of railway lines, they are the most frequent rail defects in the main Polish railroad tracks managed by PKP PLK. According to UIC 712 card [1] and PKP PLK – Catalog of Defects in Rails [2], these defects are defined as head check defect - type 2223; squat defect -type 227, respectively.

The Railway Research Institute together with PKP PLK conducts a research project implemented as part of the scientific and research project POiR (Intelligent Development Program), which aims to increase the detection of defects in rails, including head checking. In the first part of the tests, as part of the research work, samples of rails with defects were taken, followed by observations, measurements, ultrasound and microscopy. The tests were carried out on 40 rail sections. The next stage was the selection of various head-checks propagation course and their 3D observations using computed tomography [3]. CT tests allowed obtaining images and information about the defect propagation inside the rail, defect size, depth, etc. The article presents the obtained examples of head-checks observed during the tests. Based on the information provided, the 3D models were created and FEM calculations were performed to compare the same mathematical model considering material properties, rail geometry, load, support for a rail with no defects and with head-check defects. The head-check defect was modeled based on previously obtained data from CT tests and observations and measurements performed with light microscopy.

What is a squat? Squats occur in several different forms and their precise definition is still the subject of some debate. All varieties, however, share some common features. They are characterized by cracking which initiates on the rail surface and grows down to a point about 3-6mm below the surface. The cracking then spreads along and across the rail, without growing substantially deeper. The rail surface becomes depressed and a dark patch appears due to a reduced contact from train wheels. Eventually the rail surface may spall out. Figure 1 illustrates some areas of multiple squats commonly termed "squatty" rail.

Terotechnology XI
Materials Research Proceedings **17** (2020) 246-250

Materials Research Forum LLC
https://doi.org/10.21741/9781644901038-36

The head checking defect (H-C) most often appears on the inner edge of the head of the outer rail, arranged in arches. It arises in places where the largest dynamic impact occurs (centrifugal force). It also appears on the lateral edges of railways in a straight track and at crossovers. It looks like small, parallel gaps with varying intervals. The distance between the gaps varies depending on local conditions and the type of steel the rail is made of, ranging from 1 mm to several cm. Defects of this type are difficult to detect, due to the promotion of them, among others, in areas a few millimeters below the rolling surface. The short distance of the ultrasonic wave through the material and the weak reflections make detection of this type of defects difficult [4].

*Fig. 1 Headchecking - rail under operating conditions*

Microscopic observations of defects in the rail head were made on a KEYENCE VHX - 900F digital microscope. Examples of headchecking defects in the rail head are shown in Fig. 1 and Fig. 2.

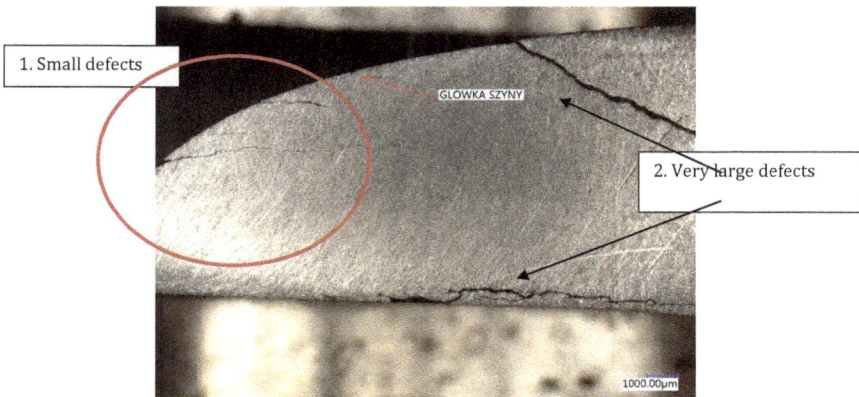

*Fig. 2 Head checking occurring in the rail head - cross section (microscopic observations).*

| Terotechnology XI | Materials Research Forum LLC |
| Materials Research Proceedings **17** (2020) 246-250 | https://doi.org/10.21741/9781644901038-36 |

Observations and measurements in a plane perpendicular to the rail cross section presented in Fig. 2, shows two cracks reaching up 10 mm, as well as four smaller ones in the surface zone of approx. 3 to 5. These photographs highlight how dangerous this type of defect is, because the complex crack net with significant propagation may lead to material decohesion in a short time.

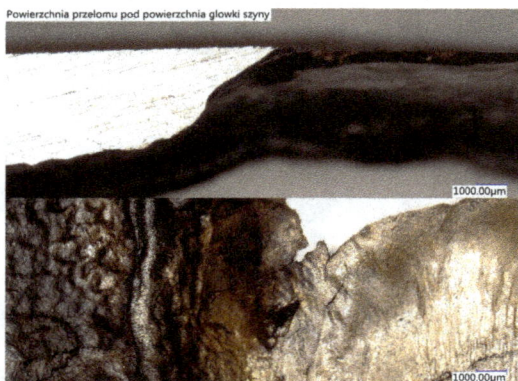

*Fig. 3 Head checking defects occurring in the rail head - cross-section (top) - destruction zone under the surface of the rolling zone approx. 2 mm (microscopic observations) - (bottom).*

**Computed tomography tests**

CT examinations were performed on GEphoenix v/tome/x m tomograph with a panel detector using a 300 kV X-ray tube. The examples of spatial images of propagation of head checking defects are presented below.

*Fig. 4. Visualization of the sample in the editor "myVGL"*

Terotechnology XI                                                Materials Research Forum LLC
Materials Research Proceedings 17 (2020) 246-250        https://doi.org/10.21741/9781644901038-36

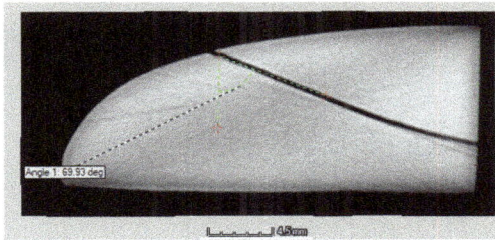

*Fig. 5. Visualization of the method of measuring the defect penetration angle.*

Based on the CT observation, the dimensions of the head-check defect on the surface and the depth of their retention were assessed. These quantities were characterized on the graph – Fig. 6.

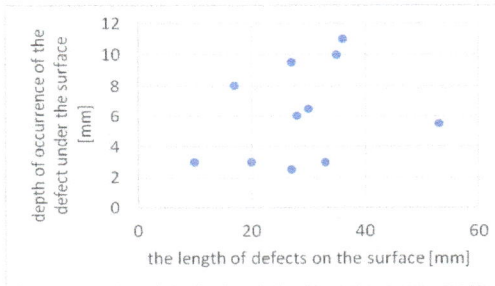

*Fig. 6. Relationship between the depth of the crack and its length on the surface [own source]*

**FEM Analysis**

On the basis of previously performed CT examinations and microscopic observations, examples of defect models in the 60E1 profile rail were created. Simulations of the propagation of ultrasonic waves in the rail material were made in the Altair Hyperworks program with Solver Optistruct [5]. Examplary simulations are shown below in Fig. 7 and Fig. 8.

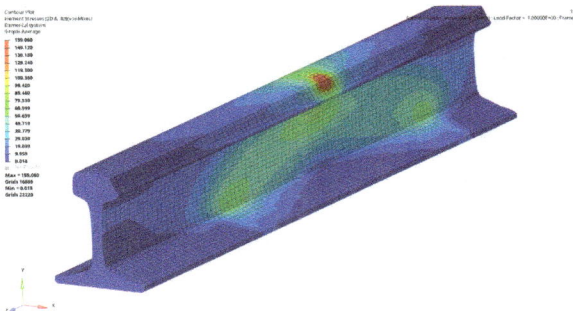

*Fig. 7 Distribution von Mises stresses in rail 60E1 rail without defects.*

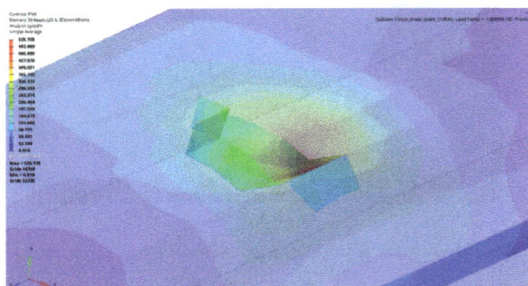

*Fig. 8 View of the 60E1 rail with modeled head-check defects (see transparency view).*

**Conclusion**

FEM analysis showed that for the same simulation assumptions, both without and after the introduction of a specific head-check defect, there is a significant increase in plastic deformation in the defect area, and thus a significant increase in stress. These calculations indicate that any contact of a freight wagon wheel (with a load close to the permissible load) causes crack propagation.

The presented photographic documentation and analysis of the results of headcheck defects show that these defects are difficult to detect due to their nature - occurrence below the surface and in the area of the rolling part of the rail, and numerous millimeter cracks. In addition, the numerous cracks net can quickly lead to material decohesion under further operating conditions.

**Acknowledgment**

The work was the result of the implementation of the research project No. POIR.04.01.01.-00-0011 / 17 financed by the National Center for Research and Development under the BRIK Project.

**References**

[1] UIC Code 712 Rail Defects / UIC International Union of Railways, Paris, 2002.

[2] PKP PLK – Katalog wad w szynach.

[3] E. Ratajczyk Tomografia komputerowa CT w zastosowaniach przemysłowych. Cz. I Ideapomiarów, główne zespoły i ich funkcje. Mechanik nr 2/2011.

[4] I.Mikłaszewicz Drogi szynowe – wady i uszkodzenia szyn, powstawanie i eliminacja, Prezentacja na Konferencji Drogi szynowe, Warszawa 2015.

[5] Altair HyperWorks 2017, OptiStruct and Examples, Altair® HyperWorks® v.2017.

Terotechnology XI
Materials Research Proceedings 17 (2020) 251-257

Materials Research Forum LLC
https://doi.org/10.21741/9781644901038-37

# The Influence of Pulse Current Frequency on Selected Aspects of Heat Transfer during GTA Welding of 321 Steel

OSTROMĘCKA Małgorzata[1a*], KOLASA Andrzej[2b] and CZARNECKI Marcin[1c]

[1]Instytut Kolejnictwa (The Railway Research Institute), Materials & Structure Laboratory, 50, Chłopicki Street, 04-275 Warsaw, Poland

[2]Warsaw University of Technology, 85, Narbutta Street, 02-524 Warsaw, Poland

[a]mostromecka@ikolej.pl, [b]a.kolasa@wip.pw.edu.pl, [c]marcinczarnecki@ikolej.pl

Keywords: GTAW, Heat Input, Welding Current, Arc Voltage, Pulsation Frequency

**Abstract:** Heat input is one of the parameters describing the arc welding process shown in the welding procedure specification. However, there is a strong need to precise its determination particularly when welding with arc pulsation. The paper presents an experiment illustrating the influence of the pulsed current frequency on heat transfer results when welding 321 steel with the GTAW method.

**Introduction**

The significance of knowledge about heat input in arc welding is well understood. But the issue of heat input representing the energy delivered to the workpiece in relation to the unit length causes many controversies. In recent years, several studies have been developed to underline the strong need to precise this technological parameter, which is widely used in industry and mentioned in the welding procedure specification (WPS). Welding standards, recommendations, and procedures usually determine ranges that the predicted heat input must fall within. But there are differences in the recommendations and formulas for heat input calculations depending on the country. Besides, the ranges of recommended heat input are sometimes very wide. The measurement and calculation aspect in relation to pulsed current processes [1, 2] is particularly controversial. Some welding equipment companies offer devices for measuring the heat input of pulsed current welding process in accordance with the newest interpretation of US standards. Unfortunately, the usage of these devices on the Polish market is still rare. In the case of pulsed current GTA welding at various pulsation frequencies, different results can be obtained for the same calculated value of heat input. These differences arise regardless of the calculation or measurement methodology used. Therefore, one can ask the question about the usefulness of a process parameter which at constant value cannot guarantee the same welding result.

**Experimental procedure**

In order to determinate the influence of the pulse current frequency on heat transfer during GTAW welding, the following experiment was performed:

1. mechanized pulsed current tungsten arc welding,
2. oscilloscope measurements of welding current and arc voltage and calculation of the heat input value,
3. macro- and microscopic observations of the structures of welds.

Workpieces of 321 steel [3] with dimensions of 250 x 50 x 3 mm were fixed on a copper back plate. Due to the possibility of precise setting of welding current pulsation parameters, Fronius TIG MagicWave 2500 DC power was used. The torch with the 2.4 mm 1.5% lanthanated

Materials Research Forum LLC
https://doi.org/10.21741/9781644901038-37

tungsten electrode was mounted on the line positioner. The process was performed with no filler material. As a shielding gas pure argon (9.8 l/min. flow) was used.

The measurements were performed with the Siglent SDS 1072CML oscilloscope and an intermediary KWR1 casette containing the necessary current and voltage measuring transducers [4]. The oscilloscope has two independent measuring channels with a range of 70 MHz. The screen displayed voltage and current waveforms as a function of time and measured true RMS and average values for these parameters.

*Table 1 Scheduled process parameters*

| Waveform color | Pulse current $I_p$ | Base current $I_b$ | Average current $I_{av}$ | RMS current $I_{RMS}$ | Duty cycle ratio $r_i$ | Current ratio $r_e$ | Welding speed v [mm/s] |
|---|---|---|---|---|---|---|---|
| blue | 150 | 20 | 85 | 107 | 0.5 | 0.13 | 2.33 |
| red | 130 | 40 | 85 | 96 | 0.5 | 0.31 | 2.33 |
| green | 110 | 60 | 85 | 88.6 | 0.5 | 0.55 | 2.33 |

The current and voltage measurements were made at the process stabilization and included an average calculated from a minimum of 5 pulse cycles. For microstructure observation Olympus BX 51M microscope was used. The etching was performed with the Mi16Fe reagent. Standard metallographic procedures were adopted for examining the microstructure of the weldments. The experiment included the execution of the waveforms shown in Fig.1 for three pulsation frequencies: 5.20 and 100 Hz. Settings on the device and calculated values of average and RMS current at 50% pulse duty ratio are given in Table 1.

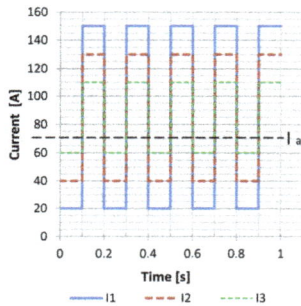

*Fig. 1 The waveforms representing the settings on the device for f = 5 Hz*

Considering the rectangular waveform, it is possible to analyze the electrical aspect of the arc's operation in terms of average or RMS current values. The average pulse intensity can be calculated using the following formula (1):

$$I_{av} = \frac{1}{T} \int_0^T I \, dt = \frac{(I_p \cdot t_p) + (I_b \cdot t_b)}{(t_p + t_b)} \tag{1}$$

Terotechnology XI
Materials Research Proceedings **17** (2020) 251-257

Materials Research Forum LLC
https://doi.org/10.21741/9781644901038-37

where: $T = t_p + t_b$ (period); $t_p$ - duration of the pulse; $t_b$- duration of the base current; $I_p$ and $I_b$ - the current of the pulse and the base current, respectively

When calculating the RMS current value, the following formula should be used (2):

$$I_{RMS} = \frac{1}{T}\left[\int_0^T I^2\, dt\right]^{\frac{1}{2}} = \left[\frac{(I_p{}^2 \cdot t_p) + (I_b{}^2 \cdot t_b)}{(t_p + t_b)}\right]^{\frac{1}{2}} \tag{2}$$

$r_e$, $r_i$ are defined:

$$r_e = \frac{I_b}{I_p} \qquad\qquad r_i = \frac{t_p}{T}$$

current ratio                    duty cycle ratio

The relation between the average and RMS current value of the current is as follows(3) [5]:

$$\frac{I_{av}}{I_{RMS}} = \frac{r_i + r_e \cdot (1 - r_i)}{(r_i + r_e^2 \cdot (1 - r_i))^{\frac{1}{2}}} \tag{3}$$

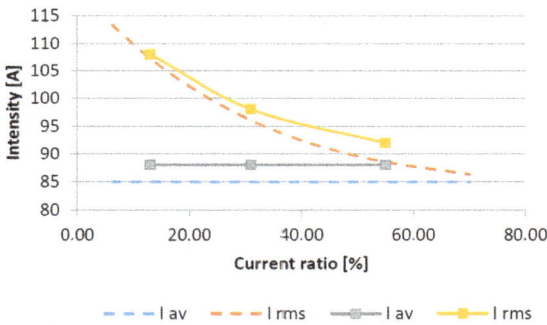

*Fig. 2 Current intensity and current ratio dependence:*
*full lines – experimental results; dashed lines – calculations from formulas (1.2).*

The above dependence indicates that the average current intensity is lower than the RMS one and the difference between $I_{RMS}$ and $I_{av}$ is the higher the lower the $r_e$ and $r_i$ values are, which is confirmed both at the theoretical level (TABLE 1) and during measurements and calculations made as part of this work (TABLE 2, Fig. 2). This leads to the conclusion that the use of average values in the calculation of pulse waveforms will cause large deviations from the real values, and for heat input it could be up to 30 % [2].

**Results and discussion**

It was observed that according to the change in the current ratio there is a slight change in voltage for both average and RMS values (Table 2). The average intensity is constant in all cases, whereas the RMS one decreases with the increase of the current ratio.

*TABLE 2. Actual parameter values measured using an oscilloscope and HI calculation*

| Spec. | Settings | | | Measurements | | | | Calculations | |
|---|---|---|---|---|---|---|---|---|---|
| | Pulse/base current | f [Hz] | $I_{av}$ [A] | $I_{RMS}$ [A] | $U_{av}$ [V] | $U_{RMS}$ [V] | | $HI_{av}$ [kJ/mm] | $HI_{RMS}$ [kJ/mm] |
| 1 | 150/20 | 5 | 88 | 108 | 14.4 | 15.2 | | 0.33 | 0.42 |
| 2 | 150/20 | 20 | 88 | 108 | 14.4 | 15.2 | | 0.33 | 0.42 |
| 3 | 150/20 | 100 | 88 | 108 | 14.4 | 15.2 | | 0.33 | 0.42 |
| 4 | 130/40 | 5 | 88 | 98 | 12.8 | 13.6 | | 0.29 | 0.34 |
| 5 | 130/40 | 20 | 90 | 98 | 12.8 | 14.4 | | 0.30 | 0.36 |
| 6 | 130/40 | 100 | 90 | 100 | 12.8 | 13.6 | | 0.30 | 0.35 |
| 7 | 110/60 | 5 | 88 | 92 | 12.8 | 14.4 | | 0.29 | 0.34 |
| 8 | 110/60 | 20 | 88 | 92 | 12.8 | 14.4 | | 0.29 | 0.34 |
| 9 | 110/60 | 100 | 90 | 94 | 13.6 | 14.4 | | 0.32 | 0.35 |

The frequency of current pulsation has no influence on the voltage and current in the given range of parameters. Instantaneous voltage changes can be interpreted by the weld pool oscillation during operation of the heat source, which leads to short-term changes in the arc length [5, 6]. Measured values were used to calculate heat input based on the following dependence (4) [7]:

$$HI = k \cdot \frac{U \cdot I}{v} \cdot 10^{-3} \tag{4}$$

where: HI – [kJ/mm], k - arc efficiency, U - arc voltage, I - welding current, v - welding speed

According to the conventional approach, arc efficiency coefficient k = 0.6 was assumed. Heat input values were calculated based on average and RMS values of voltage and current and the results were presented in TABLE 2.

The highest values of heat input (both for average and RMS parameters) were obtained for the process at the current ratio of 0.13 (samples 1 to 3 in Table 3). In the case of processes performed at 0.31 and 0.55 current ratios, the heat input values were similar. Heat input values calculated on the basis of RMS values were higher than those calculated from average ones and decreased according to the increase of the current ratio. In the formula (4), the pulsation frequency is not considered. Therefore, it would be necessary to exclude its influence on the calculated heat input of the process.

In macrostructure of the obtained weldments no physical surface defects like arc strike, cracks, and undercut were observed. Changes in the cross-sectional area of fusion zone can be considered as small (TABLE 3). However, comparison of the cross-sectional areas of the welds can provide information on the differences in the thermal efficiency. This information is of a general nature and its value is limited to a relative term, in which case the volume of molten material represented by the cross-section of the weld is greater than that of the other samples. The relationship between the amount of heat input and weld cross-section area is currently the subject of investigators' interest [8, 9] but it is an important part of the analysis of the heat input as a process parameter.

The quantitative aspect of the dependence between the melted area and the heat input can be described by formula (5) [10]:

$$A = \frac{k' \cdot k \cdot I \cdot U}{H \cdot v} \tag{5}$$

Terotechnology XI
Materials Research Proceedings **17** (2020) 251-257

Materials Research Forum LLC
https://doi.org/10.21741/9781644901038-37

where: A - cross-sectional area [mm$^3$], k' and k melting and arc efficiency, respectively, I - welding current intensity [A], U - arc voltage [V], v - welding speed [mm/s] H - the theoretical amount of heat necessary to melt the volume unit of metal J/mm$^3$ (melting enthalpy).

The highest cross-sectional value (9 mm$^2$) was obtained for specimen 3 welded with the process parameters: $I_p$ = 150A, $I_b$ = 20 A and f = 100 Hz. The lowest cross-sectional area (4.95 mm$^2$) was obtained for sample 9 with the parameters: $I_p$ = 110 A, $I_b$ = 60 A and f = 100 Hz. With reference to formula (5), the conclusion arises that the change in the pulsation frequency affects the melting efficiency and/or thermal efficiency. The volume of molten metal for the current ratio of 0.13 increases with increasing frequency. It should be noted that for different current ratio, the frequency of pulsations exhibits different tendencies of such changes (Fig. 3, Fig. 4)

*TABLE 3. Weld cross-section areas in dependence on process parameters*

| Spec. | 1 | 2 | 3 | 4 | 5 | 6 | 7 | 8 | 9 |
|---|---|---|---|---|---|---|---|---|---|
| Pulse/base current; frequency | 150/20A 5 Hz | 150/20A 20 Hz | 130/40A 100 Hz | 130/40A 5 Hz | 130/40A 20 Hz | 110/60A 100 Hz | 110/60A 5 Hz | 110/60A 20 Hz | 110/60A 100 Hz |
| A [mm$^2$] | 7.36 | 8.47 | 9.0 | 6.44 | 5.82 | 7.24 | 6.39 | 6.75 | 4.95 |

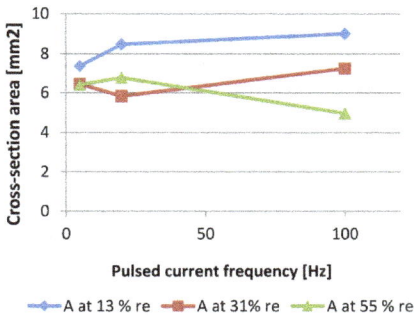

*Fig. 3 Cross-sectional area and pulsation frequency dependence at various current ratios*

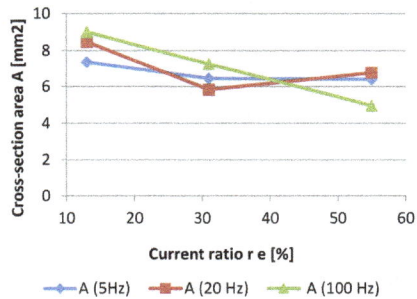

*Fig. 4 Cross-sectional area and current ratio dependence at various pulsation frequencies*

Terotechnology XI
Materials Research Proceedings **17** (2020) 251-257

Materials Research Forum LLC
https://doi.org/10.21741/9781644901038-37

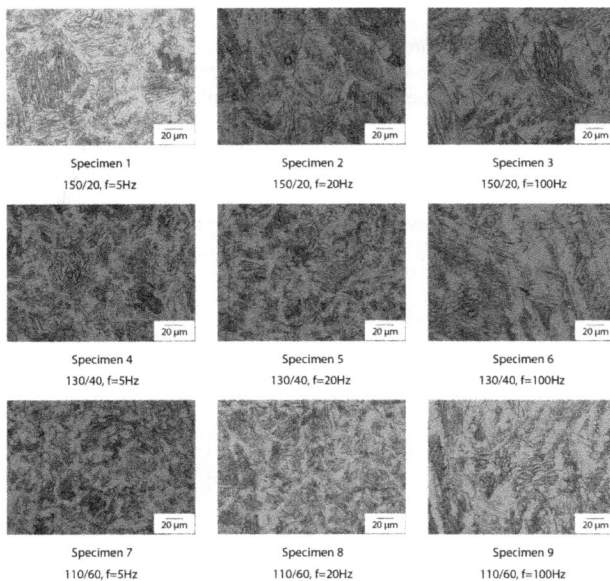

| | | |
|---|---|---|
| Specimen 1 | Specimen 2 | Specimen 3 |
| 150/20, f=5Hz | 150/20, f=20Hz | 150/20, f=100Hz |
| Specimen 4 | Specimen 5 | Specimen 6 |
| 130/40, f=5Hz | 130/40, f=20Hz | 130/40, f=100Hz |
| Specimen 7 | Specimen 8 | Specimen 9 |
| 110/60, f=5Hz | 110/60, f=20Hz | 110/60, f=100Hz |

*Fig. 5 Microstructures of weld metal at x500 magnification*

Weld metal microstructures contained austenitic matrix with delta ferrite in the form of elongated grids and plates (Fig. 5). In the structure of the fusion, heat affected zones and base metal, titanium carbide crystals were present. Closer to the weld face, in the axis of fusion zone, heterogeneous solidification of austenite dendrites could be observed, which may indicate relatively high degree of constitutional supercooling. Delta ferrite was also observed in the HAZ in the form of long stringers arranged according to the direction of steel rolling. The presence of delta ferrite in the structure of such steel is usually explained by the low stability of the austenitic structure. The comparison of microstructures obtained at different current ratios does not provide definitive conclusions, although it may seem that with the increase of the current ratio, the structure became more fine-grained. On the other hand, according to the frequency increase, the structure undergoes a significant change as the austenite grains grow. Such a change must be connected with the extension of the cooling time, and thus the reduction of the cooling rate. The observation of the microstructure led to the conclusion that there are some differences in heat transport in dependence of pulsation frequency. The highest heat input was transferred into the material when melting the sample No. 3 at the settings $I_p = 150$ A, $I_b = 20$ A f = 100 Hz. The influence of frequency on the microstructure is visible in all three current ratio settings. Heat input calculations based on formula (4) show no value differences in relation to the pulsation frequency, and therefore are not consistent with the observations of real macro- and microstructures. The results of macroscopic and microscopic investigations lead to the conclusion that they represent different aspects of heat transport to the material. Similar volumes of molten metal can differ significantly in microstructure.

Terotechnology XI                                         Materials Research Forum LLC
Materials Research Proceedings **17** (2020) 251-257        https://doi.org/10.21741/9781644901038-37

## Conclusions

The frequency of current pulsation does not affect significantly the intensity or arc voltage in the tested frequency range.

1. The calculated heat input values are constant regardless of changes of pulsing frequency. It results from the omitting the effect of the pulsation frequency in the commonly used formula. In fact, the change in the pulsation frequency causes the formation of welds with different cross-sectional area and with different microstructure. The volume of molten metal and microstructure inform about different aspects of heat transport.
2. An increase in the frequency of current pulsation does not necessarily lead to an increase in the depth of penetration.
3. The suitability of the formula for heat input in the context of pulsed TIG welding should be reviewed.

## References

[1] P. Cegielski, A. Kolasa, M. Kuczyński, R. Rostkowska, Some aspects of monitoring and measurements during arc welding, Welding Technology Rev. 88 (12) (2016) 43-50. https://doi.org/10.26628/ps.v88i12.719

[2] LEB2/2010 Lincoln Electric

[3] Welding Handbook vol.4. Materials and applications - Part 2.1998. Chapter 5 pp. 233-332

[4] P. Cegielski, Ł. Bugyi, Selected aspects of welding defects identification in MIG/MAG arc welding, Welding Technology Rev. 89 (6) (2017) 30-35. https://doi.org/10.26628/ps.v89i6.784

[5] W. H. Kim, S. J. Na, Heat and fluid flow in pulsed current GTA weld pool, Int. J. Heat Transf. 41 (1998) 3213-3227. https://doi.org/10.1016/S0017-9310(98)00052-0

[6] H. G. Fan, S-J. Na and Y. W. Shiz, Mathematical model of arc in pulsed current gas tungsten arc welding, J. Phys. D: Appl. Phys. 30 (1997) 94–102. https://doi.org/10.1088/0022-3727/30/1/012

[7] PN-EN 1011-1: 2009 Spawanie – Zalecenia dotyczące spawania metali – Część 1: Ogólne wytyczne dotyczące spawania łukowego.

[8] K. Wojsyk, M. Macherzyński, Determination of Welding Linear Energy by Measuring Cross-Sectional Areas of Welds, Biul. Inst. Spaw. 60 (5) (2016) 83-89. https://doi.org/10.17729/ebis.2016.5/11

[9] K. Wojsyk, M. Macherzyński, R. Lis, Evaluation of the amount of heat introduced into the welds and padding welds by means of their transverse fields measurement in conventional and hybrid welding processes, Welding Technology Rev.89 (10) (2017) 67-82.

[10] A. Klimpel, Technologia spawania i cięcia metali, Gliwice, Politechnika Śląska, 1997.

Terotechnology XI
Materials Research Proceedings **17** (2020) 258-263

Materials Research Forum LLC
https://doi.org/10.21741/9781644901038-38

# Tests and FEM Calculations for the Screw Coupling 1MN

KOWALCZYK Dariusz[1, a *] and RAGUS Izabela[1,b,]

[1] Instytut Kolejnictwa (The Railway Research Institute), Materials & Structure Laboratory, 50 Chlopicki Street, 04-275 Warszawa, Poland

[a] dkowalczyk@ikolej.pl [b] iragus@ikolej.pl

**Keywords:** FEM Analysis, Screw Coupling, Crack

**Abstract** The article describes mandatory requirements for screw couplings in the service release. Additionally, the test results for the coupling and their comparison with FEM calculations are presented.

**Introduction**
Screw couplings are universal connections for railway vehicles. The safety of the integrity of the train composition in operating conditions depends on these devices, understood as a combination of various types of wagons and locomotives. Not only are these devices subjected to very high loads, but also must be resistant to long-term influence of changing weather conditions. Approval of this type of equipment for operation requires carrying out tests in the scope described in current European standards and the following directives: EN 15566: 2016 [1]; EN15566: 2011[2]; Rolling Stock TSI - Freight Wagons WAG / Commission Regulation (EU) No. 321/2013[3]; TSI - Rolling stock - Locomotives and passenger rolling stock / Commission Regulation (EU) No. 1302/2014 [4].

The EN 15566 standard defines the range of forces to be applied for screw couplings operating in load ranges of 1MN; 1.2 MN; 1.5 MN, depending on the time of use (20 or 30 years).

*Table 1. Condition of dynamic tests for screw coupling [1]*

| Operational requirements | Range of forces to be applied | | |
|---|---|---|---|
| Lifecycle in years | Designation | Step 1 | Step 2 |
| | 1 MN | $\Delta F1 = 170$ kN | $\Delta F2 = 575$ kN |
| | 1.2 MN | $\Delta F1 = 205$ kN | $\Delta F2 = 690$ kN |
| | 1.5 MN | $\Delta F1 = 270$ kN | $\Delta F2 = 910$ kN |
| | | $N_1$ in cycles | $N_2$ in cycles |
| 20 | all | $10^6$ | $1.45 \times 10^3$ |
| 30 | all | $1.5 \times 10^6$ | $2.15 \times 10^3$ |

The area (surface) of the shackle-hook interaction is the crucial place in the screw coupling. In this area, there is a surface contact of the coupling and hook elements, as well as force transmission, including high surface pressure. The following are the results of tests for 1MN screw coupling in the 30-year operation variant.

Terotechnology XI                                             Materials Research Forum LLC
Materials Research Proceedings **17** (2020) 258-263          https://doi.org/10.21741/9781644901038-38

## Fatigue test

Fatigue tests of screw coupling 1MN (part of the required tests) were carried out on a LFV testing machine with a maximum tensile force of 2.5 MN Fig.2. An example of a fragmentary graph of fatigue history is presented in Fig.3. The maximum tensile forces for the coupling were 575 kN during this study. The number of the highest loads F = 575 kN in the entire fatigue tests was 2150 cycles.

*Fig. 1. Screw coupling 1MN on the testing machine during tests*

After the completion of fatigue tests, the screw coupling was subjected to destructive tests, using a magnetic powder method to assess the surface condition of the bow (screw coupling) in the area of surface contact with the hook. The NDT tests revealed cracks on the shackle of the 1MN screw coupling. The results of the tests (Fig.4) are presented below. Regarding the requirements and assessment criteria described in standard EN 15566:2016, discontinuities / fractures up to 20 mm are allowed.

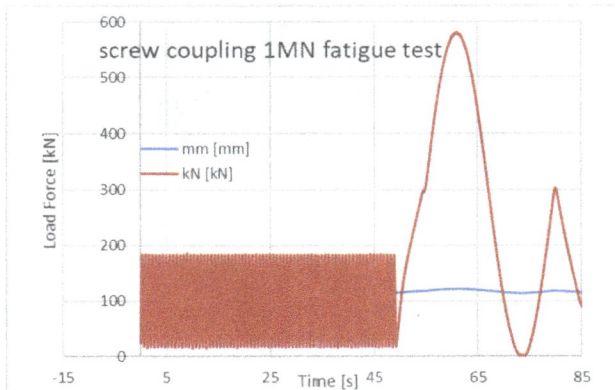

*Fig. 2. The course of fatigue loads for the screw coupling*

Terotechnology XI

Materials Research Proceedings **17** (2020) 258-263

Materials Research Forum LLC

https://doi.org/10.21741/9781644901038-38

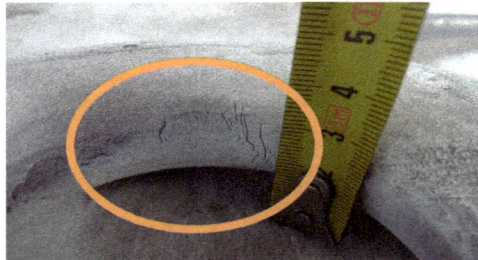

*Fig. 3 Non-destructive testing NDT - magnetic-powder method, observations of the shackle screw coupling 1MN after fatigue tests. Visible cracks in the marked area.*

In order to accurately assess cracks in the shackle screw coupling 1MN, it was subjected to further examinations.

**Computed tomography tests**

CT examinations were performed on GEphoenix v/tome/x m tomograph with a panel detector using a 300 kV X-ray tube. The examples of spatial images of cracks propagation of shackle screw coupling 1MN are presented in Fig.4.

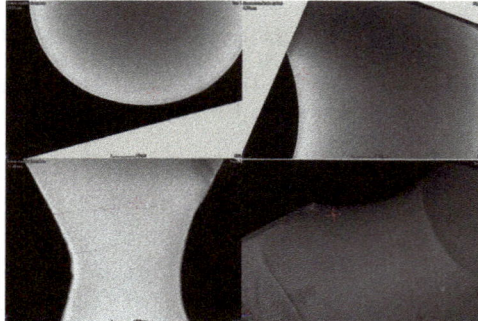

*Fig. 4 The cracks propagation in the shackle screw coupling 1MN*

The obtained CT images allow for observing the propagation course of cracks in 3D and determining their actual length and depth of occurrence below the surface.

Currently, the requirements of this standard only determine the surface length of the defect. According to the observations made, it was found that cracks could propagate to 5 mm even below the 1MN screw connection of the contact surface.

**FEM Analyses**

The shackle of 1MN screw coupling was modelled in the SolidWorks program. Boundary

Terotechnology XI
Materials Research Proceedings **17** (2020) 258-263

Materials Research Forum LLC
https://doi.org/10.21741/9781644901038-38

conditions of FEM model were prepared in the HyperMesh program. The Altair Optistruct program was the calculation solver [5],[6].

FEM calculations are often a good support in the analysis of various events, damage and help to understand the causes of damage and destruction [7].

Literature data was used for 40CrMo4 steel [8], from which the shackle was made. The load was applied considering plastic deformations.

Sample results of FEM load simulation for the shackle of 1MN screw coupling model are shown in Fig.5, Fig.6 and Fig.7.

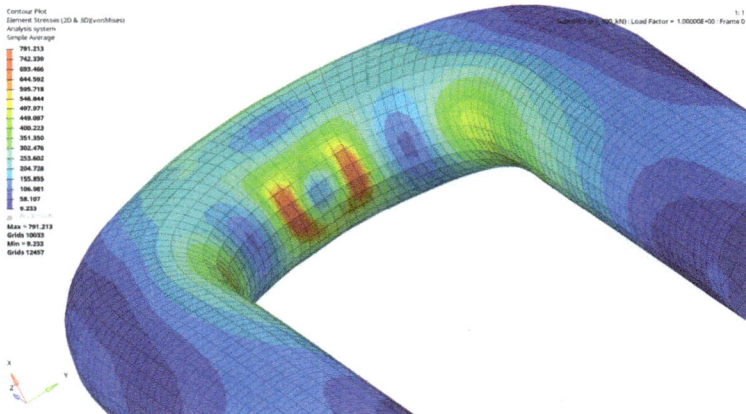

*Fig. 5. Stress according to von Mises for a load of 575kN.*

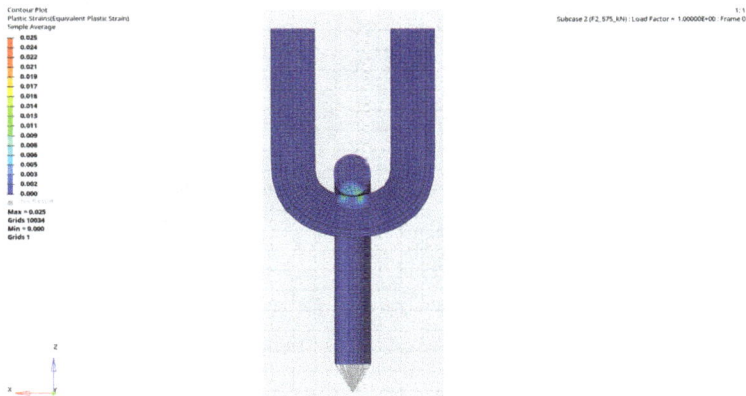

*Fig. 6. Plastic strains for a load of 575kN.*

Terotechnology XI                                           Materials Research Forum LLC
Materials Research Proceedings **17** (2020) 258-263        https://doi.org/10.21741/9781644901038-38

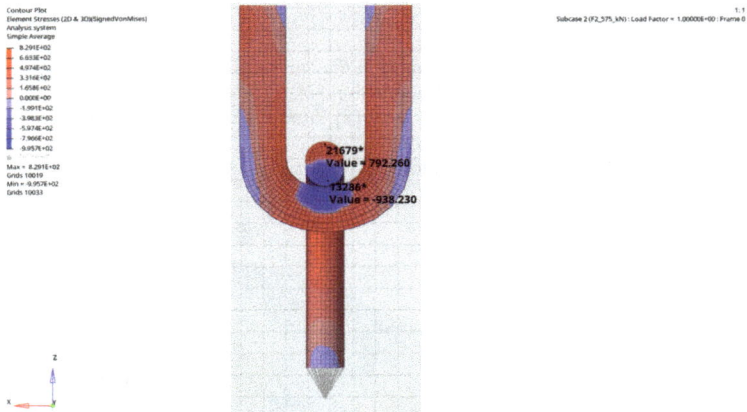

*Fig. 7. Graphic representation of stress distribution in the shackle of screw coupling. Tensile and compressive stresses according to Signed/von Mises for a load of 575kN.*

**FEM calculation results**

As it results from FEM calculations, plastic deformation may already occur at the load of 575 kN at the highest load in fatigue tests repeated up to 2150. These deformations may occur in the area of contact and near-surface areas of shackle of 1MN crew coupling shown in Fig.6.

In the contact hook/shackle area, mainly compressive stresses (blue colour) occur, which is shown in Fig. 7.

**Summary**

The NDT (non-destructive testing) and CT (computed tomography) tests performed after fatigue loads of 1M screw coupling confirm that the contact surface of the shackle / hook is a critical area of crack occurrence. CT studies show that in the analyzed case the crack propagated up to 5 mm below the surface of the bow. As it results from FEM calculations, plastic deformation may already occur at the load of 575 kN.

**References**

[1]  PN-EN 15566: 2016 Railway applications. Railway rolling stock. Draw gear and screw coupling

[2]  PN-EN 15566: 2011 Railway applications. Railway rolling stock. Draw gear and screw coupling

[3]  Commission Regulation (EU) No. 321/2013 of 13 March 2013 concerning the technical specification for interoperability relating to the subsystem 'rolling stock — freight wagons' of the rail system in the European Union and repealing Decision 2006/861/EC.

[4]  TSI - Rolling stock - Locomotives and passenger rolling stock / Commission Regulation (EU) No. 1302/2014.

[5]  Altair Engineering, Inc. RADIOSS AND OPTISTRUCT THEORY MANUAL 11.0 Version Jan, Large Displacement Finite Element Analysis (2011).

Terotechnology XI
Materials Research Proceedings **17** (2020) 258-263

Materials Research Forum LLC
https://doi.org/10.21741/9781644901038-38

[6] Altair Engineering, Inc. OptiStruct 13.0 User's Guide.

[7] D. Kowalczyk, R. Bińkowski, Causes of wheelset damage. FEM analysis., Railway Reports 61 (2017) 47-52. https://doi.org/10.36137/1756p

[8] Materials -Mechanical properties of steel   42CrMo4 (1.7225)

Terotechnology XI
Materials Research Proceedings **17** (2020) 264-269

Materials Research Forum LLC
https://doi.org/10.21741/9781644901038-39

# Laser Welding of Girth Joint – Numerical Simulation and Experimental Investigation

DANIELEWSKI Hubert[1,a]*, SKRZYPCZYK Andrzej[1,b], PAŁA Tadeusz[1,c], FURMAŃCZYK Piotr[1,d], TOFIL Szymon[1,e] and WITKOWSKI Grzegorz[1,f]

[1]Kielce University of Technology, Faculty of Mechatronics and Mechanical Engineering, Al. 1000-lecia P.P. 7, 25-314 Kielce, Poland

[a]*hdanielewski@tu.kielce.pl, [b]tmaask@tu.kielce.pl, [c]tpala@tu.kielce.pl, [d]pfurmanczyk@tu.kielce.pl, [e]tofil@tu.kielce.pl, [f]gwitkowski@tu.kielce.pl

**Keywords:** Laser Welding, Numerical Simulation, Girth Joints, Circumferential Trial Joint

**Abstract.** Laser welding of similar girth joints is a commonly used method of permanent joining process. Parameters can be easily analytically estimated. Nevertheless, performing more advanced types of joints, such as girth joints, is more complex. Estimating process parameters using numerical simulation can be performed. Assuming laser beam energy absorption in material heat expansion isotherms of a melting point can be calculated. Performing girth joint requires adequate angle process head orientation and welding parameters. Laser welding process is an unconventional method and does not require any additional material. The arc welding of a flange joint requires a multi run welding, and a single run welding can be performed using a laser beam. The article presents the possibility of numerical modeling laser welding of girth joints using the SimufactWelding software. Thermal and thermo-mechanical analysis were performed. The parameters estimated in the laser welding simulation are the following: output power, feed rate, efficiency and intensity distribution - Gaussian parameter. Low carbon construction steel S235JR is the material used for the process simulation. Double cylindrical simulated laser beam absorption and keyhole effect are the heat source used for the laser welding simulation. Thermal and thermo-mechanical properties such as fusion zone, distortion, displacement and hardness distribution were calculated [1, 2]. Experimental laser welding of lap joints was performed based on estimating parameters. Welding process using 6 kW $CO_2$ laser with simulated parameters in order to compare numerical and experimental results was performed.

**Introduction**
Laser welding uses concentrated photons beam for melting material. Heat source moves and starts the solidification and crystallization process of a material. Using the concentrated energy of focused electrons or photons, high energy density can be obtained. High velocity of the welding process combined with high energy density affects in relative low thermal energy absorbed in the material. Welding parameter investigations for application like girth joints are problematic, require many technological trials and some knowledge of the welding process. Simple calculations of welding based on solving conduction equation which are proposed by Rosenthal (1) can be used [3, 4].

$$T - T_0 = \frac{q}{2\pi k r} e^{\frac{-v(r-x)}{2\alpha}} \tag{1}$$

Terotechnology XI                                          Materials Research Forum LLC
Materials Research Proceedings **17** (2020) 264-269          https://doi.org/10.21741/9781644901038-39

The evolution of computation is based on mathematic description of heat sources evolution. A resolving moving heat sources equation shape of weld can be estimated. In the article, the possibility of a numerical simulation of girth joint using SimufactWelding software are presented.

## Numerical simulation of laser girth joints welding

An estimation of weld dimension by solving moving heat source equation can be performed. Material phase transformation affects material properties. Using the numerical analysis of a thermo- mechanical simulation, a simplified material metallographic structure can be estimated. A numerical simulation using a volumetric cylindrical heat source model was calculated. For estimating the laser radiation absorption surface disc and for the simulated keyhole effect, conical cylindrical heat source is used [5-8]. Dimensions of volumetric heat sources were defined. In the simulation of a welding process fusion zone and properties of welded joint were calculated. In boundary conditions, rigid restraint for welded elements are used. A disc-shaped sheet plate with a thickness of 4mm and pipe shape cylindrical elements were meshed using a ringmesh with hexahedral finite elements. The weld trajectory across the contact line is determined. A simulation of welding is conducted for S235JR steel, and complete multi-phase material library is used. A numerical simulation of complete joint penetration was performed (Fig 1.). To perform a phase transformation, the adequate cooling time was established as 30s.

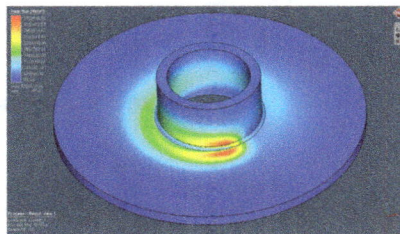

*Fig. 1. Numerical simulation of laser welding girth joint - heat expansion in closing of the weld.*

A welding simulation for estimated parameters of laser complete penetration was performed. Simulations with constant speed rate equal to 1m/min, efficiency of 0.95 and Gaussian parameter of disc and conical heat source equal to 3 were programmed. Laser output power was changing from 3 to 6kW with the step of 500W. Complete welding penetration was obtained at 6kW (Fig.2).

Welding of circumferential girth joints is problematic, as the forced position of welding head orientation gives some limitation and has great impact on the process. Thermo-mechanical analysis with phase transformation gives realistic results of the welding process with the convex face of a weld and material deformation. Fillet joints such as girth joints have considerably high welding stress and deformation, some applications require additional heat treatment, nevertheless a stress-strain analysis is required. The results of numerical analysis of stress and strain indicated in welded material (Fig. 5) showed concentration in HAZ and fusion zone. The crystallographic structure of a welded material (Fig. 3, 4, 5) is related to phase transformations during the laser welding process.

Terotechnology XI                                                    Materials Research Forum LLC
Materials Research Proceedings **17** (2020) 264-269          https://doi.org/10.21741/9781644901038-39

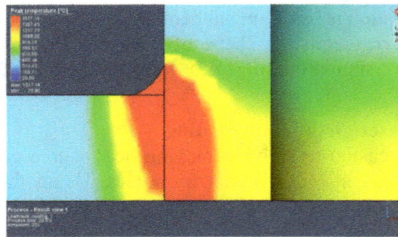

*Fig. 2. Simulated results of fusion zone geometry in cross section.*

The numerical simulation showed phase transformation of fusion zone, primary a ferritic-pearlitic crystallographic structure change. The results of numerical simulation showed a transformation into a bainitic structure with ferritic-pearlitic inclusions (Fig. 3). Low carbon construction steel S235JR used in numerical simulation is a typical ferritic material with precipitation of pearlite.

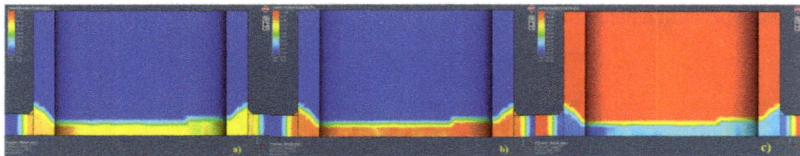

*Fig. 3. Distribution of a) bainite, b) pearlite and ferrite in simulated girth joint.*

The structure of a material defines material properties. Strength characteristic can be tested using tensile tests and hardness tests. Strength analysis requires destructive methods, nevertheless hardness based on phase transformation and thermal cycles in material can be calculated (Fig. 4).

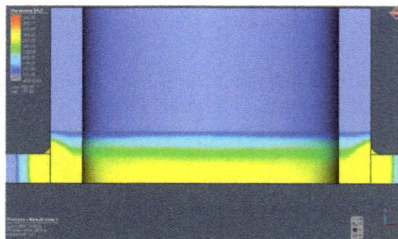

*Fig. 4. Distribution of hardness in a simulated laser welded girth joint.*

The phase transformation of welded material and heat expansion generated thermal stress. State of stress can be reduced by an additional heat treatment. Due to defined value of occurring stresses, an experimental strain gauge measurement or numerical simulation can be performed (Fig. 5). If stress exceeds maximum acceptable value defined by element usage, stress relaxation is recommended.

The value of stress calculated in the numerical simulation does not exceed 340 MPa. The greatest stress concentration occurred in the weld and HAZ. Stress in base materials is relatively low. Due to high energy density in the laser welding process, stress and strain in material are lower comparing to conventional welding methods [9-12]. Nevertheless, stress state in welded joint needs to be controlled.

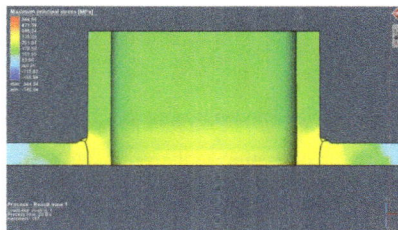

*Fig. 5. Stress distribution in simulated girth joint.*

**Laser welding of the test girth joint**
Process parameters of the laser girth joint welding using numerical simulation were estimated. Laser welding trial girth joint was performed using a high power $CO_2$ laser Trumpf TruFlow 6000 with 6 axis work centre LaserCell 1005. Estimated parameters for trial joint to obtain complete penetration were performed [13-15]. 6kW output power with speed ratio equal to 1m/min and welding heat with a focal length of 270 was used. To reduce the plasma effect, helium as a shielding with a coaxial flow rate of 15l/min gas was used.

The results of trial welding girth circumferential joint were observed (Fig.6). A complete penetration was obtained; nevertheless, due to the forced position, the length of edge fusion zone was obtained only partially [16, 17].

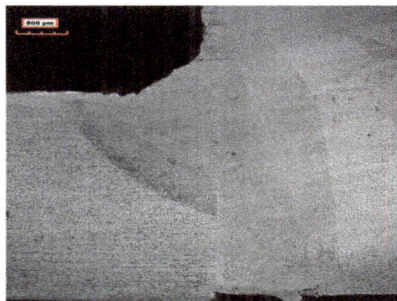

*Fig. 6. Macrostructures of partial lap joint weld in cross section.*

**Hardness test of laser welded lap joint**
Welding process affects the crystallographic structure of welded materials. To confirm simulation results, joint hardness test using Innovatest Nexus 4304 was performed (Fig 7). The hardness test was carried out according to PN-EN ISO 6507-1 [18].

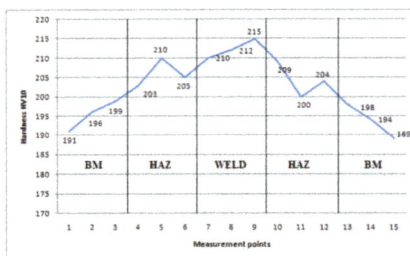

*Fig. 7. Hardness distribution in weld cross-section*

Hardness test results showed strengthening in the weld and HAZ. The measured hardness in the weld has the value of approximately 210HV10, and does not exceed 215HV10, HAZ are lower and take value from 200 to 210HV10. According to PN-EN ISO 15614-11, maximum allowable limit of Vickers hardness HV10 after welding process is 350. No additional post weld heat treatment is required.

**Summary**
Numerical simulation of laser welding allows for estimating process parameters of full penetration. A simulation of girth joint processes to assume welding parameters was performed. The results of simulation and experimental trial joint showed similarity. Caused by force welding head position, no full edge fusion zone in the welded material was performed. To define properties of the weld, hardness test was performed. The results showed good strength characteristics. Stress analysis based on numerical simulation showed low stress concentration, located only in the weld and HAZ. The hardness test showed hardening in the weld and HAZ, but no additional heat treatment was required.

**Funding**
Research carried out in the NCBiR (Narodowe Centrum Badań i Rozwoju) project nr LIDER/31/0173/L-8/16/NCBR/2017 „Technology of manufacturing sealed weld joints for gas installation by using concentrated energy source".

**References**
[1] O. Andersson, N. Budak, A. Melander, N. Palmquist, Experimental measurements and numerical simulations of distortions of overlap laser-welded thin sheet steel beam structures, Welding in the World 61 (2017) 927-934. https://doi.org/10.1007/s40194-017-0496-z

[2] B. Kogo, B. Wang, L. Wrobel, M. Chizari, Experimental and Numerical Simulation of Girth Welded Joints of Dissimilar Metals in Clad Pipes, Int. J. Offshore Polar 28 (2018) 380-386. https://doi.org/10.17736/ijope.2018.oa22

[3] T. Kik, J. Górka, Numerical Simulations of Laser and Hybrid S700MC T-Joint Welding, Materials 12 (2019) art. 516. https://doi.org/10.3390/ma12030516

[4] W. Piekarska, M. Kubiak, A. Bokota, Numerical simulation of thermal phenomena and phase transformations in laser-arc hybrid welded joints, Arch. Metall. Mater. 56 (2011) 409-421. https://doi.org/10.2478/v10172-011-0044-6

Terotechnology XI                                     Materials Research Forum LLC
Materials Research Proceedings **17** (2020) 264-269          https://doi.org/10.21741/9781644901038-39

[5] H. Danielewski, Laser welding of pipe stubs made from super 304 steel, Numerical simulation and weld properties, Technical Transactions 116 (1) (2019) 167-176. https://doi.org/10.4467/2353737XCT.19.011.10051

[7] N. Radek, J. Pietraszek, A. Goroshko, The impact of laser welding parameters on the mechanical properties of the weld, AIP Conf. Proc. 2017 (2018) art. 020025. https://doi.org/10.1063/1.5056288

[8] E. D. Derakhshan, N. Yazdian, B. Craft, S. Smith, R. Kovacevic, Numerical simulation and experimental validation of residual stress and welding distortion induced by laser-based welding processes of thin structural steel plates in butt joint configuration, Optics & Laser Technology 104 (2018) 170-182. https://doi.org/10.1016/j.optlastec.2018.02.026

[9] N. Radek, K. Bartkowiak, Laser treatment of electro-spark coatings deposited in the carbon steel substrate with using nanostructured WC-Cu electrodes, Physics Procedia 39 (2012) 295-301. https://doi.org/10.1016/j.phpro.2012.10.041

[10] N. Radek, J. Konstanty, Cermet ESD coatings modified by laser treatment, Arch. Metall. Mater. 57 (2012) 665-670. https://doi.org/10.2478/v10172-012-0071-y

[11] M. Węglowski, J. Niagaj, J. Rykała, E. Turyk, P. Sędek, Mechanical properties and metallographic characteristics of girth welded joints made by the arc welding processes on pipe steel grade api 5l x70, Adv. Manuf. Sci. Tech. 41 (4) (2017) 51-62.

[12] F. Coste, T. Azeroual, Double sided laser girth welding of high thickness pipe-line, ICALEO 2016, 1801. https://doi.org/10.2351/1.5118553

[13] J. Pietraszek, N. Radek, K. Bartkowiak, Advanced statistical refinement of surface layer's discretization in the case of electro-spark deposited carbide-ceramic coatings modified by a laser beam, Solid State Phenom. 197 (2013) 198-202. https://doi.org/10.4028/www.scientific.net/SSP.197.198

[14] A. Gądek-Moszczak, N. Radek, S. Wroński, J. Tarasiuk, Application the 3D image analysis techniques for assessment the quality of material surface layer before and after laser treatment. Advanced Materials Research 874 (2014) 133-138. https://doi.org/10.4028/www.scientific.net/AMR.874.133

[15] N. Radek, J. Pietraszek, B. Antoszewski, The average friction coefficient of laser textured surfaces of silicon carbide identified by RSM methodology, Advanced Materials Research 874 (2014) 29-34. https://doi.org/10.4028/www.scientific.net/AMR.874.29

[16] M. Scendo, J. Trela, N. Radek, Influence of laser power on the corrosive resistance of WC-Cu coating, Surface & Coatings Technology 259 (2014) 401-407. https://doi.org/10.1016/j.surfcoat.2014.10.062

[17] N. Radek, K. Bartkowiak, Laser treatment of Cu-Mo electro-spark deposited coatings, Physics Procedia 12 (2011) 499-505. https://doi.org/10.1016/j.phpro.2011.03.061

[18] PN-EN ISO 6507-1:Metale – Pomiar twardości sposobem Vickersa - Część 1: Metoda badań

Terotechnology XI
Materials Research Proceedings 17 (2020) 270-

Materials Research Forum LLC
https://doi.org/10.21741/9781644901038-40

# Improvement of the Surface of the Combustion Chamber of a Piston using Selected Techniques of Production Organization

CZERWIŃSKA Karolina[1,a] *, DWORNICKA Renata[2,c] and PACANA Andrzej[1,c]

[1]Rzeszow University of Technology, The Faculty of Mechanical Engineering and Aeronautics
Poland

[2]Cracow University of Technology, Faculty of Mechanical Engineering, Kraków, Poland

[a]k.czerwinska@prz.edu.pl, [b]renata.dwornicka@mech.pk.edu.pl, [c]app@prz.edu.pl

**Keywords:** Mechanical Engineering, Piston, Combustion Chamber, Eddy Current Testing

**Abstract.** The durability of aluminum pistons is determined by the resistance of the piston crown (combustion chamber) on pressure and temperature. There is a significant risk of fatigue cracking of the edges and the base of the chamber. Hence the need for continuous monitoring of the surface condition of the combustion chamber and prevention of possible incompatibilities. In the paper, the method of eddy currents in the quality control of the surface of combustion chamber of diesel engine pistons used in light vehicles was applied. The aim of the study was to determine, using traditional quality management tools, the sources of nonconformity of castings found in eddy current testing. Ultimately, the aim of the analysis was to reduce the number of non-compliant products or to eliminate them completely. In the analyzed batch of pistons, surface discontinuity in the form of cracks in the combustion chamber occurs most frequently. Based on the Pareto-Lorenza analysis, it can be concluded that this type of non-compliance generates the largest amount of losses (e. g. economic losses). The pistons in which the presented nonconformity has been identified will not be repaired. Therefore, in order to identify and eliminate the causes of non-compliance in the combustion chamber, the Ishikawa diagram in the 6M system was used to analyze the causes of non-compliance in the combustion chamber. On the basis of the in-depth analysis, it was concluded that insufficient feeding during the mold flooding was the key reason for the non-compliance.

## Introduction

Internal combustion engines are the primary source of propulsion for vehicles. Despite intensive work on alternative sources of energy (fuel cells, electric motors), none of them is sufficiently developed to compete with combustion engines (referring to such aspects of its operation as traction characteristics, versatility, comfort of use taking into account the universality of fuel/energy, operating costs, service network, spare parts and finally the habits of vehicle users) [1, 2]. An important role is played by ensuring the desired quality of engine components, which determines the need to eliminate products with sub-surface and surface incompatibilities. [3, 4, 5, 6]. The application of non-destructive testing in production quality control systems may be an effective method of detecting nonconformity in a finished product and its non-application to service. Tests qualified in the electromagnetic tests category are a type of non-destructive tests particularly suitable for diagnostics in the production process or during the product operation. [7, 8]. An example of such detection can be the eddy current method, during which all changes in the analyzed material, such as the change of structure, change of hardness and discontinuities affect the value of electromagnetic parameters, the value of eddy current intensity and induced magnetic field. Diagnostics of the values of electromagnetic field changes and amplitude as well as phase shift of voltage and intensity create the possibility to assess the condition of the examined product area. [9, 10, 11].

One of the main objectives of the production plants is to lead to a situation in which the production of defective products is significantly reduced. The large number of parameters and the complexity of the production process contribute to the fact that the determination of the causes and effects of a problem is not a simple task which requires systematization. [12, 13, 14]. For this reason, quality management tools are used that enable analysis, monitoring and immediate impact on the process throughout the product's production cycle. The Ishikawa diagram is one of the most popular and frequently used quality management tools. This diagram makes it possible to identify the most frequently occurring inconsistencies and to determine the reasons for their occurrence. [15, 16, 17].

The presented approach may be useful in other industrial and research activities that have problems with the occurrence of defects, e.g. biotechnology [18, 19], heat flows [20], hydraulic equipment of heavy duty machines [21], tools production [22] and improving their characteristics, in particular the surface layer [23-25]. It should be taken into account in related analysis e.g. process analysis [26, 27], image analysis [28], stereology analysis [29] and structure analysis [30].

## Experimental

**Purpose and scope of the research.** The aim of the research was to diagnose the surface condition of the combustion chamber in a diesel piston in between operational and quality control using the eddy current method. It was also to identify the reasons for the occurrence of non-compliant castings where appropriate corrective and preventive action could ultimately contribute to reducing the number of non-compliant castings.

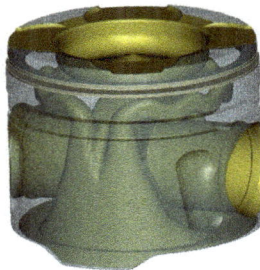

*Fig. 1. Subject of the study - 3D model of a diesel piston.*

The conducted research concerned the batch of products manufactured in the 1st quarter of 2019 in one of the production companies located in the southern part of Poland. The scope of casting control included verification of the surface of the casting combustion chamber, marking the place of non-compliance and precise determination of the type of identified non-compliance. The quality control also included verification of the correctness of the casting marking. Verification of the surface condition of the piston combustion chamber is carried out using visual

Terotechnology XI                                     Materials Research Forum LLC
Materials Research Proceedings **17** (2020) 258-263          https://doi.org/10.21741/9781644901038-38

inspection, dimensional inspection and eddy current method. Quality control was performed in accordance with the internal procedure of the company according to each production order.

**Subject of the research.** In order to assess the possibility of detecting internal inconsistencies in the material of the product, experimental tests were conducted. The subject of research was a piston designed for a diesel engine - a diesel engine used in passenger cars, manufactured by Toyota. The 3D model of the diesel piston is shown in Figure 1.

Pistons are cast from B2 alloy (designation functioning in the company), which is a eutectic aluminum and silicon alloy designed for the production of petrol and diesel pistons used in light vehicles. B2 alloy has no international or national equivalents.

**Methodology of the research.** Detection of discrepancies in the piston crown is carried out with the use of the Foerster Statograph Ds 6. 440 system using appropriate handling systems. Built-in systems control the Kuka manipulator arm (equipped with Siemens Sinumerick D640/i software), which positions and rotates the tested product in relation to the diagnostic probe. Thanks to the rotating system, the probe rotates close to the surface of the piston combustion chamber. With each subsequent rotation of the probe, a new track is scanned. During the test, the results of the test are visualized. The device used is equipped with a system that controls the eddy current method according to EN 12084.

**Results and analysis**
The obtained results of tests carried out with the use of eddy current method in the combustion chamber of the diesel engine piston are presented in Figure 2a. The result of the tests showed the presence of an unacceptable material discontinuity in the combustion chamber of the piston. As a result, as part of the in-depth quality control and analysis, the identified discontinuity area was sampled and metallographed. The results of microscopic observations are presented in Figure 2b.

a)                            b)

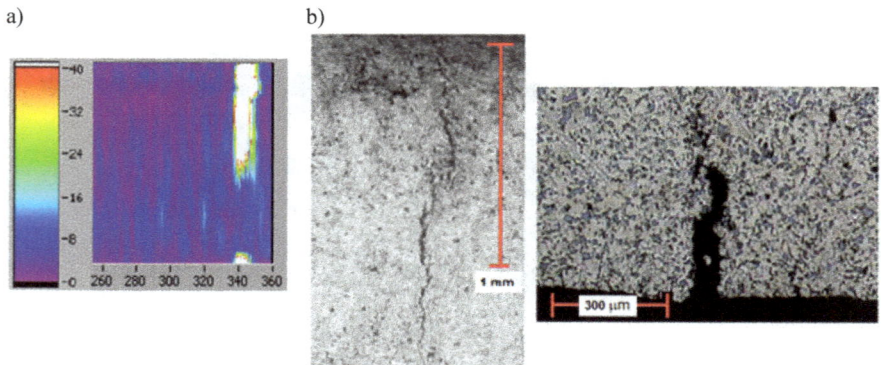

*Fig. 2. a) The result of the eddy current testing of the piston crown with the indication of the detected unacceptable surface discontinuity; b) The result of metallographic surveys from the discontinuity area.*

The presence of observed discontinuity - fracture of the chamber base results in qualitative disqualification of the piston. In the analyzed batch of products, the discontinuity of the combustion chamber surface in the form of cracks occurs most frequently. On the basis of the

Terotechnology XI                                                    Materials Research Forum LLC
Materials Research Proceedings **17** (2020) 258-263         https://doi.org/10.21741/9781644901038-38

Pareto-Lorenza analysis, it can be concluded that this type of non-compliance contributes to the generation of a greatest number of losses (e. g. economic). Products in which the presented type of non-compliance has been identified are not subject to repair. Therefore, in order to identify and eliminate the causes of surface cracks in the combustion chamber, the Ishikawa diagram in the 6M system was used for analysis (Figure 3).

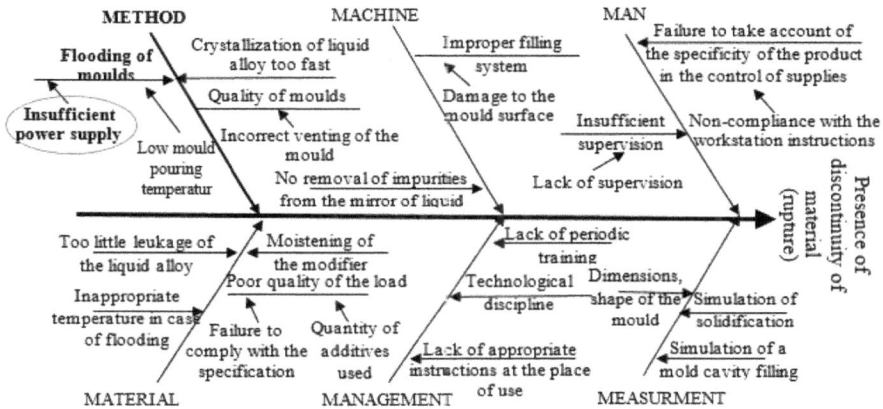

*Fig. 3. Ishikawa diagram of the causes of cracks in the combustion chamber of pistons*

Figure 3 presents factors influencing the formation of one of the most important pistons incompatibility for the enterprise- surface cracks in the combustion chamber. The most important factor influencing the occurrence of incompatibilities in the discussed series of products was distinguished in the scope of the method. In this group, the most important factor was insufficient power supply during the pouring of mould.

**Conclusion**

The paper presents an analysis with a diagnostic test of the combustion chamber of a diesel engine piston using the eddy current method. The aim of the test was to control the quality of the batch of products and to check the usefulness of the control and diagnostic test in the production process.

The non-destructive testing with the eddy current method located the discontinuity in the area of the combustion chamber – a casting defect (crack in the surface of the piston crown) and confirmed its presence by metallographic testing. The occurrence of discontinuities disqualifies the crown. Within the framework of the Pareto-Lorenz analysis, it was diagnosed that cracks in the combustion chamber are the most frequent inconsistencies in the tested batch of products. In order to prioritize the potential causes of non-compliance, a diagram of Ishikawa has been compiled. According to the diagram, the key reason for the non-compliance was insufficient power supply during the flooding of the mould.

The tool used in combination with the quality management method is largely complementary. The proposed combination may be a component of methods supporting quality management processes.

## References

[1]  A. Olczyk, Analiza możliwości zwiększania mocy tłokowych silników spalinowych, Zeszyty Naukowe. Cieplne Maszyny Przepływowe – Turbomachinery, Politechnika Łódzka, 2012, nr 121.

[2]  D. Malindzak et al., An effective model for the quality of logistics and improvement of environmental protection in a cement plant, Przemysl Chemiczny 96(9) (2019) 1958-1962.

[3]  E.K. Vukelja, I. Duplančič, B. Lela, Continuous roll casting of aluminium alloys – casting parameters analysis, Metalurgija 49(2) (2010) 115-118.

[4]  A. Pacana, A. Radoń-Cholewa, J. Pacana: The study of stickiness of packaging film by Shainin method, Przemysl Chemiczny 94(8) (2015) 1334-1336.

[5]  M. Hajkowski, Ł. Bernat, J. Hajkowski, Mechanical Properties of Al-Si-Mg Alloy Castings as a Function of Structure Refinement and Porosity Fraction, Archives of Foundry Engineering 12(4) (2012) 57-64. https://doi.org/10.2478/v10266-012-0107-9

[6]  A. Sanz, New coatings for continuous casting rolls, Surface and Coatings Technology 177-178 (2004) 1-11. https://doi.org/10.1016/j.surfcoat.2003.06.024

[7]  A. Lewińska-Romicka, Badania nieniszczące. Podstawy defektoskopii. Warszawa, WNT, 2001.

[8]  A. Pacana, L. Bednárová, I. Liberko et al., Effect of selected production factors of the stretch film on its extensibility, Przemysl Chemiczny 93(7) (2014) 1139-1140.

[9]  N. Gil, G. Konovalov, A. Mayorov, Devices for non-destructive testing of adhesion quality of a ni-resist insert in diesel engine pistons, Previous Experience and Current Innovations in Non-Destructive Testing, Slovenia, 2001.

[10] J. Senkara, Współczesne stale karoseryjne dla Przemyslu motoryzacyjnego i wytyczne technologiczne ich zgrzewania, Przegląd Spawalnictwa 81 (11) (2009) 3-7.

[11] L. Tian, Y. Guo, J. Li, J. Wang, H. Duan, F. Xia, M. Liang, Elevated re-aging of a piston aluminium alloy and effect on the microstructure and mechanical properties. Materials Science and Engineering A 738 (2018) 375-379. https://doi.org/10.1016/j.msea.2018.09.078

[12] B. Slusarczyk, M. Szajt, Globalizacja jako element wzrostu konkurencyjności, Zeszyty Naukowe Politechniki Częstochowskiej. Zarządzanie 10 (2013) 98-110.

[13] A. Ebenzer, S. R. Daradasn, Total failure mode and effects analysis in tea industry: A theoretical treatise. Total Quality Management & Business Excellence 22 (2011) 1353-1369. https://doi.org/10.1080/14783363.2011.625188

[14] R. Ulewicz, Quality Control System in Production of the Castings from Spheroid Cast Iron, Metalurgija 42(1) (2003) 61-63.

[15]  A. Pacana, Praca zespołowa i liderzy, Rzeszów, Oficyna Wydawnicza Politechniki Rzeszowskiej, 2017.

[16] A. Gwiazda, Koncepcja ważonego wykresu Ishikawy, Problemy Jakości 4 (2005) 13-17.

[17] J. Sęp, R. Perłowski, A. Pacana, Techniki wspomagania zarządzania jakością. Reszów, Oficyna Wydawnicza Politechniki Rzeszowskiej, 2006.

[18] E. Skrzypczak-Pietraszek, A. Hensel, Polysaccharides from Melittis melissophyllum L. herb and callus. Pharmazie 55 (2000) 768-771.

[19] E. Skrzypczak-Pietraszek, A. Urbanska, P. Zmudzki, J. Pietraszek, Elicitation with methyl jasmonate combined with cultivation in the Plantform™ temporary immersion bioreactor highly increases the accumulation of selected centellosides and phenolics in Centella asiatica (L.) Urban shoot culture. Engineering in Life Sciences. 19 (2019) 931-943. https://doi.org/10.1002/elsc.201900051

[20] L. J. Orman, Boiling heat transfer on meshed surfaces of different aperture. AIP Conf. Proc. 1608 (2014) 169-172. https://doi.org/10.1063/1.4892728

[21] M. Domagala, H. Momeni, J. Domagala-Fabis, G. Filo, M. Krawczyk, J. Rajda, Simulation of particle erosion in a hydraulic valve. Materials Research Proceedings 5 (2018) 17-24. 10.21741/9781945291814-4

[22] M. Mazur, K. Mikova, Impact resistance of high strength steels. Materials Today-Proceedings 3 (2016) 1060-1063. https://doi.org/10.1016/j.matpr.2016.03.048

[23] S. Wojciechowski, P. Twardowski, T. Chwalczuk, Surface Roughness Analysis after Machining of Direct Laser Deposited Tungsten Carbide, Met & Props 2013, 14th Int. Conf. on Metrology and Properties of Eng. Surf., Journal cf Physics Conference Series 483 (2014) art. 012018. https://doi.org/10.1088/1742-6596/483/1/012018

[24] D. Przestacki, M. Kuklinski, A. Bartkowska, Influence of laser heat treatment on microstructure and properties of surface layer of Waspaloy aimed for laser-assisted machining. Int. J. Adv. Manuf. Technol. 93 (2017) 3111-3123. https://doi.org/10.1007/s00170-017-0775-2

[25] P. Kieruj, M. Kuklinski, Tool life of diamond inserts after laser assisted turning of cemented carbides. MATEC Web of Conf. 121 (2017) art. UNSP 03011. https://doi.org/10.1051/matecconf/201712103011

[26] J. Pietraszek, E. Skrzypczak-Pietraszek, The optimization of the technological process with the fuzzy regression. Adv. Mater. Res-Switz. 874 (2014) 151-155. https://doi.org/10.4028/www.scientific.net/AMR.874.151

[27] J. Pietraszek, A. Gadek-Moszczak, T. Torunski, Modeling of Errors Counting System for PCB Soldered in the Wave Soldering Technology. Advanced Materials Research 874 (2014) 139-143. https://doi.org/10.4028/www.scientific.net/AMR.874.139

[28] A. Szczotok, D. Karpisz, Application of two non-commercial programmes to image processing and extraction of selected features occurring in material microstructure. METAL 2019: 28th Int. Conf. on Metallurgy and Materials, Ostrava, TANGER, 1721-1725. https://doi.org/10.37904/metal.2019.971

[29] L. Wojnar, A. Gadek-Moszczak, J. Pietraszek, On the role of histomorphometric (stereological) microstructure parameters in the prediction of vertebrae compression strength. Image Analysis and Stereology 38 (2019) 63-73. https://doi.org/10.5566/ias.2028

[30] J. Pietraszek, A. Gadek-Moszczak, The Smooth Bootstrap Approach to the Distribution of a Shape in the Ferritic Stainless Steel AISI 434L Powders. Solid State Phenomena 197 (2012) 162-167. https://doi.org/10.4028/www.scientific.net/SSP.197.162

# Keyword Index

# About the Editor

**Agnieszka SZCZOTOK, M.Eng., Ph.D.**

is an assistant professor at the Silesian University of Technology, The Faculty of Materials Engineering, Katowice, Poland.

She has scientific interests in: modern methods of materials characterization, an image analysis application, a quantitative description of the materials structures and statistics in materials science analysis.